T0279837

ORGANIC, INORGANIC, AND HYBRID SOLAR CELLS

IEEE Press
445 Hoes Lane
Piscataway, NJ 08855

ORGANIC, INORGANIC, AND HYBRID SOLAR CELLS

Principles and Practice

Ching-Fuh Lin
Wei-Fang Su
Chih-I Wu
I-Chun Cheng

IEEE PRESS

A JOHN WILEY & SONS, INC., PUBLICATION

For general information on our other products and services please contact our Customer Care Department within the United States at (800) 762-2974, outside the United States at (317) 572-3993 or fax (317) 572-4002.

Wiley also publishes its books in a variety of electronic formats. Some content that appears in print, however, may not be available in electronic formats. For more information about Wiley products, visit our web site at www.wiley.com.

Library of Congress Cataloging-in-Publication Data:

Organic, inorganic, and hybrid solar cells : principles and practice / Ching-Fuh Li ... [et al.].
p. cm.
Includes bibliographical references.
ISBN 978-1-118-16853-0 (hardback)
1. Solar cells—Materials. 2. Thin films. 3. Solar cells—Design and construction. I. Lin, Ching-Fuh.
TK8322.O735 2012
621.31'244—dc23
2012005089

10 9 8 7 6 5 4 3 2 1

CONTENTS

FOREWORD

Woodrow W. Clark II

The importance of this book at this time in world history is very significant. Today, nations in Asia and the European Union (EU) are deeply involved in the Green Industrial Revolution (GIR). However, the United States has not gotten there yet. As the book notes from its beginning, the Second Industrial Revolution (SIR) started over two centuries ago, but became intense and the dominant technical and economic paradigm during the twentieth century. The reason for this was that the invention of the combustion engines for transportation and infrastructure systems demanded fossil fuels which western industrialization became very dependent upon. The results of this at the end of the twentieth century were wars in the Middle East and constant conflicts, including terrorism, around the world.

By the start of the twenty-first century, the Third Industrial Revolution (3IR) had started. Jeremy Rifkin referred to the 3IR in his book, *The European Dream* (2004). Then in 2011, he wrote a book about the Third Industrial Revolution. However, in both books, the focus was on a 3IR in the EU, especially Germany. There was no mention of what had gone on in Asia since the 1980s, especially in Japan, South Korea, Taiwan, and China.

Hence, my book, *Global Energy Innovation* (2011) highlighted the Green Industrial Revolution. The main point is that many Asian countries have had a key role in developing renewable energy systems, and are now leading the world in terms of efficiencies and cost reductions. The Green Industrial Revolution is more than just renewable energy systems and storage technologies. It combines and integrates these systems in order to be more conservational and efficient as well as reducing the carbon and emissions in the global atmosphere.

What Asian nations realized and did something about early in the 3IR is see that the fossil fuel era is at an end with its dependency on climate-changing fossil fuels that damage the environment and cause weather problems around the world. Asian nations put research funds and corporate development as prime national goals for renewable energy systems, specifically as they apply to basic needs like water, waste treatment, and transportation. Hence, companies developed water saving systems and IT communication to reduce the use of wood and plastic products for manufactured goods. The development of all-electric and hydrogen fuel cell cars came from Asian nations.

More significantly, the Asian countries saw the need for renewable energy systems that use the smart grid and IT controls to achieve cheaper electricity. Above all, Asian nations knew before the Fukushima earthquake and nuclear power plant explosion in Japan that dependency on nuclear power was both questionable and not the answer to their energy needs, aside from addressing or mitigating climate change. The global competition for fuels would continue with conflicts and wars in which no nation, in the end, ever won. The conflicts over fuel supplies would continue. The basic issue was that renewable energy systems, like oceans, air, and sun power were inexhaustible and readily available to people and nations around the world.

This book presents the historical background of solar cells in both a technological and economic context. Even more significantly, the book provides a road map to the near (3–5 years) and distant (6–10 years) future of solar cells. With a discussion on the kinds of solar cells that exist today, to a discussion of organic and inorganic materials and applications, the scientific research results are significant. Clearly, one of the key outcomes learned from the book is the shift in science and engineering from mechanical to chemical and electrical systems. What this means for education, innovation, entrepreneurship, and job growth has not been determined yet.

The future is here now and world institutions need to take notice or be left behind.

PREFACE

Human beings have enjoyed the benefits of fossil energy since the Industrial Revolution began with the invention of steam engine in 1765 by James Watt. Unfortunately, over two centuries of consumption of fossil fuels, in particular after the twentieth century, we are now facing three major crises: depletion of fossil fuels, global warming, and dramatic changes in weather. Although there are some debates on those issues, most people have realized the importance of alternative energies. Some people deem nuclear power a good solution because it involves almost no emission of greenhouse gases. Nonetheless, nuclear fuel is not renewable and the threat of nuclear contamination is probably even worse than the problems caused by fossil-fuel energy. On March 11, 2011, an earthquake-induced giant tsunami in Japan destroyed the Fukushima nuclear power plant and caused nuclear pollutants to spread over a very large area. Therefore, clean renewable, noncontaminating energies should be the major trend for developing future energy needs. Among them, solar energy is a very good choice because it is clean, has almost no limit in supply, and is available everywhere on earth.

Although solar energy has the above advantages and Si solar cells have gained increasing popularity over recent years, it still cannot compete with fossil energy due to the high cost of production and implementation. Therefore, the reduction of cost for solar energy is the key issue. However, how do we achieve the goal of low-cost solar cells that are competitive with fossil fuels? This book is aimed at providing knowledge on various types of solar cells, including inorganic, organic, and hybrid solar cells, in the hope of offering a good foundation to readers, so they can evaluate the possible technologies in solar cells that have the potential to compete with fossil energy. Hence, in the first chapter of this book, in addition to the discussions on the crises resulting from fossil energy, we particularly address the cost issue and energy economics.

Chapter 2 discusses the properties of light and its interaction with materials. Because solar cells involve light–matter interaction, this chapter gives readers the necessary background of physics for understanding the working principles of solar cells. In Chapter 3, the properties of inorganic materials are introduced. The focus is particularly on the most common materials for inorganic solar cells—semiconductors. In Chapter 4, we switch the focus to organic materials, which have become very popular nowadays for their potential of very low processing cost and tailored properties. We will particularly emphasize their electrical and optical properties because of their possible applications in solar cells. On the other hand, in order to realize the hybrid solar cells that employ both organic and inorganic materials, the interface between them needs to be understood. Chapter 5 discusses the issues that arise when the organic and inorganic materials are brought together, including compatibility of deposition, adhesion problems, formation of surface states, and band-level realignment.

After the basic knowledge of inorganic and organic semiconductors as well as their interface is established, we are ready to examine the device characteristics. Therefore, Chapter 6 will start with the discussions on inorganic semiconductor solar cells, which are the most common devices used commercially nowadays. This chapter discusses the functional principles of those solar cells and also provides the basic knowledge for comparison with organic solar cells and hybrid solar cells. The reasons for their high production cost will also be addressed. Chapter 7 focuses on the discussion of organic solar cells, which have the potential to produce very low cost solar electricity (<US$0.5/Wp). At present, there are four kinds of organic solar cells under development. They are discussed in detail.

With the knowledge of both inorganic and organic solar cells, we are prepared to understand the organic–inorganic hybrid solar cells, which are described in Chapter 8. Here, the fundamental concepts of forming such solar cells will be introduced. The technique of the solution process will be particularly addressed because it has the potential of very low production cost. The sandwiched structure formed through the solution process is also discussed because it gives very good stability to solar cells consisting of organic materials. Furthermore, two alternatives of forming hybrid solar cells are described in this chapter: (1) using organic materials for light absorption and inorganic materials for carrier transportation, and (2) using inorganic semiconductors for light absorption and organic polymers for carrier transportation. Finally, Chapter 9 will examine the outlook for solar cells. We have also included exercises at the end of each chapter in this book so that it can serve as a textbook for solar-cell technologies to teach the future generation of solar engineers.

To resolve the crises caused by fossil energy, clean and renewable energies have to be developed at a very fast pace. With good knowledge of solar cells, scientists, engineers, and investors, as well as general citizens, can better assess the future development of new energies and have good faith in new technologies based on a solid foundation.

Although there are already many books on solar cells, most of them are focused only on certain types, either only semiconductor solar cells or just organic solar cells. We hope that this book brings together the knowledge on various types of solar cells so readers can closely compare their similarities and differences, including their working principles and production technologies, as well as the costs involved therein. We also hope that the close comparison of various solar cells will speed up the development of solar-cell technologies and industries that provide very cheap and clean solar electricity.

Finally, Ching-Fuh Lin, Chih-I Wu, and I-Chun Cheng would like to acknowledge those that help and provide relevant information for this book. They are Jing-Shun Huang, Shu-Chia Shiu, Hong-Jhang Syu, Ming-Yi Lin, Yu-Hong Lin, Po-Ching Yang, Shiang Lan, Shang-Hong Lin, Chen-Yu Chou, Jen-Yu Sun, Kuai-Yu Chien, Chi-Hsing Hsu, Shih-Jieh Lin, Ping-Yi Ho, Shao-Hsuan Kao, Tzu-Ching Lin, Jin-Lin Pan, Chia-Lin Chuang, Wei-Hsuang Tseng, Po-Sheng Wang, I-Wen Wu, Jung-Hung Chang, Jan-

Kai Chang, Chia-Wei-Liu, I-Chung Chiu, Yun-Shiuan Li, Bo-Shiung Wang, Chin-Cheng Chiang, Yang-Shieng Lin, Ching-Wen Hsu, Po-Yuan Chen, Hsiao-Wei Liu, Hsin-Hua Hou, I-Feng Lu, and Chih-Hung Tsai. Also, Wei-Fang Su would like to thank her husband, Cheng-Hong, for encouragement and support, and her Ph.D. students, Jhin-Fong Lin and Hsueh Chung Liao, for final editing and proofreading

Ching-Fuh Lin, Wei-Fang Su, Chih-I Wu, and I-Chun Cheng
Innovative Photonics Advanced Research Center (i-PARC)
National Taiwan University
Taipei, Taiwan
January, 2012

ABOUT THE AUTHORS

Professor **Ching-Fuh Lin** obtained his B.S. degree from National Taiwan University in 1983, and the M.S. and Ph.D. degrees from Cornell University, Ithaca, NY, in 1989 and 1993, respectively, all in electrical engineering. He is now the Director of the Innovative Photonics Advanced Research Center (i-PARC), the Chairman of Graduate Institute of Photonics and Optoelectronics, and a joint professor in the Graduate Institute of Photonics and Optoelectronics, Graduate Institute of Electronics Engineering, and Department of Electrical Engineering at National Taiwan University. He is also a Fellow of IEEE, a Fellow of SPIE, Member of Asia-Pacific Academy of Materials, and a member of OSA. He had obtained many awards, including the Distinguished Research Award and Class A Research Awards from National Science Council of Taiwan, the Outstanding Electrical Engineering Professor Award from the Chinese Institute of Electrical Engineering, Acer Research Golden Award, and Acer Research Excellent Awards. He has published over 140 journal papers, 360 conference papers, two scientific books, and been awarded 80 patents.

Professor **Wei-Fang Su** specializes in polymeric materials. She obtained her Ph.D. from University of Massachusetts and did postdoctoral research at Northwestern University. Then she joined the Westinghouse Research Center and worked on materials for electrical insulation and electronic devices for 16 years, during which she won six outstanding-researcher awards. She did nonlinear optical materials research at the Research Center of Mitsubishi Electric Company in 1990. She joined National Taiwan University as a full professor in 1996 and was promoted to distinguished professor in 2009. Her research is focused on innovative materials, including conducting polymers, liquid crystalline polymer, and nanocomposites for electronic and solar cell applications. She was named 2010 Outstanding Professor of the Chinese Engineer Society and 2011 Outstanding Researcher of National Science Council of Taiwan. She has published 136 SCI papers, 165 international conference papers, one book, chapters for six books, and was awarded 21 U.S. patents and 15 Taiwan patents.

Professor **Chih-I Wu** joined the Graduate Institute of Photonics and Optoelectronics and the Department of Electrical Engineering of National Taiwan University in 2004. His research is on optical-electronic devices and materials, organic light-emitting materials, metal–semiconductor interfaces, and heterojunctions in electronic devices. Prior to joining NTU, he worked at the Component Research Lab of Intel in the United States from 2000 to 2004. Professor Wu got his B.S. degree from National Taiwan University and M.S. degree from Northwestern University, both in Physics. He received his Ph.D. from the Department of Electrical Engineering at Princeton University in 1999. At Princeton, he worked on the electronic structures of optical-electronic semiconductors, including nitride-based semiconductors and organic thin films for light-emitting diodes. Professor Wu has published more than 100 journal and conference papers and holds five U.S. patents. His journal papers have been cited more than 1700 times.

Professor **I-Chun Cheng** received B.S. and M.S. degrees in mechanical engineering from National Taiwan University, Taipei, Taiwan, in 1996 and 1998, respectively, and a Ph.D. degree in electrical engineering from Princeton University, Princeton, NJ, in 2004. Following her degree, she became a research associate at Princeton University. She joined the faculty of National Taiwan University, Taipei, Taiwan, in 2007, where she is currently an Associate Professor in the Department of Electrical Engineering and Graduate Institute of Photonics and Optoelectronics. She has primarily worked in the field of novel silicon thin-film technology, metal–oxide thin-film technology, and flexible large-area electronics.

1

INTRODUCTION—WHY SOLAR ENERGY?

Ching-Fuh Lin

1.1 THE ERA OF FOSSIL ENERGY

The Industrial Revolution that began with the invention of the steam engine in 1765 by James Watt reshaped the role of labor in human history. From the 18th to the 21st centuries, machines gradually replaced human labor and animal power. As a result, energy consumption due to the increased types and numbers of machines continues to grow, so fossil fuels, including coal, oil, and natural gas, have become the major supplies of energy. The growth of energy consumption became even more significant in the 20th century. From 1900 to 2000, energy consumption grew nearly 40 times and this trend will not stop in future decades.

As it became known that the use of fossil fuels would result in emission of carbon dioxide, gasoline engines that emitted less carbon dioxide became dominant in 20th century. Oil began to be used around 1900 and natural gas around 1950. The pollutant factors for electricity generation from coal, oil, and natural gases are 322.8, 258.5, and 178, respectively [1]. However, the emission of carbon dioxide continued to grow dramatically because of the huge demands for fossil energy in general. Natural gas has a pollutant factor about half that of coal but this reduction in carbon dioxide emission is still not significant enough. Therefore, although the recovery of the global economy was slow in 2010, emissions of carbon dioxide still reached the historical record of 30,600 million tons, whereas it was less than 15,000 million tons in 1970 [2, 3]. The United States and the OECD countries used to generate most of the carbon dioxide but China surpassed United States in the emission of carbon dioxide in 2007 [2]. Other non-OECD countries also generated more carbon dioxide than OECD countries after 2004 [2]. The fact that those newly developing countries consume even more fossil fuels than economically developed countries certainly worsens the emission of carbon dioxide.

1.1.1 Possible Depletion of Fossil Fuels

After centuries of consuming fossil fuels, people are now facing three major problems, although some controversies over the causes exist. First, some predict that reserves of petroleum will last for only 40 years and natural gases for only 60 years. Such a time span seems long for people who are over 40 years old but could become a serious issue for young people. In addition, there are two hidden problems in such a time span. First, mining technologies are becoming more difficult and challenging because some fossil fuels are buried deep in the ground. Mining cost will consequently increase. Second, as the foreseeable depletion of fossil energy approaches, countries will compete for the limited resources. Then large-scale wars might occur unless other huge resources of energy supplies become ready for use.

1.1.2 Global Warming

The second problem is global warming due to the emission of greenhouse gases such as carbon dioxide. According to a recent report from NASA [4], global temperature has obviously been increasing since 1980. As shown in Figure 1.1, the average temperature has increased 0.5–0.6 °C, compared to the average temperature between 1951 and 1980. The increase of temperature in the northern hemisphere is even worse for the past ten years: 0.5–2.5 °C. As indicated in Figure 1.2, the worst area is around the Arctic Ocean. The significant increase of global temperature has caused the disappearance of glaciers in many areas, such as North America, South America, Africa, Europe, and

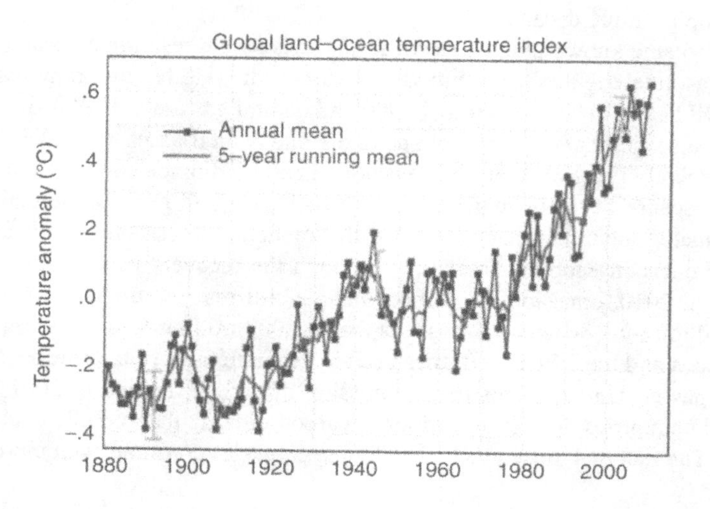

Figure 1.1. Relative variation of global temperature in recent decades [4].

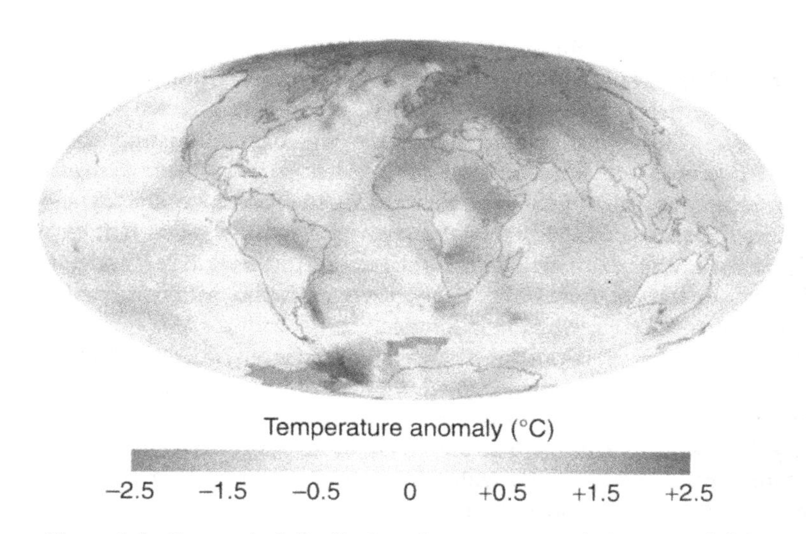

Figure 1.2. Geometrical distribution of temperature variation on earth [4].

Asia. For example, the area of the Tsho Rolpha Glacier Lake in Nepal increased from 0.23 km^2 before 1960 to 1.65 km^2 in 1997 because glacier ice melted and flowed into this lake. Also, Arctic Ocean ice has been decreasing at a rate of 9% per decade as the ice of frozen freshwater in glaciers on Greenland melts and dilutes the salinity of the Arctic Ocean, significantly influencing the conveyed ocean currents.

1.1.3 Dramatic Changes in Weather

Global warming has led to dramatic changes in weather. In Australia, record or near-record temperatures at the sea surface were recorded off the Queensland coast in late 2010. Then a series of floods hit Queensland, beginning in December 2010 and lasting until January 2011. The flood took over about three-quarters of the state and affected over 70 towns and 200,000 people. Thirty-five deaths were reported due to these floods. The state rainfall of 209.45 mm was the highest since 1900. Kevin Trenberth attributed the flood to a half-degree Celsius rise in ocean temperature around Australia as a result of global warming. [5] Thus, extra water vapor was produced and intensified the rainfall. Some scientists do not agree with this point of view, but such global debate cannot be verified by experiments in the laboratory [6]

Like Australia, Taiwan was also hit by a huge rainfall in summer 2008. The mountain area of south Taiwan experienced a record of about 3000 mm rainfall in just two days when the center of Typhoon Morakot landed in the north part of Taiwan. The huge amount of rainfall further caused a dramatic mudslide. An 80-meter layer of dirt and stone slid off the mountain and buried the entire Shiao-Lin Village in south Taiwan. More than 500 people were killed. From 2000 to 2010, Taiwan has experienced over four times the number of huge rainfalls that used to occur only once every 50 or 100 years.

Even the most powerful country, the United States, is not able to escape from the flood. In April and May 2011, the Mississippi River had catastrophic floods as two major storms gave rise to record levels of rainfall. The floods affected several states, including Missouri, Illinois, Kentucky, Tennessee, Arkansas, Mississippi, and Louisiana. At least 383 people were killed in these seven states. In 2011, there were also most destructive tornadoes. In 2011, 546 people were killed (counted to July 28) by tornadoes in the United States, compared to 564 deaths over the past ten years. The disasters were mainly due to several extremely large tornadoes in April and May. A huge and intense multiple-vortex tornado, rated EF5, badly damaged Joplin, Missouri in May 22. Over 100 people were killed by this single tornado.

It is still controversial to directly link the disastrous floods and tornadoes with global warming. The global change in weather cannot be experimentally verified in the laboratory but the coincidence between the temperature rise in the recent decade and recently detrimental weather should alert us to make efforts to reduce global warming. Therefore, replacing fossil fuels with other energy resources that do not emit greenhouse gases should be considered seriously.

1.2 RENEWABLE ENERGIES

Nuclear power had been thought of as a good alternative to replace fossil fuels. For example, Japan has 54 nuclear power plants that generate 30% of its electricity. In the beginning of 2011, Japan planned to build another 14 nuclear power plants by 2030 and hoped to have 50% of its electricity generated by nuclear power. Nevertheless, the earthquake on March 11, 2011 induced a giant tsunami that destroyed several Fukushima nuclear power reactors, causing nuclear pollutants to spread over a very large area. This disaster made Japan abandon its plans and stopped the operation of several other nuclear power plants. Other countries also reconsidered their plans to build new nuclear power plants and have given more thought to renewable energies that are more environmentally friendly.

The renewable energies include solar, hydropower, ocean wave, tide, biomass, wind, and geothermal energies. Although hydropower had been well developed and currently generates 15% of global electricity, it can only be built in regions that have large rivers with steady water streams. Not much more hydropower can be developed. On the other hand, among all other renewable energies, wind and solar are the two most developed technologies and have the potential to generate significant portions of electricity worldwide. The potential wind power is about 1.3×10^{12} kW globally, which is about 3000 times the power generated by fossil fuels and approximately 850 times global power consumption, 15–16 TW (1 TW = 1×10^{12} W). However, it has two major drawbacks: (1) wind power is not stable and (2) wind is strongly influenced by regional geography. The windy areas are not equally distributed on earth. Quite a few countries do not have sufficiently strong winds.

In addition to wind power, solar power could produce much more electricity than human beings need. The solar power that the earth receives every day is 174,000 TW, which is about 11,600 times human needs. In comparison with wind power, solar power has two major advantages. First, sunlight is most intense in summer and around noon, when most electricity is needed. It well matches the daily activities of human beings. Second, the sun shines almost everywhere, and the area required to generate electricity for human needs from sunshine is small compared to the entire land area. As mentioned previously, the total power demand of human beings is 15–16 TW. Solar intensity is approximately 1 kW/m^2. With 15% efficiency of solar panels, a square meter will generate 150 W of power. Therefore, the area that is required to generate the total power needed by human beings is 16×10^{12} W/(150 W/m^2) = 1.07×10^{11} m^2 = 1.07×10^5 km^2. This area is only 0.0723% of the total land area on earth, which is 1.48×10^8 km^2. Take the United States as an example. The United States consumes about 20% of total global power, 3.2 TW, so it will need an area of 3.2×10^{12} W/(150 W/ m^2) = 2.14×10^{10} m^2 = 2.14×10^4 km^2. This area is only 0.234% of the United States land area, which is about 9.16×10^6 km^2.

On the other hand, if solar cells can be used on the roof of a house, even with an efficiency of only 10%, a regular house with 100 m^2 of roof area will be able to generate 10 kW of power capability. For 3.5-hour equivalent daily sunlight, which is common in many areas, such a house will generate about 35 kWh of electricity each day and 1050 kWh every month. This is sufficient for a regular household with usual power consumption.

It looks as if solar energy is very promising and should be a good solution to the problems caused by the fossil energy. However, solar energy is still not popular. The reason will be discussed in the following section.

1.3 SOLAR ENERGY AND ECONOMY

Four aspects are important for solar cells: cost, efficiency, lifetime, and productivity, as illustrated in Figure 1.3. In the past, efficiency has been thought of as the key factor that indicates the advancement of solar cells. However, the cells that have the best efficiency may not be practical because of high production cost. Thus, cost is the core issue of solar cells. Solar energy has to be competitive with fossil energy in order to make the solar industry self-sustainable without government subsidies. On the other hand, the importance of efficiency cannot be ignored because it influences the cost. If the efficiency is doubled while other factors remain the same, the same area of land will generate twice the electricity. It means that the cost per watt is reduced by half.

As to the aspect of lifetime, its requirement depends on the applications. If the solar cells will be used for power plants, their lifetime is expected to be 20 years, or at least 10 years. With all other factors remaining the same, 20-year lifetime costs are almost half of the 10-year lifetime, including solar-cell cost and installation cost. Only the cost of land is not increased. If solar cells are used for consumer products, the lifetime can be lowered to much less than 10 years. For productivity, it is well known that mass production will reduce the cost significantly. For example, the cost of dynamic random access memory (DRAM) becomes one quarter as its production increases one order of magnitude. Therefore, the technology of solar cells has to be compatible with mass-production techniques for the cost and deployment of solar panels to be practical.

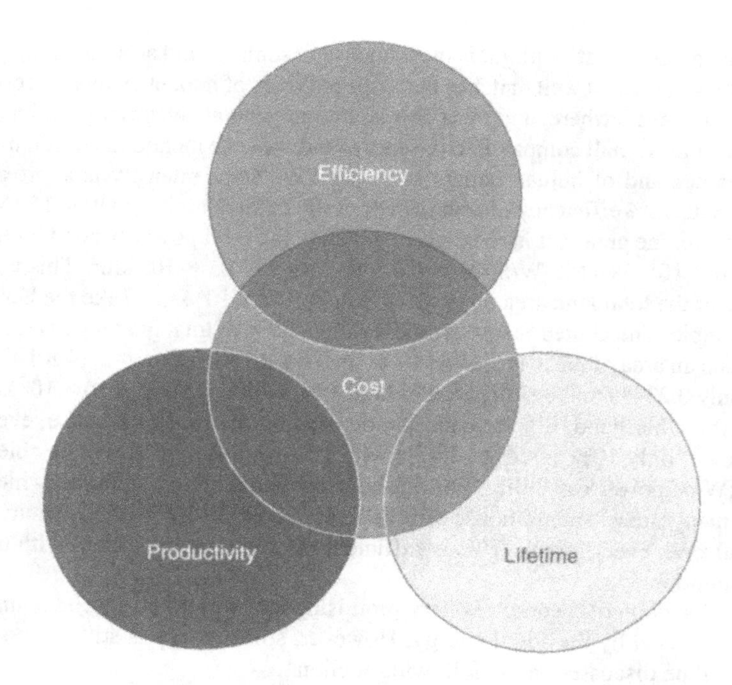

Figure 1.3. Four important aspects of solar cells.

1.3.1 Production Issue

Although the area for solar cells to generate all energy needs for human beings is only 0.0723% of the total area of earth's surface land, this area of 1.07×10^5 km^2 is still large compared to the area of integrated circuits (ICs). The IC industry produces ICs with an area of 10 km^2 per year. Therefore, if the solar panels are manufactured with IC technology, it will take 1.07×10^4 years. From the material point of view, the IC industry uses 7×10^6 kg of silicon. If silicon with a similar wafer thickness will be used for solar panels to generate all energy, the total amount will be 7.5×10^{10} kg. Even if the thickness is reduced to only one-third and the replacement of fossil energy with solar is only 20%, the total amount is still 5×10^9 kg. The large amount of material demand means that current means of material production will have to evolve very quickly or different solutions from the IC industry will be necessary to achieve the goal of using sunlight as one of the major energy supplies. Therefore, in the evaluation of solar technology, a very different scenario for the production volume has to be taken into account.

1.3.2 Types of Solar Cells

Solar energy had been applied as early as the seventh century B.C., but the photovoltaic effect was first recognized in 1839 by the French physicist A. E. Becquerel. The first solar cell was built much later. In 1883, Charles Fritts coated selenium with a very thin layer of gold to form junctions and obtained about 1% of power conversion efficiency from sunlight to electricity. Only after Albert Einstein explained the photoelectric effect in 1905 were human beings able to gradually realize the working principle of power conversion from light to electricity. More progress was made after the modern photovoltaic cell was developed at Bell Laboratories in 1954. Since then, many types of solar cells have been developed, including single-crystalline Si solar cells, multicrystalline Si solar cells, single-junction III-V solar cells, multijunction III-V solar cells, and several types of thin-film solar cells.

The III-V single-junction and multijunction solar cells are made of III-V crystals, which are much more expensive than other materials. Thus, they are usually combined with concentrators, which are lenses to focus sunlight to a small spot. As a consequence, a large area of sunlight can be collected and concentrated to a III-V solar cell with a much smaller area, so much less III-V materials are required. The single-junction III-V solar cells with concentrated sunlight have efficiency of nearly 30%. Experimentally, the multijunction III-V solar cells with concentrated sunlight are able to convert 43.5% of sunlight to electricity [7]. In principle, the multijunction III-V solar cells can have power conversion efficiency of more than 60% [8], but the fabrication of solar cells with over three junctions is very difficult and is still under development.

The crystalline Si solar cells are the most used ones commercially. The best Si solar cell has a power conversion efficiency of 27.6% [9]. Because cost is the major concern, the commercial solar cells are mainly made of multicrystalline Si that is slightly cheaper than the single-crystalline Si. However, the cost of multicrystalline Si solar cells is still too high and cannot compete with fossil fuels for electricity generation. Government subsidies are necessary to keep this industry alive.

In comparison, thin-film solar cells that consume much less materials are considered to offer the hope of future development without the necessity of government subsidies. The thin-film solar cells include amorphous Si solar cells, nanocrystalline or microcrystalline Si solar cells, $Cu(In, Ga)Se_2$ (CIGS) solar cells, CdTe solar cells, dye-sensitized solar cells, organic solar cells, and organic–inorganic hybrid solar cells. As mentioned before, if solar cells will be used as a major energy supply, the total area required will be very large, so the production capability and material consumption have to be considered. Roll-to-roll production and ease of conveyance will be important issues for future production and deployment. From this point of view, those that can be fabricated with solution processes or under atmospheric pressure will have an advantage over those that need high-vacuum apparatus. From the material point of view, those that use abundant chemical elements will be more beneficial than those that use rare-earth chemical elements.

1.3.3 Cost Analysis—Grid Parity

To make solar electricity attractive, the first step is to make it comparable in price to what people pay to power companies, which varies among countries, from 3.05 US cents/kWh (Ukraine) to 42.89 US cents/kWh (Denmark). Table 1.1 lists the prices of grid electricity in many countries. Within each country, the price also varies depending

TABLE 1.1. Prices of grid electricity in different countries

Country	Price, US cents/kWh
Argentina	5.74
Australia	18.55
Belgium	11.43
Canada	6.18
Denmark	42.89
France	19.25
Germany	30.66
Italy	37.23
Netherlands	34.70
Russia	9.49
Singapore	20.69
Spain	19.69
South Africa	17.10
Sweden	27.34
Taiwan	8.80
United Kingdom	18.59
Ukraine	3.05
United States	11.20

on the region and the amount of consumption. For example, the average price in the US is 11.2 US cents/kWh, but the highest price is 31.04 US cents/kWh in Hawaii and the lowest price is 7.31 US cents/kWh in North Dakota. Both New York and California have higher prices than the national average, 17.45 cents/kWh and 14.83 cents/kWh, respectively. Thus, each country or each state will vary in the degree to which the cost of solar electricity is comparable with the price of regular electricity.

In addition, the total amount of solar energy received in different regions also varies. For the same power output of solar panels, a region that has the equivalent of 6 hours of sunlight each day generates twice the electricity of a region that has 3 hours of sunlight each day. The following formula can be used to calculate the cost of solar electricity for grid parity, meaning that the cost is equal to the price of regular grid electricity:

$$\text{Cost of solar electricity } \$/\text{kWh} = \frac{\text{Cost of solar system } (\$/\text{Wp}) \times \text{Capacity of plant (W)}}{\text{Average daily electricity generated (kWh)} \times \text{lifetime of system (days)}}$$

$$= \frac{\text{Cost of solar system } (\$/\text{Wp})}{\text{lifetime of system (years)}} \times \frac{\text{Capacity of plant (W)}}{\text{Average daily electricity generated (kWh)} \times 365 \text{ (days)}} \quad (1\text{-}1)$$

From Eq. (1-1), the cost of solar electricity clearly depends on the average electricity generated. For example, the UK has only 800 kWh of solar energy annually per kW of solar plant, whereas Australia receives 1500 kWh of solar energy per kW.

Both countries have similar prices of grid electricity: 18.59 US cents/kWh for UK and 18.55 US cents/kWh for Australia. To make the cost of solar electricity have grid parity, that is, the cost of solar electricity equals the price of grid electricity, the cost of the solar system has to be as low as US\$2.94/W_p for UK, whereas it can be as high as US\$5.56/W_p for Australia, assuming that the solar system will be used for 20 years. This means that a high-price solar system can be more easily adapted by Australia than the UK. Figure 1.4 shows the relation of system cost to average price of grid electricity in different countries or regions; capital interests are not taken into account.

In this figure, the annual solar energy yield equals the effective hours of sunshine (1 kW/m^2) in one year times 1 kW of solar system power. If the annual solar energy yield is 1000 kWh/kW, the effective hours per year is 1000, so each day has 2.74 hours of average effective sunshine. For countries with low annual solar energy yield, the curve at constant system cost moves up, indicating that a lower system cost has to be achieved to make the price of solar electricity competitive with grid electricity. Most countries in Europe are located in the region of low annual solar energy yield.

From Figure 1.4, the cost of solar electricity (\$/kWh) can also be obtained for a constant system cost according to the annual solar energy yield or the average effective hours of sunshine per day. For example, if the system cost is US\$3.00/W_p and is expected to be used for 20 years, the cost of solar electricity is US\$0.1644/kWh if the average effective sunshine is 2.5 hours per day and reduces to US\$0.0685/kWh for 6 hours of average effective sunshine per day. Table 1.2 shows the relation between the cost of solar electricity and the effective hours of sunshine per day.

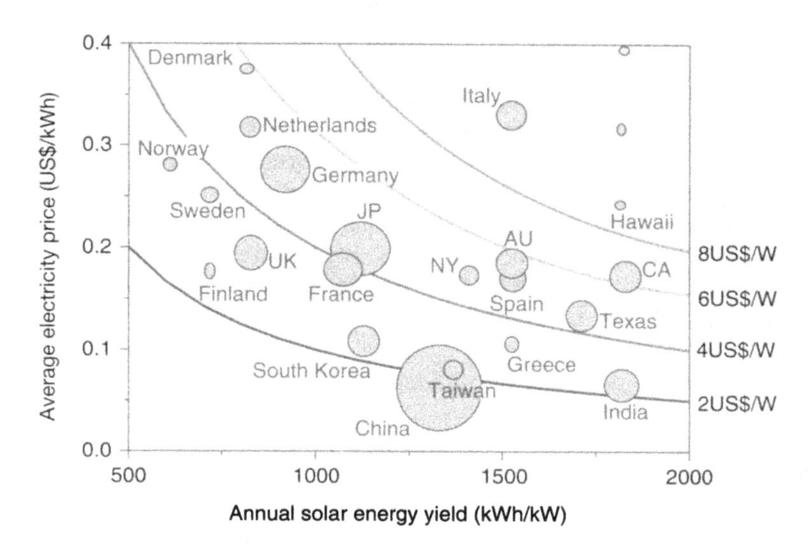

Figure 1.4. Relation of system cost with the average price of grid electricity (modified from [10]).

TABLE 1.2. Cost of solar electricity versus effective hours of sunshine per day for system cost of US$3.00/Wp used for 20 years

Effective hours of sunshine per day (hours)	Cost of solar electricity (US$)
2.5	0.1644
3.0	0.1370
3.5	0.1174
4.0	0.1027
4.5	0.0913
5.0	0.0822
6.0	0.0685

1.3.4 Cost Analysis—Breakdown of System Cost

The cost of a solar system includes the raw material, wafer formation, cell fabrication, lamination of modules, electrical accessories, and the balance of the system. Here we take crystalline-Si solar cells as an example for evaluation, which can be easily compared with other types of solar cells. The cost of raw materials for Si is typically no lower than US$25.0/kg. With 180 μm of the wafer thickness and 130 μm thickness of cutting loss, each wafer will consume 310-μm-thick material. Then the material cost converts to US$34.6/m^2. Wafer formation is US$0.70/piece of 6-inch wafer, which converts to US$30.0/m^2. Cell fabrication is about US$0.80/piece of 6-inch wafer, equivalent to US$34.7/m^2. The cost of lamination for modules is approximately US$70.0/m^2. Therefore, the above costs add up to give the overall module cost of US$169.3/m^2, named cost I here.

The electrical accessories include the inverter, mounting cable, meter, and transformer, which add up to give about US$1.0/$W_p$, named cost II. The balance of the system consists of labor cost, EPC (engineering, procurement, and construction) cost, fencing, and land cost, which sums to US$50/m^2 approximately, named cost III. Cost I and cost III are given in units of USD/m^2. They depend on the efficiency after being converted to USD/W_p. For efficiency of η, each square meter will generate p W of power if the intensity of sunlight is 1 kW/m^2, where $p = 1000\ \text{W} \times \eta \times 100\%$. The final cost in units of USD/W_p is hence given by

$$(\text{cost I} + \text{cost III})/p + \text{cost II}\ (= 1.0) \tag{1-2}$$

If the module efficiency is 16%, each square meter will generate 160 W of power, so the total cost in units of USD/W_p will be (cost I + cost III)/160 + 1.0 = 2.37. Table 1-3 lists the total cost of solar system in units of USD/W_p at different module efficiencies for crystalline-Si solar cells.

For other types of solar cells, cost I could be significantly different. For example, the module cost of the amorphous-Si (a-Si) solar cells is expected to be no more than

TABLE 1.3. Cost of solar system at different module efficiencies for crystalline-Si solar cells

Efficiency (%)	Cost of solar system (US$/$W_p$)
10	3.19
11	2.99
12	2.83
13	2.69
14	2.57
15	2.46
16	2.37
20	2.10

US$65/$m^2$. Then the cost of the solar system could be further reduced. With 12% efficiency of the a-Si solar-cell module, (cost I + cost III)/p + 1.0 = 1.958, and the cost of solar systems will be less than US$2.00/$W_p$. For printed solar cells such as organic or organic–inorganic hybrid solar cells, cost I will be further reduced because much lower material and equipment costs are expected, less than US$35.0/$m^2$. If the solar-cell module has 10% efficiency, (cost I + cost III)/p + cost II = 1.85. Thus, the cost of the solar system is expected to be less than US$1.85/$W_p$. According to Figure 1.4, this cost will make the solar electricity cheaper than the grid electricity in most countries.

In addition to cost I, cost II should be expected to further decrease as mass production of the inverters, mounting cables, meters, and transformers is achieved and cheaper ways of fabrication are invented. Probably, only cost III will not change much with time.

1.3.5 Forecast and Practical Trends

Solar-cell development is actually progressing faster than had been forecast. Before 2007, most people agreed with the prediction of McKinsey & Company that solar electricity in Japan would have the same price as grid electricity in 2020 [10]. However, the progress has moved ahead by about 5–10 years. Now, more people believe that the goal, predicted by Fuji Keizai, for the price of solar electricity to be comparable to the grid price can be achieved by 2015 [11]. Figure 1.5 shows that the price of a solar-cell module dropped significantly during 2006–2010, so the original prediction has changed. The movement is 5–10 years ahead of the prior forecast. CdTe, CIGS, a-Si, and multicrystalline Si solar cells all have their module prices falling very fast. Organic and organic–inorganic hybrid solar cells are not included in Figure 1.5, but their module price is expected to be lower than US$0.35/$W_p$ in the future.

Grid parity can be achieved first in areas that have abundant sunshine and high prices of grid electricity, such as California and Hawaii. Grid parity has been reached in Hawaii because of the high price of grid electricity. The United States had set 2015 as the year for grid parity [12, 13]. The Chief Engineer of General Electric also predict-

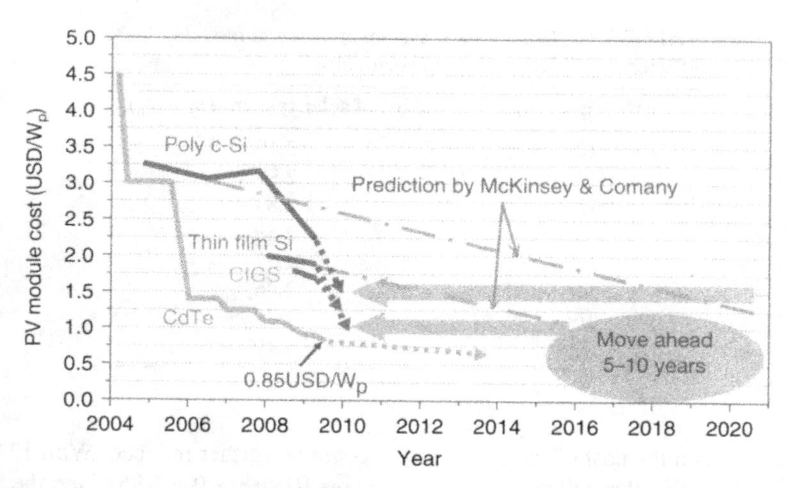

Figure 1.5. Cost of solar cells is decreasing faster than predicted. The progress has moved ahead by 5–10 years. The figure shows the trend of the fabrication cost of PV modules. CdTe solar cells defeated others by 2009, due to their low cost [11].

ed grid parity without government subsidies in sunny regions of the United States by 2015 approximately, as shown in Figure 1.6. Because the cost of solar electricity decreases very fast, while the price of grid electricity gradually increases, both values will meet around 2015 to give grid parity [14]. In brief, the progress of solar energy is moving forward at a fast pace. We should be optimistic about the replacement of fossil energy by solar energy in the near future.

1.4 MOVE TOWARD THIN-FILM SOLAR CELLS

From the above analysis, we know that the cost of crystalline-Si solar cells is higher than that of thin-film solar cells. The main reason is that the material consumption of crystalline-Si solar cells is large. Many types of thin-film solar cells have been developed. They can be categorized into two areas: inorganic semiconductor and organic ones.

1.4.1 Inorganic Versus Organic

The inorganic thin-film solar cells include the most popular a-Si solar cells, nanocrystalline or microcrystalline Si solar cells, and other compound semiconductor thin-film solar cells that combine several chemical elements, such as group I elements (Cu, Ag, Au), group III elements (Al, Ga, In), and group VI elements (S, Se, Te). Recently, CdTe thin-film solar cells have also proved to exhibit high efficiency and low-cost production. No matter how many chemical elements are involved in the inorganic semiconductor solar cells, a p–n junction is usually required. Their working principles will

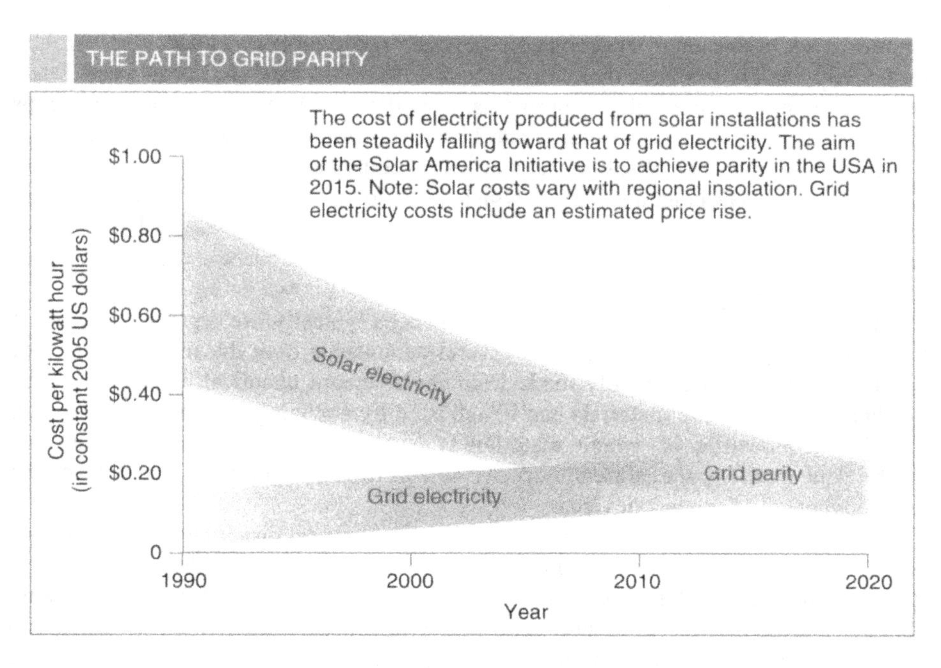

Figure 1.6. Cost of solar electricity will decrease rapidly and the price of grid electricity will gradually increase. Both values will meet around 2015 to achieve grid parity [14].

be described in later chapters. Briefly, the semiconductor thin-film solar cells operate similarly to the crystalline-Si solar cells except that materials with direct-bandgap properties are usually chosen because direct-bandgap materials have a larger absorption coefficient than indirect-bandgap ones like Si. However, the thin-film semiconductors are usually not single crystals. Many grain boundaries and voids are formed, leading to significant defects and surface states, so there are lots of recombination centers to reduce the extraction of carriers to external electrodes. It is always a challenging task to reduce the grain boundaries and voids.

Most of the thin-film semiconductors are deposited in vacuum chambers. In particular, the commercial modules of thin-film semiconductor solar cells are fabricated with entire large panels placed in vacuum chambers, which are thus very large. Solution processes with printing capability are also under development for easy fabrication and possibility of roll-to-roll production. Currently, CIGS thin-film solar cells and CdTe thin-film solar cells have attracted significant attention due to their high efficiencies and low costs. However, they will face the challenge of material supply as the solar power exceeds 100 GW [15]. Both In and Te are rare-earth elements. Shortages will develop as their demand continues to grow.

The organic solar cells are mainly divided into two types: dye-sensitized solar cells (DSSCs) and conventional organic solar cells. The dye-sensitized solar cells have proved to exhibit high efficiency (more than 10%) and low-cost production. In particular, they work in low-light conditions, so they are very suitable for cloudy weather. Unfortunately, the DSSCs have three major drawbacks. First, they use liquid electrolytes, which have temperature stability problems. In particular, when the temperature drops to 0°C, the liquid electrolyte freezes and becomes solid-state. Then the cell does not work and the expansion of solid-state electrolyte could cause damage to the cell. Second, the electrolyte contains iodine, which is toxic, so the cell has to be carefully sealed, leading to increased cost. Third, the most efficient DSSC uses a dye that contains ruthenium (Ru), a rare-earth element. When solar cells become the major supply of electricity, shortages of Ru will develop.

In comparison, conventional organic solar cells typically use organic materials in the light-absorption layer. They have several advantages over the inorganic ones or DSSCs: (1) they can be mostly made from the abundant chemical elements, (2) the bandgap of the organic materials can be adjusted by changing the chemical formula, and (3) the deposition of organic materials is usually done by a solution process and can be done in air, so the fabrication cost is lower. Roll-to-roll fabrication is deemed a great advantage of organic devices.

On the other hand, organic solar cells also have several disadvantages: (1) the organic semiconductors have low mobility, so most of organic materials with a film thickness more than 100 nm may lead to high series resistance, significantly decreasing the device performance; (2) organic semiconductors have a short exciton-diffusion distance, so complicated nanomorphology is required to assure efficient separation of electrons and holes, making the processing conditions very stringent and challenging; (3) organic materials usually degrade rapidly in air, so rigorous encapsulation is required, which may give rise to additional cost.

According to the above discussion, both inorganic semiconductor thin-film solar cells and organic solar cells have their own advantages and disadvantages. It would be great if both advantages could be utilized simultaneously while avoiding the disadvantages. Therefore, this book will particularly address the organic-inorganic hybrid solar cells.

With the combination of organic and inorganic semiconductors, the hybrid solar cells are expected to have the following major benefits: (1) improved stability, much better than organic solar cells; (2) solution process or good flexibility, so future development of roll-to-roll production can be easily adapted; (3) good conduction paths for collected carriers; (4) both organic and inorganic materials could be used for light absorption, increasing the choice of materials for light harvesting.

1.4.2 More Possible Applications

In the discussion in Section 1.3, we mainly focused on the possibility of replacing fossil energy with the solar energy from the viewpoint of power plants. However, there are also applications of solar cells that are not directly used in power plants. A few possibilities are listed below:

1. Consumer products such as calculators, watches, toys, cell-phone chargers, and notebook chargers

2. Power systems for remote villages and rural electrification
3. Telecommunications in remote areas and remote monitoring systems
4. Remote lighting lamps for houses
5. Electric fences
6. Water pumping powered by solar cells
7. Power systems for fish ponds
8. Emergency power systems
9. Portable power supplies for camping
10. Satellites and space vehicles

The above listed are niche applications that do not need very cheap solar cells. As the cost becomes very low, more applications will emerge. For example, one possible application is to place the solar panels on the outside wall of a building, like tile decoration. The entire wall area could be easily over 100 m^2. Even if the efficiency is only 3%, the overall power generated could be around 3 kW, which will give about 10 kWh of energy each day and 300 kWh each month. Another example could be use as a curtain or wallpaper if the solar cells are flexible. The overall area could also be easily more than 100 m^2. Then, even with low efficiency, probably over 10% of household electricity could be generated in this way. The key points are low cost and easy installation, so people would put up the solar cells like a curtain or wallpaper. If a city has 100,000 homes and offices generating solar electricity in this way, there will be 300 MW of solar power, which is equivalent to the power from a power plant. Therefore, a city that used to consume lots of energy could become a plant generating environmentally friendly power. Here we only consider 3% efficiency. When solar cells of higher efficiency and lower cost are developed, the generated power will be even more significant.

1.5 OUTLINE OF THIS BOOK

After the introduction in this chapter, Chapter 2, entitled "Light and Its Interaction with Matter," will discuss light properties and its interaction with materials. The different aspects of light will be introduced—rays, waves, and particles—followed by the physics of black-body radiation and the characteristics of solar light. Then the basic physics of light–matter interaction is described and discussed in detail, including reflection, transmission, dispersion, anisotropy, scattering, nonlinear optics, and light absorption. The theoretical derivation of some phenomena will also be given.

In Chapter 3, "Inorganic Materials," the properties of inorganic materials are introduced. The focus will be on semiconductors which are the most common materials for inorganic solar cells. The physics of band gap, conduction band and valence band, p-type semiconductors, n-type semiconductors, the formation of the p–n junction, carrier diffusion and drift, and light absorption in semiconductors will be discussed.

In Chapter 4, "Organic Materials," we will discuss the formation of small organic molecules as well as large molecules and how their chemical structures are related to their physical properties, with emphasis on the electrical and optical properties of the organic materials.

In Chapter 5, "Interface between Organic and Inorganic Materials," some issues in combining organic and inorganic materials are discussed, including compatibility of deposition, adhesion problems, formation of surface states, and band-level realignment.

Once the basic knowledge of inorganic semiconductors and organic semiconductors as well as their interface is established, we are ready to examine the device characteristics. Therefore, Chapter 6 will start with a discussion of inorganic semiconductor solar cells. Inorganic semiconductor solar cells are the most common devices used commercially nowadays. Therefore, this chapter will introduce the functioning principles of those solar cells and also provide the basic knowledge for comparison with organic solar cells and organic–inorganic hybrid solar cells. The reasons for their high production cost will also be addressed, so readers can realize why the future trend will shift toward organic solar cells and organic-inorganic hybrid, thin-film solar cells.

Chapter 7 will address organic solar cells. The driving force for the organic solar cell is the prospect of having very low cost solar cells ($<$US\$0.5/W_p) due to the use of less materials compared to silicon solar cells (submicron thickness versus 100 micron thickness) and the ease of using low energy consumption fabrication processes such as printing and dip coating. At present, there are four kinds of organic solar cells under development: (1) dye-sensitized, (2) organic–molecule, (3) polymer–fullerene, and (4) polymer–semiconductor nanoparticle. The principle and performance of each type of organic solar cell is not the same and will be discussed in detail in this chapter.

With knowledge of both inorganic and organic solar cells, we are now prepared to explore the organic–inorganic hybrid solar cells, which are described in Chapter 8. The fundamental concepts of forming organic–inorganic hybrid solar cells will first be introduced. Then focus will be brought to the solution-processed sandwiched structure, in which the organic layer is protected by two inorganic oxide layers. The technique to overcome deposition difficulty of the inorganic layer on the organic layer using the solution process will be addressed. Detailed investigation of the deposited oxides will be provided. Device characterization and performance of such thin-film solar cells will be described. In addition to using organic materials for light absorption and inorganic materials for carrier transportation, an alternative to using inorganic semiconductors as the light-absorption materials and organic polymers as the carrier–transportation layer will be discussed.

Finally, Chapter 9 will discuss how the future technology of solar cells will be developed and what can be expected of electricity generation from solar cells in the future.

REFERENCES

1. K. Ito (2000), "Prospect of Fossil Fuels," in *Energy for the 21st Century*, 3rd Annual Meeting Report on Nuclear Fusion.

2. IEA (2010), *CO_2 Emissions from Fuel Combustion: Highlights*, 2010 Edition, International Energy Agency (IEA), Paris.
3. M. Boselli (30 May 2011), "IEA Sees Record CO_2 Emissions in 2010," Reuters.
4. "2009 Ends Warmest Decade on Record," NASA Earth Observatory Image of the Day, January 22, 2010.
5. D. Fogalty (12 January 2011), "Scientists See Climate Change Link to Australian Floods," Reuters.
6. H. Chanson (2011), "The 2010–2011 Floods in Queensland (Australia): Observations, First Comments, and Personal Experience," *La Houille Blanche* (Paris: Societe Hydrotechnique de France) Vol. 1, 5–11, ISSN 0018-6368.
7. M. A. Green, K. Emery, Y. Hishikawa, W. Warta, and E. D. Dunlop (2011), "Solar Cell Efficiency Tables (Version 38)," *Prog. Photovolt: Res. Appl.,* Vol. 19, 565–572.
8. M. A. Green (2003), *Third Generation Photovoltaics: Advanced Solar Energy Conversion*, Springer-Verlag, Berlin–Heidelberg.
9. A. Slade and V. Garboushian (2005), in "27.6% Efficient Silicon Concentrator Cell for Mass Production," *Technical Digest, 15th International Photovoltaic Science and Engineering Conference,* Shanghai, October, p. 701.
10. P. Lorenz, D. Pinner, and T. Seitz (2008), "The Economics of Solar Power," *McKinsey Quarterly,* June.
11. *Nikkei Electronics,* (2010), Taiwan Edition, April, pp. 11–18.
12. BP Global—Reports and Publications, "Going for Grid Parity," Bp.com, http://www.bp.com/genericarticle.do?categoryId=9013609&contentId=7005395. Retrieved on August 2, 2011.
13. BP Global—Reports and Publications (2011), "Gaining on the Grid," Bp.com., http://www.bp.com/sectiongenericarticle.do?categoryId=9019305&contentId=7035199. Retrieved on August 2, 2011.
14. The Path to Grid Parity (Graphic) (2011), http://www.bp.com/popupimage.do?img_path=liveassets/bp_internet/globalbp/globalbp_uk_english/reports_and_publications/frontiers/STAGING/local_assets/images/fr19solar_parity570x417.jpg%20&alt_tag=Graphic%20about%20grid%20parity,%20when%20the%20cost%20of%20solar%20energy%20equals%20that%20of%20grid%20electricity. Retrieved on August 2, 2011.
15. M. Konagai (2011), "Thin-Film Photovoltaics: An Overview," in *26th European Photovoltaic Solar Energy Conference and Exhibition*, Hamburg, Germany, September, 3CP.1.1.

EXERCISES

1. What problems did fossil fuels bring about after the Industrial Revolution?
2. What event caused Japan to give up the plan of building future nuclear power plants?
3. How much power does the earth receive from the sun each day? How much power do human beings consume each day?
4. To supply the overall power demand of United States, how much area of the land is required? (Assume that the United States consumes one-fifth of the total power

worldwide and that solar cells have 15% efficiency. The intensity of sunlight is considered to be 1 kW/m^2.)

5. List the types of solar cells that you know of.
6. What issues are important for solar cells?
7. How does efficiency influence the cost of solar cells?
8. How does lifetime influence the cost of solar cells?
9. List the costs of crystalline-Si solar cells.
10. What cost will be the same for most types of solar cells when they are used in a solar power plant?
11. What is grid parity?
12. To have grid parity in California, what is the cost of a solar system?
13. Compare Italy and Netherlands. Which country will more possibly use solar electricity? Why? (Provide scientific data.)
14. In a country with 5 effective hours of sunshine per day, what is the cost of solar electricity (in US$/kWh) if this country establishes a power plant costing US$4.00/$W_p$?
15. What are the advantages and disadvantages of inorganic solar cells?
16. What are the advantages and disadvantages of organic solar cells?
17. What are the major drawbacks of dye-sensitized solar cells?
18. What benefits are expected from hybrid solar cells?
19. Write down all the possible applications of solar cells that you can imagine.
20. For small projects: (1) estimate the area required to supply 50% of power consumption from solar-cell power plants in your country; (2) evaluate the cost of solar systems in your country to achieve grid parity, in the units of US$/$W_p$; (3) estimate how many years are required to achieve grid parity in your country or your state/province.
21. Explain why solar energy will be needed or not needed.

2

LIGHT AND ITS INTERACTION WITH MATTER

Ching-Fu Lin

2.1 WHAT IS LIGHT?

Light, air, and water are necessary for life to exist. However, it took human beings thousands of years to understand what light is. After many years of controversial arguments among philosophers and scientists, now there are generally three ways to interpret light: light ray, light wave, and photon.

2.1.1 Light Ray

Although a light ray is not real, it provides a good way to simplify the mathematical treatment of light. Many optical systems can be designed using ray-tracing techniques, in which a trace of light is evaluated using the simple rules of light rays. Those rules include: (1) the light ray is a straight line if the light is in a homogeneous medium; (2) if the light ray hits an abrupt interface, it will either be reflected or transmitted, following the laws of reflection and transmission. Therefore, the traces can be predicted mostly from the geometrical shapes of the interfaces between objects.

2.1.2 Light as a Wave

The view of light as a wave indicates that light has oscillating properties. As a matter of fact, light is an electromagnetic wave, in which the electric field and magnetic field are oscillating with time. To be more specific, the electric field and magnetic field obey Maxwell's equations, which consist of four equations: Faraday's law, Ampere's

Organic, Inorganic, and Hybrid Solar Cells. By C.-F. Lin, W.-F. Su, C.-I Wu, and I-C. Cheng

law, Gauss's law for electric fields, and Gauss's law for magnetic fields. They are given in the following formulae:

$$\text{Faraday's law: } \nabla \times \vec{E} + \frac{\partial \vec{B}}{\partial t} = 0 \tag{2-1a}$$

$$\text{Ampere's law: } \nabla \times \vec{H} - \frac{\partial \vec{D}}{\partial t} = \vec{J} \tag{2-1b}$$

$$\text{Gauss's law for electric fields: } \nabla \bullet \vec{D} = \rho \tag{2-1c}$$

$$\text{Gauss's law for magnetic fields: } \nabla \bullet \vec{B} = 0 \tag{2-1d}$$

Faraday's law says that the variation of magnetic field $\vec{B}$ over time leads to the generation of electric field $\vec{E}$. Ampere's law describes the fact that a flow of current results in magnetic field $\vec{H}$. In addition, the variation of electric field $\vec{D}$ over time can generate magnetic field $\vec{H}$. Gauss's law for electric fields states that an electrical charge ρ generates electric field $\vec{D}$ around it. For reasons of symmetry, Gauss's law for electric fields may also be written for magnetic fields. Because there is no magnetic charge or monopole, the right-hand side of Equation (2-1d) is equal to zero. This means that the existence of the magnetic field is not due to a charge, but due to the current flow or variation of the electric field, as given by Ampere's law.

The two electric fields $\vec{D}$ and $\vec{E}$ are related by the following equation:

$$\vec{D} = \varepsilon \vec{E} \tag{2-2}$$

where ε is called the dielectric constant or permittivity.

The two magnetic fields $\vec{B}$ and $\vec{H}$ are likewise related by a similar equation:

$$\vec{B} = \mu \vec{H} \tag{2-3}$$

where μ is called magnetic permeability.

The reason for having two types of electric fields and magnetic fields is due to the interaction of light (as an electromagnetic wave) with materials. The details of light–matter interaction will be explained later in this chapter. In a vacuum, there is no interaction between light and material, so the relation between the two types of fields is very simple. Field $\vec{D}$ is proportional to field $\vec{E}$ and field $\vec{B}$ is also proportional to field $\vec{H}$. They are related by the following equations:

$$\vec{D} = \varepsilon_0 \vec{E} \tag{2-4}$$

$$\vec{B} = \mu_0 \vec{H} \tag{2-5}$$

ε_0 and μ_0 in a vacuum are just constants. Their values are given below.

$$\varepsilon_0 = \frac{1}{36\pi} \times 10^{-9} \frac{A \bullet s}{V \bullet m}$$

$$\mu_0 = 4\pi \times 10^{-7} \frac{V \bullet s}{A \bullet m}$$

With some mathematical manipulation, a wave equation can be derived from Maxwell's equations. In a homogeneous medium, the wave equation is very simple, given below for the electric filed and magnetic field, respectively.

$$\nabla^2 \vec{E} - \mu\varepsilon \frac{\partial^2 \vec{E}}{\partial t^2} = 0 \tag{2-6}$$

$$\nabla^2 \vec{H} - \mu\varepsilon \frac{\partial^2 \vec{H}}{\partial t^2} = 0 \tag{2-7}$$

2.1.3 Plane-Wave Solution of the Wave Equation

Because the above two equations have the exact same form, their solutions also have the same mathematical formula. Many mathematical functions can satisfy the wave equations (2-6) and (2-7). One of the simplest forms is as follows:

$$\phi(\vec{r}, t) = Ae^{j(\omega t - \vec{r} \bullet \vec{k})} \tag{2-8}$$

Where $\varphi(\vec{r}, t)$represents the magnitude of either the electric field $\vec{E}$ or the magnetic field $\vec{H}$, $\vec{r}$ is the spatial coordinate, and $\vec{k}$ is the wave vector with magnitude given by

$$|\vec{k}| \equiv k = \omega\sqrt{\mu\varepsilon} \tag{2-9}$$

Being a wave, it has two important properties: oscillating frequency and wavelength. The oscillating frequency and wavelength are usually represented by ν and λ, respectively. ν is related to ω by $\omega = 2\pi\nu$. The product of ν and λ is the speed of light:

$$\nu \times \lambda = \frac{1}{\sqrt{\mu\varepsilon}} = \frac{c}{n} \tag{2-10}$$

Because the speed of light in a medium is constant, the larger the frequency is, the shorter the wavelength is.

In a vacuum, the speed is equal to $1/\sqrt{\mu_0\varepsilon_0}$, which has a constant value of 3×10^{10} cm/s. When transmitted in other media, the propagation speed, $1/\sqrt{\mu\varepsilon}$, is less than $1/\sqrt{\mu_0\varepsilon_0}$. The ratio of light speed in a vacuum to light speed in a material is defined as the refractive index of this material:

n (refractive index) = light speed in vacuum/light speed in material (2-11)

$$= \frac{\sqrt{\mu\varepsilon}}{\sqrt{\mu_0\varepsilon_0}}$$

The refractive indices of some materials are given in Table 2-1.

2.1.4 Light as a Particle

In the beginning of the twentieth century, Einstein and Compton discovered that light also exhibits the property of a particle. That is, light has behaviors similar to a particle. When behaving like a particle, light is called a photon. A photon has an energy E equal to $h\nu$, where h is the Plank constant and ν is the frequency of light. Because the product of frequency and wavelength is equal to the speed of light, the energy of a photon is related to the wavelength by the following equation:

$$E = h\frac{c}{\lambda} \tag{2-12}$$

Substituting the values of the Plank constant h and the speed of light c into the above equation, we have the following simple formula to calculate the photon energy directly from the wavelength of light:

$$E = \frac{1.24}{\lambda}(eV) \tag{2-13}$$

The wavelength λ for the above formula is in the unit of μm. For example, if the wavelength is 1.55 μm, the photon energy is 0.8 eV. (1 eV = 1.6×10^{-19} joule) If the wavelength is 0.65 μm, then the photon energy is 1.91 eV. It is very clear that the shorter the wavelength of the light, the larger the energy the photon has.

From the viewpoint of the particle nature of light, the energy of light does not vary continuously. Instead, the energy is a multiple of a single photon. That is, $E = nh\nu$,

TABLE 2-1. Refractive index of some materials.

Air	n = 1.000278
Water	n =1.33
Fused silica	n =1.46
Silcon	n =3.5
GaAs	n =3.5
Crystal quartz	n =1.55
Optical glass	n =1.51–1.81
Sapphire	n =1.77
Diamond	n =2.43

where n is a positive integer. For example, assuming that the total energy of light is 10 eV, this light could consist of 10 photons. Each photon has 1 eV of energy. This light could also consist of a single photon with energy of 10 eV. However, it is not possible that this light consists of a half photon of 20 eV. In brief, the particle nature of light tells us that the appearance or disappearance of light is in units of photons.

The particle point of view is particularly important for light absorption into or emission from materials. If the energy of light is absorbed by the material, this energy is always a multiple of photon energy ($h\nu$). Similarly, when materials emit light (like lasers or other light sources), the emitted energy is also a multiple of photon energy ($h\nu$).

2.1.5 Blackbody Radiation and Solar Spectrum

The concept of the photon is very important. It well explains the light–matter interaction with energy exchange, as will be detailed later in this chapter. One of the most significant achievements of the photon concept is the prediction of blackbody radiation. In thermal equilibrium, a blackbody sends out an amount of energy equal to the amount of energy it absorbs, so it remains at a constant temperature. The thermal radiation is the most common form that the blackbody sends out at any temperature greater than absolute zero. This thermal radiation represents the energy that the blackbody converts to electromagnetic radiation. In 1901, Max Planck assumed that the energy of the oscillators in a cavity were quantized. The quantization of the energy of the electromagnetic wave is now known as photons. Applying the photon energy of $E = h\nu$ and the photon statistics, Planck's law of blackbody radiation can be derived as

$$I(\nu) = \frac{2h\nu^3}{c^2}\frac{1}{e^{h\nu/kT}-1} \tag{2-14a}$$

where $I(\nu)$ is the power radiated per unit area of emitting surface in the normal direction per unit solid angle per unit frequency by a blackbody at temperature T, h is the Planck constant, c is the speed of light in a vacuum, k is the Boltzmann constant, ν is the frequency of the electromagnetic radiation, and T is the temperature of the body in kelvins. The formula in Eq. (2-14a) is expressed in terms of frequency. It can be transformed to another one in terms of wavelength:

$$I(\lambda) = \frac{2\pi hc^2}{\lambda^5}\frac{1}{e^{hc/\lambda kT}-1} \tag{2-14b}$$

where $I(\lambda)$ is the power per unit area of emitting surface in the normal direction and per unit wavelength. The calculation will be straightforward if MKS units are used. Eq. (2-14b) is more frequently used for spectral evaluation.

The frequency of maximum radiative power shifts to higher frequencies or shorter wavelengths as the temperature of the blackbody increases. At temperature of a few hundred degrees Celsius, the radiation enters the regime of visible wavelengths. It starts as red light. When the temperature further increases, it gradually changes to orange, yellow, and then white. When the blackbody appears visually white, it is emitting a substantial fraction of ultraviolet (UV) radiation. For temperature above 7000°C, most of the radiation is in the wavelength of less than 450 nm, so the radiation appears blue and the UV radiation increases.

The sunlight in our solar system has its spectrum very close to the blackbody radiation at 5800 K, which is deemed as the sun's surface temperature. The peak wavelength is about 500 nm. Because the major part of the solar emission is in the visible wavelengths, sunlight appears white. Figure 2.1 shows the blackbody radiation at different temperatures. The one at the 5800 K is most similar to the solar radiation spectrum. When sunlight passes through the atmosphere of the earth, some portion is absorbed by gases. Thus, the measured solar spectrum exhibits disappeared bands, which correspond to the absorption spectra of those gases. The most significant part of the absorption is caused by water vapor. The gases of oxygen and carbon dioxide also contribute to the absorption.

2.1.6 The Brightness and Intensity of Sunlight

Because the spectrum of sunlight covers a broad range of wavelengths, not all of them are visible. As a result, it has a luminous efficacy of about 93 lumens/W of radiant flux. If the sunlight is not blocked by clouds, the brightness of the sunlight is about 100,000 lumens per square meter or 100,000 lux at the earth surface. With clouds, the sunlight will experience multiple scatterings by the water droplets, so it is experienced as diffused light instead of direct sunshine.

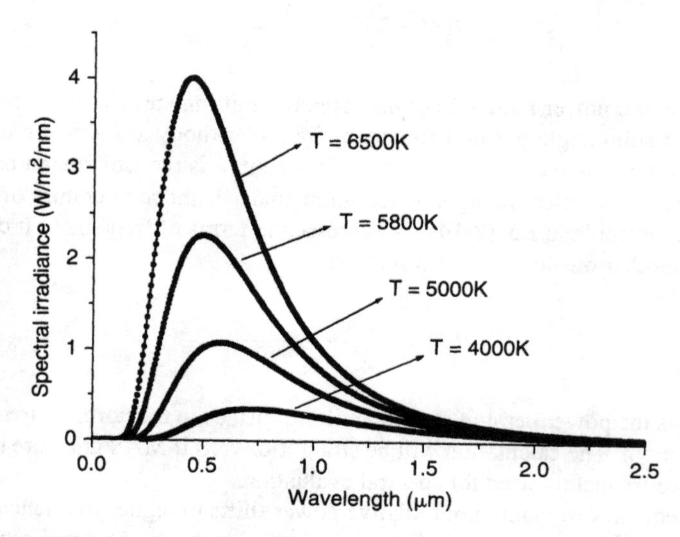

Figure 2.1. Blackbody radiation at different temperatures.

Because the sun is the radiation source, the radiation intensity being experienced by an observer is inversely proportional to the square of the distance from the source. Therefore, the intensity of sunlight experienced on Venus is greater than on the earth. On Venus, the intensity of sunlight is between 2576 and 2647 W/m^2. On the earth, the intensity is between 1321 and 1413 W/m^2. This intensity does not take into account the absorption and scattering by the atmosphere. Typically, the solar intensity right outside of the earth's atmosphere when the sun is at its mean distance from the earth is taken as 1370 W/m^2. The atmosphere influences sunlight through the processes of scattering, absorption, and reflection, so the intensity at the sea level on earth reduces to a value of about 1000 W/m^2. In good weather under a very clear sky, the intensity of direct sunshine could be slightly larger than 1000 W/m^2. In a wet climate, the intensity might be less than 200 W/m^2.

2.2 FUNDAMENTALS OF INTERACTION BETWEEN LIGHT AND MATTER

The interaction of light with materials can be divided into two types: with energy exchange and without energy exchange. We will discuss the interaction without energy exchange first. Light is an electromagnetic wave with wavelength near or within the visible spectrum, whereas an electromagnetic wave is the periodic variation of electromagnetic fields with time and space. According to Maxwell's equations, electromagnetic fields interact with charged particles. Because the materials are composed of atoms, which consist of positively charged nuclei and negatively charged electrons, the electromagnetic wave certainly interacts with materials. For example, a positive charge q under an electric field will experience a force

$$\vec{F} = q\vec{E} \tag{2-15}$$

As a result, this positive charge will be forced to move in the direction of the electric field if it is initially at a static position. Similarly, a negative charge will move in an opposite direction. All materials consist of charged particles, so they all interact with electromagnetic waves.

The interactions between light and matter without energy exchange can be categorized into two types and described using the following two equations:

$$\vec{D} = \varepsilon\vec{E} = \varepsilon_0\vec{E} + \vec{P} \tag{2-16}$$

$$\vec{B} = \mu\vec{H} = \mu_0\vec{H} + \vec{M} \tag{2-17}$$

The above equations are called material equations or constitutive equations. They will be explained separately below.

2.2.1 Interaction of Electric Fields with Dielectrics

This interaction is described by $\vec{D} = \varepsilon\vec{E} = \varepsilon_0\vec{E} + \vec{P}$, where ε and ε_0 are dielectric constants in the material or a vacuum, respectively. Here $\vec{P}$ is called polarization, which can be further described by

$$\vec{P} = \varepsilon_0\chi_e\vec{E} \tag{2-18}$$

where χ_e is the electric susceptibility. Equation (2-18) implies that the externally applied electric field $\vec{E}$ causes the polarization. Polarization is caused by the separation of electrical charges with opposite signs. The materials are typically composed of atoms, which consist of positively charged nuclei and negatively charged electrons. Without the external field, the center of negative charges is located at the same position as the positively charged nuclei. As a result, there is no effective separation of the positive and negative charges. When material is under an external field, the applied electric field pulls positive and negative components in opposite directions. Then, as shown in Figure 2.2, the positive nuclei and the center of negative charges are not at the same location. Consequently, the local separation of positive and negative charges forms a dipole. Polarization is a collection of many such dipoles in the material. This dipole then generates an additional electric field. Thus, $\vec{D}$ field is the sum of the externally applied field $\varepsilon_0\vec{E}$ and the resulting polarization field $\vec{P}$.

Further substituting Eq. (2-18) into Eq. (2-16), we obtain

$$\vec{D} = \varepsilon\vec{E} = \varepsilon_0\vec{E} + \varepsilon_0\chi_e\vec{E} = \varepsilon_0(1+\chi_e)\vec{E} \tag{2-19}$$

Thus,

$$\varepsilon = \varepsilon_0(1+\chi_e) \tag{2-20}$$

Equation (2-20) reveals the relation between the dielectric constants ε and ε_0, and the electric susceptibility χ_e.

2.2.2 Interaction of Light with Magnetic Materials

This interaction is described by $\vec{B} = \mu\vec{H} = \mu_0\vec{H} + \vec{M}$, where μ and μ_0 are magnetic permeability in the material or a vacuum, respectively. Similar to the interaction of mater-

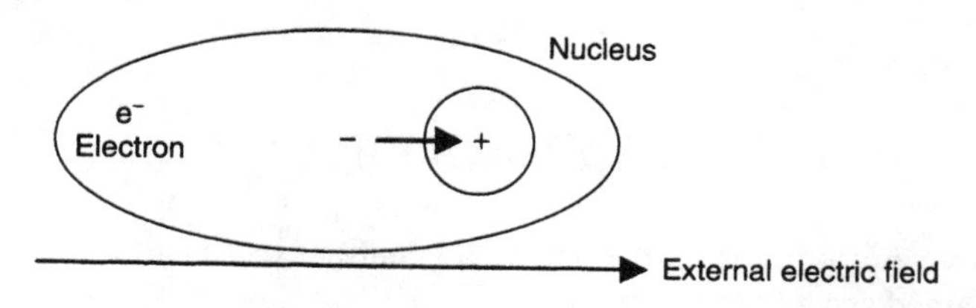

Figure 2.2. A schematic of an atom under an applied electric field. A dipole is formed.

ial with an electric field, some materials interact with magnetic fields. A magnetic field can cause the magnetization of some materials, which is described by the symbol $\vec{M}$. $\vec{M}$ can be further described by $\vec{M} = \chi_m \vec{H}$, where χ_m is the magnetic susceptibility. If $\chi_m = 0$, the material is not magnetic, meaning that it cannot interact with the magnetic field or its property cannot be changed by the applied magnetic field. If $\chi_m \neq 0$, then the material is magnetic and its property can be changed by the applied magnetic field.

2.2.3 Summary of Light–Matter Interaction Without Energy Exchange

Most materials have $\chi_m = 0$, so magnetic fields have no influence on them. However, all materials have nonzero electric susceptibility, that is, $\chi_e \neq 0$, so all materials have interaction with electric fields. Because an electromagnetic wave has both electric and magnetic fields, all materials will certainly interact with electromagnetic fields, primarily through the electric-field-induced polarization effect.

According to the above discussion, we know that Eq. (2-16) is the most common situation for light–matter interaction. Such interaction will lead to the change of the dielectric constant from its vacuum value of ε_0 to ε. In such interactions, there is no exchange of energy between the material and light. When the energy of light is neither absorbed nor amplified by the materials, we called such materials "transparent materials."

2.3 BASIC PROPERTIES OF TRANSPARENT MATERIALS

Even though the interaction between the transparent material and light exhibits no energy exchange, many important phenomena are still involved. Those phenomena will be discussed in the following sections.

2.3.1 Reflection and Refraction

Reflection and refraction are very common phenomena that can be seen in our daily lives. For example, when we look into the mirror, our eyes receive the light reflected from the mirror. Another common example is when we see a fish in water, we actually see the refracted light coming from the water. Reflection and refraction occur at the boundary of different media. When light propagates from one medium to another, some portion of light is reflected and some is transmitted. These phenomena can be explained using either ray optics or wave optics. Wave optics can explain not only the angles of reflection and refraction, but also the intensity ratio of reflection and transmission. Therefore, we will discuss reflection and refraction from the viewpoint of wave optics.

*2.3.1.1 **Boundary Conditions for Electric and Magnetic Fields.*** Because light waves are electromagnetic waves, they follow the boundary conditions for electric and magnetic fields. Referring to Figure 2.3, an interface exists between medium 1 and

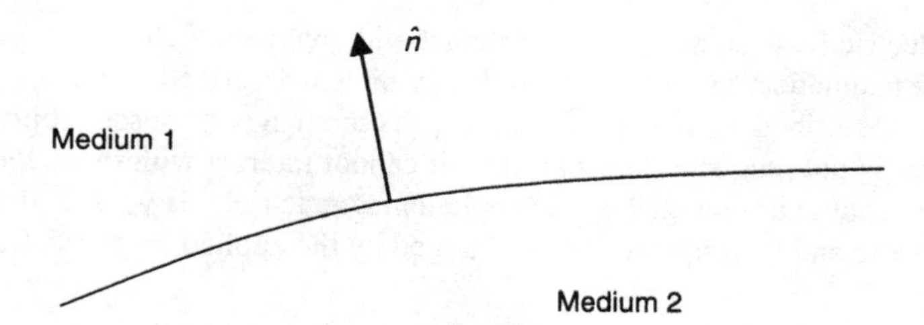

Figure 2.3. Interface of medium 1 and medium 2.

medium 2. An easy explanation follows. The fields at the interface of medium 1 and medium 2 are represented with subscripts 1 and 2, respectively. Starting from Maxwell's equations (Gauss's law for electric fields and Gauss's law for magnetic fields) and using vector analysis, we can obtain the following boundary conditions:

$$\hat{n} \bullet (\vec{B}_2 - \vec{B}_1) = 0 \qquad (2\text{-}21)$$

$$\hat{n} \bullet (\vec{D}_2 - \vec{D}_1) = \rho_s \qquad (2\text{-}22)$$

where $\hat{n}$ is the unit vector normal to the interface and ρ_s is the density of the surface charge (number of charges per unit area). For insulators, there is no surface charge, so $\rho_s = 0$. Then Eqs. (2-21) and (2-22) become

$$\hat{n} \bullet (\vec{B}_2 - \vec{B}_1) = B_{2n} - B_{1n} = 0 \qquad (2\text{-}23)$$

$$\hat{n} \bullet (\vec{D}_2 - \vec{D}_1) = D_{2n} - D_{1n} = 0 \qquad (2\text{-}24)$$

where B_{2n}, B_{1n}, D_{2n}, and D_{1n} are the projections of fields along the direction normal to the boundary. Equations (2-23) and (2-24) give the results $B_{2n} = B_{2n}$ and $D_{1n} = D_{2n}$, that is, the normal projections of the B field and D field have to be continuous even at the interface of two media if medium 1 and medium 2 are both insulators.

Not only the normal projections of the B field and D field, but also the tangential components of the E field and H field have to be continuous at the interface. This can be derived as follows. Starting from Maxwell's equations (Faraday's law and Ampere's law) and using vector analysis, we obtain the following boundary conditions:

$$\hat{n} \times (\vec{E}_2 - \vec{E}_1) = 0 \qquad (2\text{-}25)$$

$$\hat{n} \times (\vec{H}_2 - \vec{H}_1) = \vec{K} \qquad (2\text{-}26)$$

where $\vec{K}$ is the surface current density at the interface. Again, if both medium 1 and medium 2 are insulators, there is no surface current density, so $\vec{K} = 0$. Because the cross products of two vectors are perpendicular to the original two vectors, $\hat{n} \times \vec{E}_2$ points in the direction perpendicular to $\hat{n}$, which is normal to the interface boundary. Thus, $\hat{n} \times \vec{E}_2$ is the tangential direction along the interface. Then, for insulators, Eqs. (2-25) and (2-26) can be further written as

$$E_{2t} - E_{1t} = 0 \tag{2-27}$$

$$H_{2t} - H_{1t} = 0 \tag{2-28}$$

where E_{2t}, E_{1t}, H_{2t}, and H_{1t} are the tangential components of the E and H fields along the interface boundary.

The boundary conditions described by the above equations are as important as the wave equation when we are dealing with light propagation between materials. For solar cells, light has to enter the cell material from the air or vacuum, so it certainly has to follow the boundary conditions derived above.

2.3.1.2 Reflection and Transmission of Plane Waves. The solution of the plane wave equation (2-8) has the following two important characteristics. (1) It has a unique direction of propagation along the wave vector $\vec{k}$. (2) The positions of its constant phase form a plane. This constant phase plane is also called the wavefront. In other words, its phase front is a plane.

Due to the simple and clear characteristics of a plane wave, the reflection and transmission of a plane wave at the interface between two media is particularly simple. It gives the simplest and the most important concept for the understanding of light–matter interaction. The situation is illustrated in Figure 2.4.

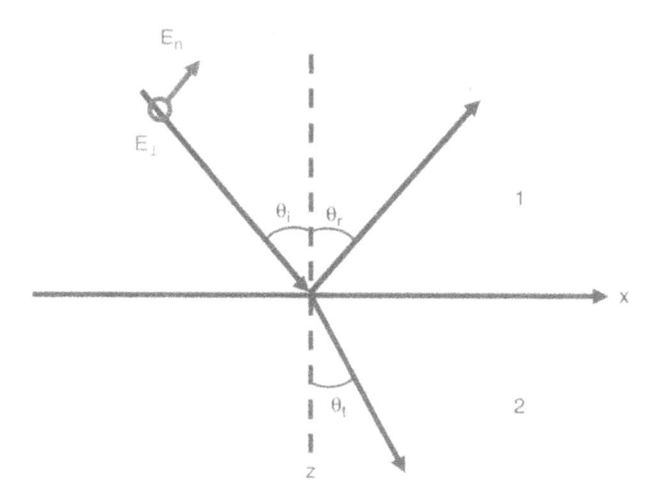

Figure 2.4. Reflection and transmission of a plane wave at a boundary located on the x–y plane.

Assume that the boundary of medium 1 and medium 2 is on the x–y plane, so the normal of the boundary is parallel to the z-axis. Figure 2.4 shows the x-axis and z-axis. The z-axis is assumed to be the normal of the boundary. A plane wave is incident on the boundary. Its direction of propagation, $\vec{k}_i$, is on the x–z plane. The incident angle, defined as the angle between the $\vec{k}_i$ vector and the normal of the boundary (z-axis), is θ_i. The electric field is perpendicular to the wave vector $\vec{k}_i$. For the convenience of mathematical manipulation, the incident electric field is decomposed into two components, $\vec{E}_{//}$ and $\vec{E}_\perp$ $\vec{E}_{//}$ is the component on the x–z plane, whereas $\vec{E}_\perp$ is the component parallel to the y-axis. Certainly, both $\vec{E}_{//}$ and $\vec{E}_\perp$ are perpendicular to the wave vector $\vec{k}_i$. This electric field can also be expressed using Cartesian coordinates, $\vec{E} = \hat{x}E_x + \hat{y}E_y + \hat{z}E_z$, with

$$E_x^{(i)} = E_{//}^{(i)} \cos\theta_i e^{j(\omega t - \vec{r}\bullet\vec{k}_i)} \tag{2-29a}$$

$$E_y^{(i)} = -E_\perp^{(i)} e^{j(\omega t - \vec{r}\bullet\vec{k}_i)} \tag{2-29b}$$

$$E_z^{(i)} = -E_{//}^{(i)} \sin\theta_i e^{j(\omega t - \vec{r}\bullet\vec{k}_i)} \tag{2-29c}$$

where the superscript (*i*) indicates the incident field. According to Maxwell's equations, (2-1a) and (2-1b), the magnetic field is related to the electric fields, so the x-, y-, and z-components of the magnetic field can be obtained:

$$H_y^{(i)} = \frac{\omega\varepsilon_1}{k} E_{//}^{(i)} e^{j(\omega t - \vec{r}\bullet\vec{k}_i)} \tag{2-30a}$$

$$H_y^{(i)} = \frac{\omega\varepsilon_1}{k} E_{//}^{(i)} e^{j(\omega t - \vec{r}\bullet\vec{k}_i)} \tag{2-30b}$$

$$H_z^{(i)} = -\frac{\omega\varepsilon_1}{k} E_\perp^{(i)} e^{j(\omega t - \vec{r}\bullet\vec{k}_i)} \tag{2-30c}$$

where ε_1 is the dielectric constant in medium 1.

Similarly, the electric field and magnetic field of the reflected wave are as follows:

$$E_x^{(r)} = E_{//}^{(r)} \cos\theta_r e^{j(\omega t - \vec{r}\bullet\vec{k}_r)} \tag{2-31a}$$

$$E_y^{(r)} = -E_\perp^{(r)} e^{j(\omega t - \vec{r}\bullet\vec{k}_r)} \tag{2-31b}$$

$$E_z^{(r)} = -E_{//}^{(r)} \sin\theta_r e^{j(\omega t - \vec{r}\bullet\vec{k}_r)} \tag{2-31c}$$

$$H_x^{(r)} = \frac{\omega\varepsilon_1}{k} E_\perp^{(r)} \cos\theta_i e^{j(\omega t - \bar{r}\bullet\bar{k}_r)} \tag{2-32a}$$

$$H_y^{(r)} = \frac{\omega\varepsilon_1}{k} E_{//}^{(r)} e^{j(\omega t - \bar{r}\bullet\bar{k}_r)} \tag{2-32b}$$

$$H_z^{(r)} = -\frac{\omega\varepsilon_1}{k} E_\perp^{(r)} e^{j(\omega t - \bar{r}\bullet\bar{k}_r)} \tag{2-32c}$$

where the superscript (r) indicates the reflected field. Also, the transmitted wave is as follows:

$$E_x^{(t)} = E_{//}^{(t)} \cos\theta_t e^{j(\omega t - \bar{r}\bullet\bar{k}_t)} \tag{2-33a}$$

$$E_y^{(t)} = -E_\perp^{(t)} e^{j(\omega t - \bar{r}\bullet\bar{k}_t)} \tag{2-33b}$$

$$E_z^{(t)} = -E_{//}^{(t)} \sin\theta_t e^{j(\omega t - \bar{r}\bullet\bar{k}_t)} \tag{2-33c}$$

$$H_x^{(t)} = \frac{\omega\varepsilon_2}{k} E_\perp^{(t)} \cos\theta_i e^{j(\omega t - \bar{r}\bullet\bar{k}_t)} \tag{2-34a}$$

$$H_x^{(t)} = \frac{\omega\varepsilon_2}{k} E_\perp^{(t)} \cos\theta_i e^{j(\omega t - \bar{r}\bullet\bar{k}_t)} \tag{2-34b}$$

$$H_x^{(t)} = \frac{\omega\varepsilon_2}{k} E_\perp^{(t)} \cos\theta_i e^{j(\omega t - \bar{r}\bullet\bar{k}_t)} \tag{2-34c}$$

where the superscript (t) indicates the transmitted field and ε_2 is the dielectric constant of medium 2.

From the boundary conditions expressed by Eqs. (2-27) and (2-28), the tangential components of electric field E and magnetic field H should be continuous, so

$$E_x^{(i)} + E_x^{(r)} = E_x^{(t)} \tag{2-35a}$$

$$E_y^{(i)} + E_y^{(r)} = E_y^{(t)} \tag{2-35b}$$

$$H_x^{(i)} + H_x^{(r)} = H_x^{(t)} \tag{2-35c}$$

$$H_y^{(i)} + H_y^{(r)} = H_y^{(t)} \tag{2-35d}$$

Substituting the electric fields and magnetic fields expressed in Eqs. (2-29)–(2-34) into Eqs. (2-35a)–(2-35d), we obtain the following results:

1. $$\vec{r} \bullet \vec{k}_i = \vec{r} \bullet \vec{k}_r \Rightarrow xk_{ix} + yk_{iy} + zk_{iz} = xk_{rx} + yk_{ry} + zk_{rz} \quad (2\text{-}36a)$$

$$\vec{r} \bullet \vec{k}_i = \vec{r} \bullet \vec{k}_t \Rightarrow xk_{ix} + yk_{iy} + zk_{iz} = xk_{tx} + yk_{ty} + zk_{tz} \quad (2\text{-}36b)$$

2. $$\cos\theta_i (E_{//}^{(i)} - E_{//}^{(r)}) = \cos\theta_t E_{//}^{(t)} \quad (2\text{-}37a)$$

$$\sqrt{\varepsilon_1}(E_{//}^{(i)} + E_{//}^{(r)}) = \sqrt{\varepsilon_2} E_{//}^{(t)} \quad (2\text{-}37b)$$

$$\sqrt{\varepsilon_1}\cos\theta_i (E_{\perp}^{(i)} - E_{\perp}^{(r)}) = \sqrt{\varepsilon_2}\cos\theta_t E_{\perp}^{(t)} \quad (2\text{-}37c)$$

$$(E_{\perp}^{(i)} + E_{\perp}^{(r)}) = E_{\perp}^{(t)} \quad (2\text{-}37d)$$

Equations (2-36) and (2-37) will be used to further derive the relation between the incident angle, the reflection angle, and the refraction angle, as well as the ratios of magnitudes between the incident field, the reflection field, and the transmission field.

2.3.1.3 Laws of Reflection and Refraction. Equation (2-36a) characterizes the relation between the incident angle and the reflection angle. Equation (2-36b) characterizes the relation between the incident angle and the transmitted angle. Because the wave vector is on the x–z plane, $k_{iy} = k_{ry} = k_{ty} = 0$. Also the boundary is at the x–y plane, so $z = 0$. Thus Eqs. (2-36a) and (2-36b) reduce to $k_{ix} = k_{rx}$ and $k_{ix} = k_{tx}$, where $k_{ix} = k_i \sin\theta_i = \omega\sqrt{\mu\varepsilon_1}\sin\theta_i$, $k_{rx} = k_r \sin\theta_r = \omega\sqrt{\mu\varepsilon_1}\sin\theta_r$, and $k_{tx} = k_t \sin\theta_t = \omega\sqrt{\mu\varepsilon_2}\sin\theta_t$. Therefore, Eq. (2-36a) gives the law of reflection

$$\sin\theta_i = \sin\theta_r \quad (2\text{-}38)$$

Also, Eq. (2-36b) gives the law of refraction (transmission), $\sqrt{\varepsilon_1}\sin\theta_i = \sqrt{\varepsilon_2}\sin\theta_t$, or in a form more familiar to most of people

$$n_1 \sin\theta_i = n_2 \sin\theta_t \quad (2\text{-}39)$$

where $n_1 = \sqrt{\varepsilon_1}$ and $n_2 = \sqrt{\varepsilon_2}$ under the case of materials without magnetic interaction with the electromagnetic waves.

2.3.1.4 Reflection and Transmission Coefficients. Equations (2-37a)–(2-37d) give the relation among the amplitudes of incident wave, reflected wave, and transmitted wave. Notice that we divide the electric field into the $\vec{E}_{\parallel}$ and $\vec{E}_{\perp}$ components for mathematical convenience at the beginning. Our derivation confirms that the $\vec{E}_{\parallel}$ and $\vec{E}_{\perp}$

components are not related to one another. They actually have physics meanings, as will be further explained shortly. Equations (2-37a) and (2-37b) relate the amplitudes of incident wave, reflected wave, and transmitted wave for the $\vec{E}_{\parallel}$ component. This component is called transverse magnetic (TM) polarization because its magnetic field, along the y-axis shown in Figure 2.4, is perpendicular to the x–z plane, which is the plane that contains the wave vector and the normal of the boundary. On the other hand, Eqs. (2-37c) and (2-37d) relate the amplitudes of incident wave, reflected wave, and transmitted wave for the $\vec{E}_{\perp}$ component. It is called transverse electric (TE) polarization because its electric field is perpendicular to the plane formed by the wave vector and the normal of the boundary. In some cases, the TM component is also called a p-wave and the TE component is called an s-wave, depending on the areas of applications. The TE polarization (s-wave) and TM polarization (p-wave) can be treated as two independent modes because they do not influence each other. This is another reason that we decompose the electric field into these two components, $\vec{E}_{\parallel}$ (p-wave) and $\vec{E}_{\perp}$ (s-wave), as shown in Figure 2.4.

With proper mathematical manipulation, Eqs. (2-37a)–(2-27d) lead to the following ratios:

For TM polarization (p-wave)

$$R_{//} \equiv \frac{E_{//}^{(r)}}{E_{//}^{(i)}} = -\frac{n_2 \cos\theta_i - n_1 \cos\theta_t}{n_2 \cos\theta_i + n_1 \cos\theta_t} \tag{2-40a}$$

$$T_{//} \equiv \frac{E_{//}^{(t)}}{E_{//}^{(i)}} = \frac{2n_1 \cos\theta_i}{n_2 \cos\theta_i + n_1 \cos\theta_t} \tag{2-40b}$$

For TE polarization (s-wave)

$$R_{\perp} \equiv \frac{E_{\perp}^{(r)}}{E_{\perp}^{(i)}} = \frac{n_1 \cos\theta_i - n_2 \cos\theta_t}{n_1 \cos\theta_i + n_2 \cos\theta_t} \tag{2-41a}$$

$$T_{\perp} \equiv \frac{E_{\perp}^{(t)}}{E_{\perp}^{(i)}} = \frac{2n_1 \cos\theta_i}{n_1 \cos\theta_i + n_2 \cos\theta_t} \tag{2-41b}$$

$R_{//}$ and $R_{\perp}$ are reflection coefficients, whereas $T_{//}$ and $T_{\perp}$ are transmission coefficients. They are the ratios of amplitudes instead of intensity. For light waves, the intensity ratio (or power ratio) is more frequently used than the amplitude ratio. The intensity of light is related to the amplitude of the electric field as follows:

$$I = \frac{1}{2}\sqrt{\frac{\varepsilon}{\mu}}\left|\vec{E}\right|^2 \tag{2-42}$$

2.3.1.5 Reflectivity and Ratio of Transmitted Intensity. People more frequently consider the portions of reflected intensity and transmitted intensity rather than the amplitudes from the energy point of view, in particular for the applications in photovoltaics. The intensity reflectivity (or simply reflectivity) is given by:

For TM polarization (p-wave)

$$\mathcal{R}_{//} = |R_{//}|^2 \equiv \left|\frac{E_{//}^{(r)}}{E_{//}^{(i)}}\right|^2 = \left|\frac{n_2 \cos\theta_i - n_1 \cos\theta_t}{n_2 \cos\theta_i + n_1 \cos\theta_t}\right|^2 \tag{2-43a}$$

For TE polarization (s-wave)

$$\mathcal{R}_{\perp} = |R_{\perp}|^2 \equiv \left|\frac{E_{\perp}^{(r)}}{E_{\perp}^{(i)}}\right|^2 = \left|\frac{n_1 \cos\theta_i - n_2 \cos\theta_t}{n_1 \cos\theta_i + n_2 \cos\theta_t}\right|^2 \tag{2-43b}$$

The ratio of transmitted intensity is given for a TM polarization (p-wave) by:

$$\mathcal{T}_{//} = \frac{n_2}{n_1}|T_{//}|^2 \equiv \frac{n_2}{n_1}\left|\frac{E_{//}^{(t)}}{E_{//}^{(i)}}\right|^2 = \frac{n_2}{n_1}\left|\frac{2n_1 \cos\theta_i}{n_2 \cos\theta_i + n_1 \cos\theta_t}\right|^2 \tag{2-44a}$$

and for a TE polarization (s-wave) by:

$$\mathcal{T}_{\perp} = \frac{n_2}{n_1}|T_{\perp}|^2 \equiv \frac{n_2}{n_1}\left|\frac{E_{\perp}^{(t)}}{E_{\perp}^{(i)}}\right|^2 = \frac{n_2}{n_1}\left|\frac{2n_1 \cos\theta_i}{n_1 \cos\theta_i + n_2 \cos\theta_t}\right|^2 \tag{2-44b}$$

Note that the refractive indices have to be included in Eqs. (2-44a) and (2-44b) because the incident light and transmitted light are in different media. This can be realized if one goes through the derivation using the intensity formula given by Eq. (2-42). If both medium 1 and medium 2 have no absorption, the total power is conserved, so $\mathcal{R}_{//} + \mathcal{T}_{//} = 1$ and $\mathcal{R}_{\perp} + \mathcal{T}_{\perp} = 1$.

Glass and plastics will probably be the most common substrates that organic solar cells or organic–inorganic composite solar cells are fabricated on. The refractive indices of glass, plastics, organics, and inorganic oxides used for such solar cells are mostly between 1.3 and 2.3. One can easily estimate the amount of light to be reflected from the surface of those materials using the formula given above. Figure 2.5(a) shows an example of the intensity reflectivity versus the incident angle for the example of light incident from the air to the glass with refractive index of 1.5. If the light is incident from the glass to the air, the situation is similar except that the incident angle is different, as shown in Figure 2.5 (b). As we can see, the TE polarization mode (s-wave) has minimal reflection at the normal angle of incidence. At the small angle of incidence, the reflection is only about 4%. When the incidence angle increases from 0°

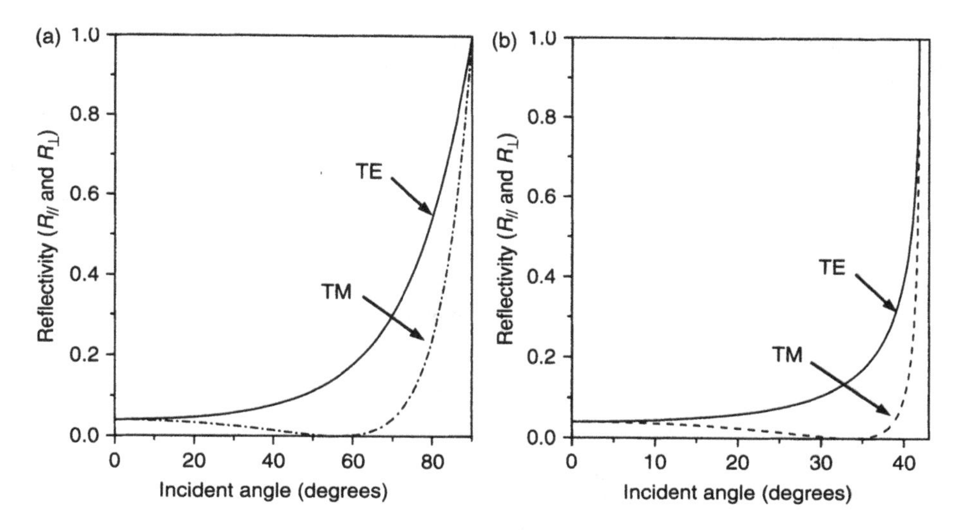

Figure 2.5. Intensity reflectivity versus incident angle. (a) Light is incident from the air ($n = 1$) to the glass ($n = 1.5$); (b) light is incident from the glass ($n = 1.5$) to the air ($n = 1$).

(normal direction) to 90°, the reflection increases to almost 100%. For the TM polarization mode (p-wave), there is an angle at which the incident light has no reflection. This angle is called the Brewster angle. Above this angle, the reflection also increases to almost 100%. The sunlight usually has no particular polarization, that is, neither p-wave nor s-wave, so the sunlight experiences no zero reflection. Its reflection typically increase with the incident angle. For conventional Si or III-V solar cells, the refractive index of Si or III-V semiconductors is as large as 3.5 or even larger. The reflection will be slightly over 30% at the normal incidence. As the incident angle increases, the reflection will mostly increase to even much more than 30% because the sunlight is not a TM-polarized wave (p-wave). As a result, a lot more than 30% of solar intensity will be reflected back to air, without being utilized by the solar cells. Therefore, additional antireflection techniques are usually required to increase light absorption by the solar cells. This implies that additional cost of fabrication will be involved for the conventional Si and III-V solar cells.

2.3.1.6 Total Reflection. Figure 2.5(a) shows that the reflection intensity approaches 100% when the incident angle is close to 90°. On the other hand, Figure 2.5(b) shows that the reflection of intensity approaches 100% as the incident angle nears 41.8°, whether the light is TE or TM polarized. Then one question arises: What happens if the incident angle is larger than 41.8°?

We can answer the above question from two points of view. First, from the law of refraction, $n_1 \sin\theta_i = n_2 \sin\theta_t$ [Eq. (2-39)], when light is incident from a medium of a larger refractive index to a medium of a smaller refractive index ($n_1 > n_2$), the incident angle is smaller than the refractive angle ($\theta_i < \theta_t$). As a result, it is possible for the refractive angle to approach 90°, but the incident angle will still be much less than 90°. The incident angle for which the refractive angle is equal to 90° is called critical angle, θ_c:

$$\sin\theta_c = \frac{n_2}{n_1}\sin 90^\circ = \frac{n_2}{n_1} \tag{2-45}$$

At this angle θ_c and beyond, there is no transmission of light. That is, the light is reflected completely. This is called total reflection and can be further understood from the second point of view. From Eq. (2-39), we have

$$\sin\theta_t = \frac{n_1}{n_2}\sin\theta_i \tag{2-46}$$

For $\theta_i > \theta_c$, $\sin\theta_t = (n_1/n_2)\sin\theta_i > (n_1/n_2)\sin\theta_c$. Using Eqs. (2-45) and (2-46), we obtain $\sin\theta_t > 1$. Because $\cos\theta_t = \sqrt{1-\sin^2\theta_t}$, $\cos\theta_t$ is a pure imaginary number. Then Eqs. (2-40a) and (2-41a) assume the form of $(X - jY/X + jY)$, where X and Y are real numbers and j represents the imaginary number, $j^2 = -1$. Therefore, the absolute value of $R_{//}$ and $R_\perp$ is of the form $(\sqrt{X^2+Y^2}/\sqrt{X^2+Y^2})$ (= 1), which makes the intensity reflectivity ($\mathcal{R}_{//}$ and $\mathcal{R}_\perp$) equal to one. It clearly explains that the incident light is reflected completely, so there is no transmission of light. The total reflection only occurs for light incident from the medium of a larger refractive index to the medium of a smaller refractive index. It does not occur in the reverse direction. The three cases for $\theta_1 < \theta_c$, $\theta_1 = \theta_c$, and $\theta_1 > \theta_c$, are schematically shown in Figure 2-6(a–c).

2.3.1.7 ***Brewster Angle.*** Figure 2-5 (a and b) shows that at a certain angle, the reflection is zero for TM polarization (p-wave). This angle is called the Brewster angle, which can be calculated from Eq. (2-40a). For zero reflection,

$$n_2\cos\theta_i - n_1\cos\theta_t = 0 \tag{2-47}$$

Combining Eq. (2-47) with the law of refraction, we obtain $\tan\theta_i = (n_2/n_1)$ and $\tan\theta_t = (n_1/n_2)$. Therefore, the Brewster angles, θ_{Bi} and θ_{Bt}, in medium 1 and medium 2, respectively, are

$$\theta_{Bi} = \tan^{-1}\left(\frac{n_2}{n_1}\right) \tag{2-48a}$$

$$\theta_{Bt} = \tan^{-1}\left(\frac{n_1}{n_2}\right) \tag{2-48b}$$

For example, when light is incident from air to glass with $n = 1.5$ or from glass to air, the Brewster angle is 56.3° in the air side and 33.7° in the glass side.

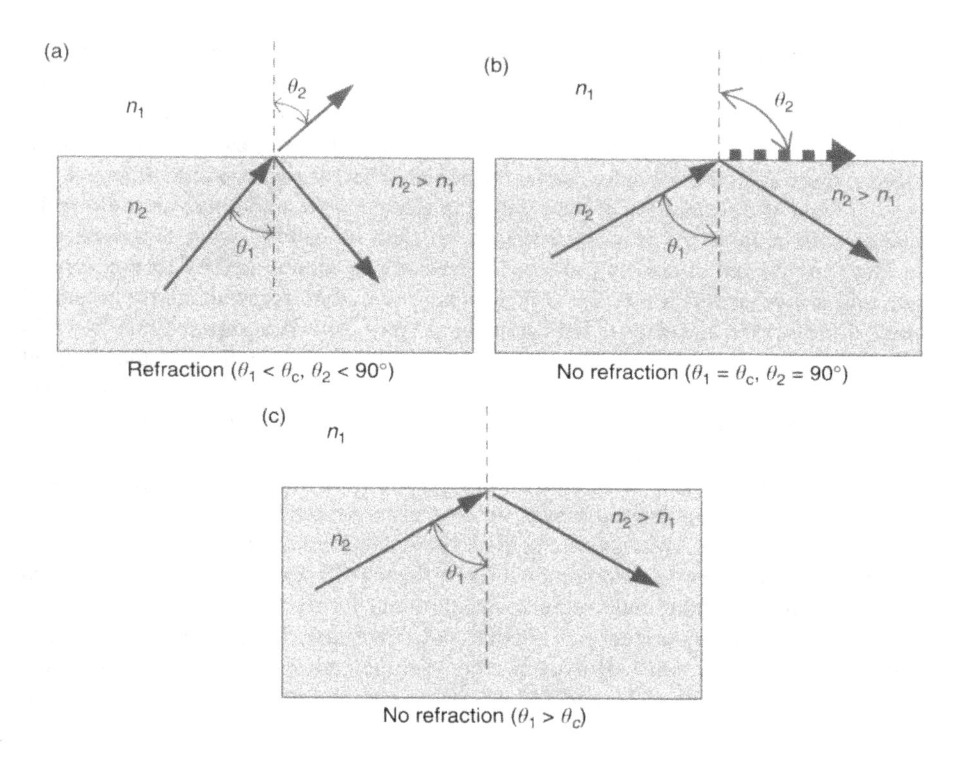

Figure 2.6. Illustration of reflection and refraction for (a) $\theta_1 < \theta_c$, (b) $\theta_1 = \theta_c$, and (c) $\theta_1 > \theta_c$.

For TE polarization (s-wave), Eq. (2-41a) gives no such angle for zero reflection, so there is no Brewster angle. Because sunlight is not particularly polarized, it will not clearly experience zero reflection at the Brewster angle.

2.3.2 Polarization

As we discussed in the previous section, the reflection and transmission of light depends on its polarization. Light may be TE-polarized (s-wave) or TM polarized (p-wave). What is the polarization of light?

The direction of the oscillating electric field in an electromagnetic wave is usually perpendicular to the direction of propagation. If the direction of the oscillating electric field is fixed and along the x-axis, we call it linearly polarized and x-polarized, assuming that the propagation direction is along z-axis. The linearly polarized wave could also be y-polarized for a wave with the direction of the electric field oscillating along the y-axis. The polarization could be along directions other than the x-axis or y-axis because there are many directions perpendicular to the propagation direction. Those directions are actually on a plane, which can be represented by two orthogonal vectors. Hence, there are only at most two independent directions of the electric field, called two independently polarized waves. It is also possible that the direction of the electric field rotates regularly on the plane perpendicular to the propagation direction. If the strength of the rotating field does not change, then this wave is circularly polarized. If the magnitude of the rotating field also varies regularly, then it is called elliptically polarized. Figure 2.7 schematically shows the above different polarization situations. On the other hand, if the direction of the oscillating electric field randomly varies along the plane perpendicular to the propagation direction, then we call this wave unpolarized.

Unpolarized and circularly polarized waves can be represented by an x-polarized wave and a y-polarized wave, $E_x\hat{x} + E_y\hat{y}$ or any two other perpendicularly polarized waves. For example, the electric field shown in Figure 2.4 is decomposed into two other perpendicular components: TM polarization and TE polarization. If the incident wave is TM polarized only, then the reflection and transmission are completely given by Eqs. (2-40a) and (2-40b). Similarly, for the TE-polarized wave (s-wave), the reflection and transmission are given by Eqs. (2-41a) and (2-41b). However, if the incident wave is unpolarized, then the reflection and transmission are the mixed behaviors of TE-polarized and TM-polarized waves. In this case, we can decompose the unpolarized wave into TE and TM waves, calculate the reflection and transmission of the TE and TM waves separately, then combine the reflected TE wave and the reflected TM wave by adding their amplitude vectors, and similarly for the transmitted wave.

For a circularly polarized and unpolarized wave, the decomposed TE and TM components should be equal. However, when the incident wave is not normal to the boundary, the reflections of TE and TM waves are not the same and, as a result, the TE

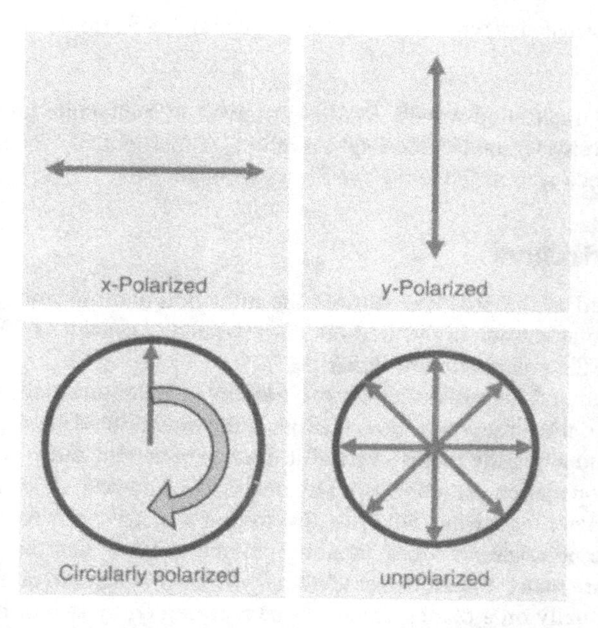

Figure 2.7. Different polarization situations.

and TM components of the reflected wave are not equal. Similarly, the transmitted wave has different amounts of TE and TM components. In this case, they are called partially polarized waves. An extreme case is when the light is incident at the Brewster angle. The TM wave will be completely transmitted, so the reflected wave is only TE-polarized, as shown in Figure 2.8. However, the transmitted wave contains both the TE and TM waves. If the incident wave has equal amounts of TE and TM components, the transmitted wave should have more TM component than TE component.

Because sunlight is unpolarized, it will have mixed behaviors of the TE-polarized wave (s-wave) and the TM-polarized wave (p-wave). For detailed and precise analysis of the reflection and transmission of sunlight from the surface of a certain medium, we can go through the same procedure: decompose the sunlight into TE-polarized waves (s-waves) and TM-polarized waves (p-waves), calculate their reflection and transmission separately, then combine the reflection for the TE-polarized waves (s-waves) and the TM-polarized waves (p-waves) as well as their transmissions summing those individual terms together, then adding the reflected TE waves and the reflected TM waves to give total reflection of the sunlight, and similarly adding the transmitted TE waves and the transmitted TM waves to give total transmission of the sunlight.

2.3.3 Dispersion

It is well known that a prism will divide white light into different colors, as shown in Figure 2.9. The reason for this phenomenon is because the refractive index of a material varies with the wavelength of the light. As a consequence, the refractive angle depends on the wavelength, as described by the following equation:

$$\frac{d\theta}{d\lambda} = \frac{d\theta}{dn}\frac{dn}{d\lambda} \tag{2-49}$$

The variation of the refractive index with wavelength is called dispersion. Usually, the refractive index decreases with the wavelength, as shown in Figure 2.10. The short-

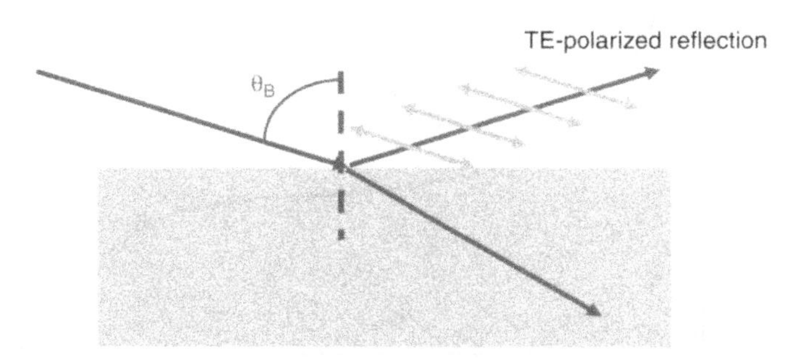

Figure 2.8. The reflected wave is TE-polarized only when the incident angle is equal to the Brewster angle θ_B).

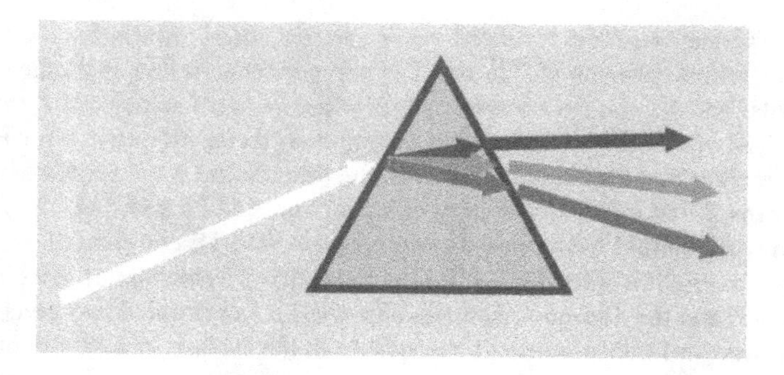

Figure 2.9. The beam of white light incident from the left side of the prism is divided into several beams of different colors.

wavelength component of light has a larger refractive angle than the long-wavelength component of light. As shown in Figure 2.9, the blue-color light (lower arrow) is bent more than the red-color light (upper arrow). Sunlight covers a very wide spectrum, extending from ultraviolet wavelength (< 300 nm) to infrared (> 3 μm). In general, the dispersion of sunlight is very significant, as most people have seen in colorful rainbows in the sky. For solar cell applications, because most of solar cells only absorb a relatively small wavelength range of sunlight, the dispersion phenomenon may not be worthy of consideration.

2.3.4 Isotropy and Anisotropy

According to Eq. (2-18), $\vec{P} = \varepsilon_0 \chi_e \vec{E}$, an external field will cause polarization, which is due to the separation of positive and negative charges. For materials with good symmetry of atomic configuration, as illustrated in Figure 2.11(a), the separation of positive

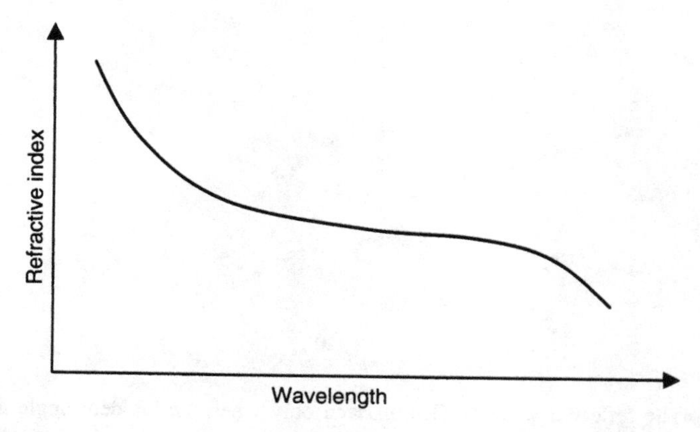

Figure 2.10. A general variation of refractive index with wavelength.

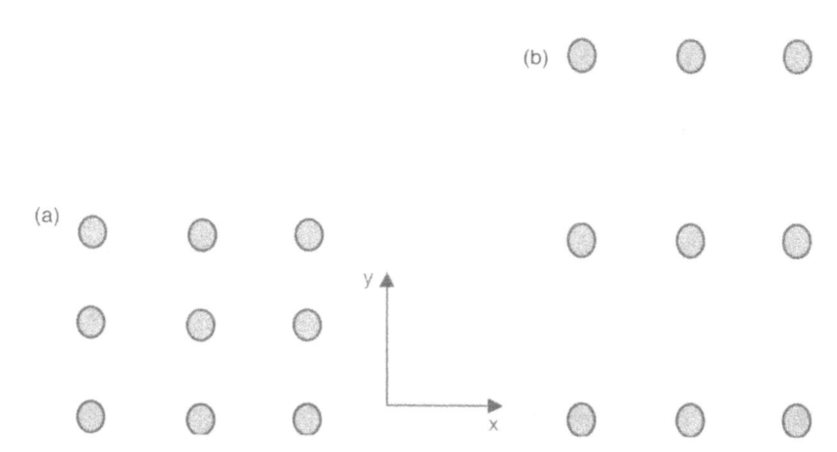

Figure 2.11. (a) Isotropic. When an electric field exists, charge separation is the same for the x and y directions due to symmetric structure along both directions. (b) Anisotropic. Because atom spacing along y-direction is larger, resulting in less bonding force for this direction, charges are easier to move along y-direction. Thus, when an electric field exists, charge separation along y-direction is larger than along x-direction.

and negative charges is independent of the direction of applied electric field. In addition, the dipole direction (the arrow line pointing from the negative charge to the positive charge) is along the direction of electric field. Then the polarization $\vec{P}$ has the same direction as the electric field $\vec{E}$. Thus, the proportional constant, electric susceptibility χ_e, is only a scalar number, so the dielectric constant, $\varepsilon = \varepsilon_0(1 + \chi_e)$, is a scalar number as well. In this case, $\vec{D} = \varepsilon\vec{E}$ also has the same direction as the $\vec{E}$ field. The material with such property is called an isotropic material.

For materials with an asymmetric arrangement of atoms, as illustrated in Figure 2.11(b), the positive and negative charges will not separate along the direction of the applied electric field due to different bonding forces caused by the atoms with different distances along the x- and y-directions. As a result, polarization depends on the direction of the incident electric field and the material is called an anisotropic material. In this case, the value of χ_e depends on the direction of the applied electric field. The dielectric constant, $\varepsilon = \varepsilon_0(1 + \chi_e)$, also depends on the direction of the applied electric field. Then the polarization $\vec{P}$, the electric field $\vec{E}$, and the field $\vec{D}$ are not necessarily parallel. In addition, because the value of ε now depends on the direction of the electric field, the refractive index $(=\sqrt{\mu\varepsilon}/\sqrt{\mu_0\varepsilon_0})$ is not a constant. The anisotropic property will result in the phenomenon of birefringence, to be described in the following section.

For organic polymers, their molecular structures usually have a longish shape, so they have anisotropic properties in general. However, because sunlight is not particularly polarized, the anisotropic property of polymers will not clearly influence the sunlight. On the other hand, for some inorganic crystals, such as ZnO, the asymmetric structure also exhibits anisotropic properties. If those inorganic crystals are not for light absorption, the polarization of sunlight may be changed after it passes through those anisotropic materials. Then the transmitted light becomes polarized. Consequently, the anisotropic polymers may preferentially interact with the transmitted light that has a certain polarization due to the influence of the transparent and anisotropic materials on top of the polymers.

2.3.5 Scattering

When light is incident on a plane surface, its direction of reflection can be easily predicted according to the law of reflection. This reflection is called specular reflection. On the other hand, as light is incident on a surface with an irregular shape, its direction of reflection becomes less trivial. This phenomenon is called scattering. The object that gives rise to the scattering of light is called the scattered center. Depending on the size of the object that the light is incident on, scattering is divided into three categories: Rayleigh scattering, Mie scattering, and geometric scattering.

When the object is much smaller than the wavelength of light, the phenomenon is called Rayleigh scattering [1]. This phenomenon was first modeled by Lord Rayleigh, who explained the blue color of sky. For the object with size less than 1/10 of the light wavelength, the scattering magnitude is proportional to $1/\lambda^4$. As a result, blue light is scattered more than the red light because the blue light has a shorter wavelength than the red light.

When the object has a size similar to the wavelength of light, the Mie theory [2, 3] is used to explain the scattering, called Mie scattering. This theory was developed by Gustav Mie, who gave a good analytical solution to Maxwell's equations for light incident on an object with a size close to the light wavelength, in particular for spherical objects. With some modification, closed form solutions can also be obtained for spheroids and ellipsoids or other simple shapes, but no general from is known for an arbitrary shape. The analytical solution allows for the calculation of the electric and magnetic fields inside and outside a spherical object, but it is generally used to calculate how much light is scattered.

For an object that has a size much larger than the wavelength, the law of reflection can be applied. If a surface consists of many particles of irregular shapes, light is reflected by many local interfaces between the air and the edges of the particles. Therefore, the reflected light can be directed in many directions, even though light is incident from only a particular direction. This will give rise to the diffusion effect. If the object has a concave surface, the reflected light may undergo another reflection, so the calculation will not be just a direct summation of a single reflection at each interface.

The phenomenon of scattering could be useful in photovoltaic applications because light will possibly experience a longer path of propagation in the media, for example, through a zigzag path or with multiple reflections. The increased light path can enhance light absorption, so the light absorption layer need not to be very thick. This will be particularly useful for solar cells using organic materials. Because organic materials do not usually have good conductivity, a thick layer of organic materials could degrade the performance of solar cells. A thin layer of organic materials, however, provides insufficient light absorption. Light scattering possibly increases light absorption because it makes light undergo a long propagation path even though the organic layer is thin.

2.3.6 Nonlinear Optics: Energy Up-Conversion and Down-Conversion

According to Eq. (2-18), $\vec{P} = \varepsilon_0 \chi_e \vec{E}$, polarization is proportional to the electric field. However, the proportional constant χ_e is not always a constant. Just as a spring's extension is sometimes not linearly proportional to the external force, the field-induced separation of positive and negative charges is also sometimes not linearly proportional to the field strength. As a result, the polarization $\vec{P}$ is related to the electric field $\vec{E}$ in a more complicated way because the electric susceptibility χ_e could also depend on the electric field $\vec{E}$:

$$\vec{P} = \varepsilon_0 [\chi_e \vec{E} + \chi_2 \vec{E}\vec{E} + \chi_3 \vec{E}\vec{E}\vec{E} + \cdots] \tag{2-50}$$

The second, third, and subsequent terms in the right hand side of the above equation are nonlinear terms that cause nonlinear phenomena.

Several nonlinear phenomena such as self-phase modulation, soliton effect, parametric process, energy up-conversion and down-conversion, Raman scattering, and Brillouin scattering, to name a few, are often observed in optical fibers. Those phenomena can result in new frequency components in addition to the input signal. Among those nonlinear phenomena, the energy up-conversion and down-conversion are particularly useful for photovoltaic applications because most of light-absorbing materials can absorb only a certain range of photon energy. For example, if the photon energy is lower than the bandgap energy of a semiconductor, such photons cannot be absorbed by the material. Energy up-conversion can then change the photon energy to above the bandgap energy and make it possible to be absorbed by the material. For organic polymers, the light absorption band is even narrower than that of the usual inorganic semiconductors. The energy up-conversion and down-conversion can convert the spectra of sunlight from outside of the absorption band to within, effectively increasing absorption of sunlight.

2.4 INTERACTION OF LIGHT AND MATTER WITH ENERGY EXCHANGE

2.4.1 Interaction of Light with Conductors

There are two major types of interaction of light and matters with energy exchange. The first type of interaction can be described by the following equation:

$$\vec{J} = \sigma \vec{E} \tag{2-51}$$

In this equation, $\vec{J}$ represents current density and σ is the conductivity of the material. This equation describes the interaction of an electric field with conductors in which σ is large. For insulators, σ is very small. The reason for large σ in conductors is because there are lots of carriers, which are charged particles like electrons, holes, or ions. Carriers move in the presence of an electric field, resulting in current. Therefore, as an electric field is applied to conductors, there is significant current flow in the conductor. In contrast, insulators have almost no carriers, so negligible current can be measured even with the existence of an electric field. For simplicity, the insulator is usually treated as a material with $\sigma = 0$. Although carriers move under the influence of an electric field, they still often collide with atoms. Then heat is generated so energy is dissipated. This is known as the ohmic heat of conducting materials and is usually described by Ohmic law, in the form of $V = IR$, where I is the current, V is the voltage, and R is the resistance. The electromagnetic waves or light can only propagate at the surface or penetrate to a very shallow depth in the conductors, called skin depth, which is on the order of the wavelength of the light.

The second type of light–matter interaction with energy exchange is the absorption of light by a material or the emission of light from a material. This type of energy exchange can be better understood from the photon point of view, which requires the concept of quantum physics. The quantum physics includes two parts. First, the energy of light is quantized. When light is quantized, it is called a photon. This viewpoint has already been explained briefly and will not be elaborated upon further here. Second, the energy of an atomic system is quantized.

2.4.2 Quantum Concept of an Atomic System

Next, we will discuss the concept of energy quantization in an atomic system and then describe the interaction of light with matter. The quantization of energy in an atomic system means that the energy is not continuous. The energy consists of many discrete levels, like the steps of a ladder. For example, the hydrogen atom has the energy levels given by the formula

$$E_n = -\frac{13.6}{n^2}\ \text{eV} \tag{2-52}$$

Figure 2.12 illustrates the energy levels of a hydrogen atom. The lowest energy level is called the ground state. The other energy levels are called excited states.

The electron of the hydrogen atom can stay on one of the energy levels. However, when the electron transits from level $n = 1$ to level $n = 2$, it has to absorb the energy ΔE:

$$\Delta E = 13.6\left(1-\frac{1}{2^2}\right) = 10.2\ \text{eV} \tag{2-53}$$

Such transition is schematically shown in Figure 2.13. Similarly, when the electron transits from level $n = 1$ to level $n = 3$, it has to absorb the energy ΔE, so

$$\Delta E = 13.6\left(1-\frac{1}{3^2}\right) = 12.09\ \text{eV}$$

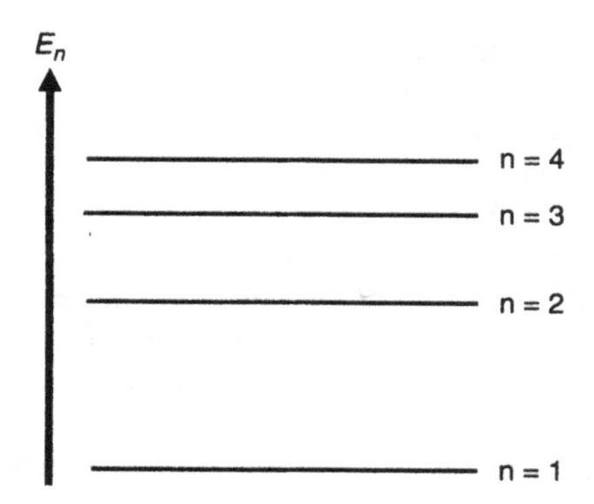

Figure 2.12. A schematic of energy levels of the hydrogen atom.

The electron could also possibly transit from level $n = 2$ to level $n = 3$. Then it absorbs the energy ΔE. In this case,

$$\Delta E = 13.6\left(\frac{1}{2^2} - \frac{1}{3^2}\right) = 1.89 \text{ eV}$$

The electron could have a downward transition from level $n = 2$ to level $n = 1$, too. Then it emits energy ΔE with ΔE given by Eq. (2-53). Solar cells are designed to absorb sunlight. They are not used for light emission, so we will not further elaborate the emission of light from downward transition of electrons from excited states to lower or ground states. In short, when an electron transits from a low energy level to a high energy level, it absorbs energy equal to the energy difference of the two levels. When an electron transits from a high energy level to a low energy level, it emits energy equal to the energy difference of the two levels.

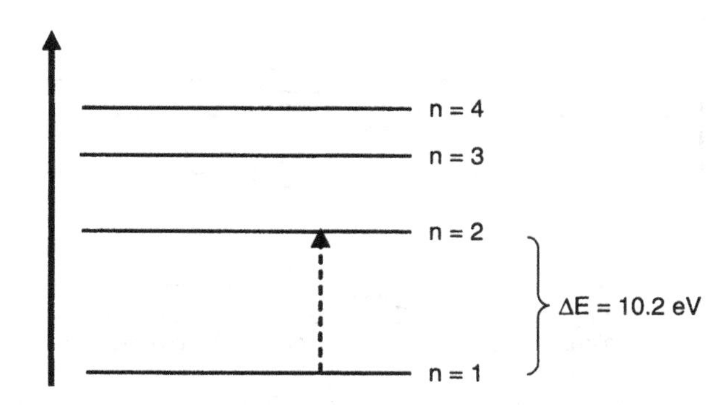

Figure 2.13. A schematic of electron transition from level $n = 1$ to level $n = 2$ in the hydrogen atom.

For most of materials, their atomic systems are more complicated than a hydrogen atom and their energy levels cannot be simply described by Eq. (2-52). For example, a hydrogen molecule is composed of two hydrogen atoms. When the two hydrogen atoms approach one another, their original energy level starts to split, as shown schematically in Figure 2.14. The general tendency is that the widely separated energy levels gradually become an energy band when more atoms are involved in forming a molecule. Nonetheless, the transition of electrons from one energy level to another behaves in the same way. Therefore, from the viewpoint of quantum physics, light absorption is caused by the transition of electrons from low energy levels to high energy levels.

An organic polymer or an inorganic crystal consists of many atoms. As a result, those energy levels become very closely spaced and form energy bands. For most cases, those energy bands even extend to the upper band and lower band, so there is no gap between bands. However, certain polymers and semiconductors have a gap between two groups of energy bands. This gap is called the energy bandgap, typically represented by *Eg*. The bands above the energy bandgap are called the conduction bands for semiconductors or the lowest unoccupied molecule orbitals (LUMO) for polymers, whereas the bands below the energy bandgap are called the valence bands for semiconductors or highest occupied molecule orbitals (HOMO) for polymers. The concepts of energy band and bandgap may differ between the inorganic semiconductors and the organic polymers. The details will be discussed in later chapters.

2.4.3 Light–Matter Interaction

The energy bandgap is an important property that characterizes the possible spectral range of light absorption. When the energy of a photon is larger than the energy bandgap, it can be absorbed by the materials because there are corresponding energy levels for electrons to transit from the low-energy band to the high-energy band, as shown schematically in Figure 2.15. If the energy of photon is less than the bandgap,

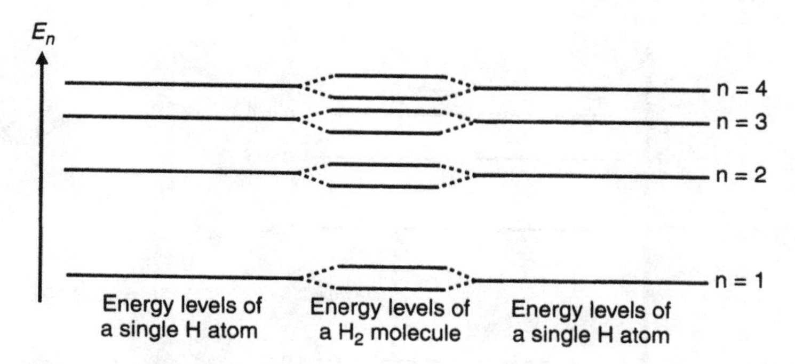

Figure 2.14. A schematic of energy levels of a hydrogen molecule (center). They are formed from the split of each individual energy level from two hydrogen atoms (two sides).

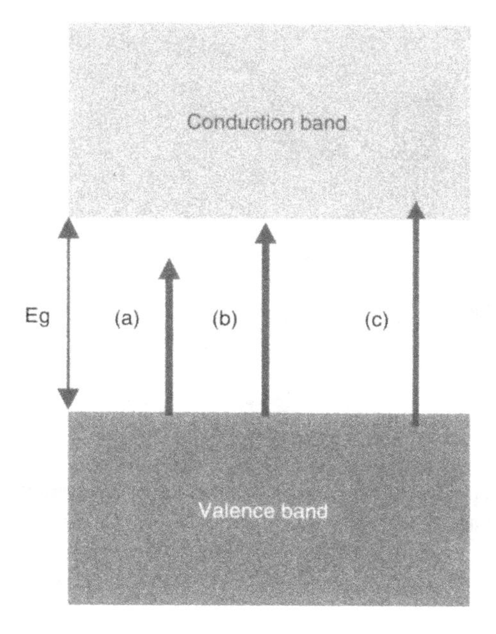

Figure 2.15. Transition of an electron from valence band to conduction band by the absorption of a photon. (a) If the energy of the photon is less than the bandgap, $h\nu < Eg$, then there is no corresponding energy level for the electron to transit. (b) If the energy of photon is just equal to the bandgap, $h\nu = Eg$, the transition is possible. (c) If the energy of photon is larger than the bandgap, $h\nu > Eg$, the transition is also possible.

$h\nu < Eg$, then there is no corresponding energy level for the electron to transit. On the other hand, if the energy of photon is equal to or larger than the bandgap, $h\nu \geqq Eg$, the transition is possible. Therefore, the light-absorption materials in the solar cells can only absorb the sunlight with energy larger than Eg. In other words, solar cells can absorb light with wavelength shorter than the value corresponding to Eg, that is,

$$\lambda < \frac{1.24}{Eg}(eV) \tag{2-54}$$

Figure 2.16 shows the spectrum of sunlight and the ratio of its energy integrated from the short wavelength to the long wavelength. The longer wavelength the bandgap corresponds to, the more solar energy can be absorbed by the materials. For example, if the material has the bandgap of 2.0 eV, which corresponds to 620 nm of wavelength, its absorption of sunlight is approximately only 20% of the total solar energy. If the bandgap is reduced to 1.24 eV, corresponding to 1000 nm of wavelength, it can absorb about 60% of the solar energy in the sunlight.

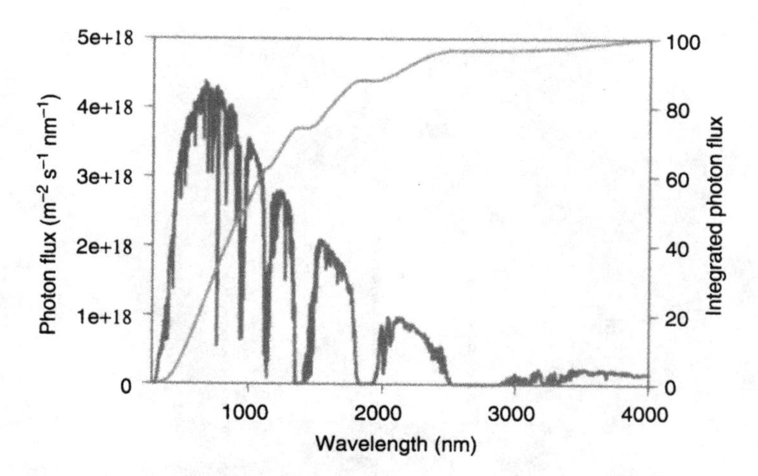

Figure 2.16. The solar spectrum and the ratio of solar energy integrated from the short wavelength to the long wavelength [4, 5].

REFERENCES

1. P. Lilienfeld, (2004), "A Blue Sky History," *Optics and Photonics News,* 15(6), 32–39.
2. C. F. Bohren and D. R. Huffmann, (2010), *Absorption and Scattering of Light by Small Particles,* Wiley.
3. G. Mie, (1908), "Beiträge zur Optik trüber Medien, Speziell Kolloidaler Metallösungen," Leipzig, *Ann. Phys.,* 330, 377–445.
4. E. Bundgaard and F. C. Krebs, (2007), "Low Band Gap Polymers for Organic Photovoltaics," *Solar Energy Materials & Solar Cells,* 91, 954–985.
5. R. Kroon, M. Lenes, J. C. Hummelen, P. W. M. Blom, and B. de Boer, (2008), "Small Bandgap Polymers for Organic Solar Cells—Polymer Material Development in the Last 5 Years," *Polymer Reviews,* 48, 531–582.

EXERCISES

1. Verify that the solution given in Eq. (2-8) satisfies Eq. (2-6) and Eq. (2-7).
2. In addition to the plane wave solution given in Eq. (2-8), find another form of solution for the wave equation of light.
3. Show that the wavelength corresponding to the maximum intensity of the blackbody radiation only depends on the temperature. [Hint: use the formula of Eq. (2-14b).]
4. Calculate the reflectivity when light is incident on the surface of a silicon surface. The refractive index of Si is 3.5.

5. Describe the difference between isotropy and anisotropy.
6. When light is incident on a glass window at the Brewster angle, will the reflected light be polarized? Why?
7. Sunlight is directed to pass through an anisotropic material. Give the conditions for the transmitted light being polarized and unpolarized.
8. A semiconductor has bandgap energy of 1.0 eV. Will the sunlight with wavelength longer than 1 μm be absorbed by this material? Explain.
9. Categorize scattering according to the size of the scattered center.
10. Explain the mechanism of material dispersion.
11. Explain why blue light has a larger refractive angle than red light when the light is incident on glass.
12. Explain why the sky has a blue color and why the sunset has a red or orange color.
13. Please explain why the sandy surface of a wall does not behave as a mirror-like surface.
14. A light beam that has total power of 1 mW is incident on a black wall. (a) If the wavelength of the light beam is 600 nm, calculate the number of photons that the wall absorbs per second. (b) Redo the above problem if the wavelength is 1000 nm.
15. A light beam that has total power of 5 mW is incident on a Si wafer. The refractive index of Si is 3.5 and the bandgap of Si is 1.12 eV. (a) If the wavelength of the light beam is 600 nm, calculate the number of photons that is absorbed by the Si wafer per second. (b) Redo the above problem if the wavelength is 1500 nm.
16. A light beam that has the same spectrum as the sunlight on the earth surface is incident on a GaAa wafer. Calculate the percentage of light being absorbed by the GaAs wafer. The refractive index of GaAs is 3.5 and the bandgap of GaAs is 1.42 eV.

3

FUNDAMENTALS OF INORGANIC SOLAR CELLS

Chih-I Wu

3.1 FROM ATOMIC BONDS TO ENERGY BANDS

As we all know from fundamental modern physics, the energy of levels of electrons in a single atom are discrete and can be classified as 1s, 2s, 2p, 3s, ..., orbital, and so on. When the two atoms of a same kind brought close to each other, the interaction of the electrons and the nuclei of the two atoms could form another stable state or chemical bond between two atoms. During the bonding process, the potential energy that the electrons in the atoms feel would change due to the existence of the other atom. As a result, the original energy level of the electron in a single atom would vary somewhat so that the same energy level of the electrons in two atoms would split. The separation of the energy of the same atomic level would depend on the strength of the chemical bonds, as shown in Figure 3.1. Each energy level can accommodate two electrons, one from each atom. Therefore, in the two split energy levels, one would be full and one would be empty. If we bring more atoms of a same kind close together, the same atomic level would keep varying a little bit, again due to the charge of surrounding charged particles, including electrons and nuclei. The more the atoms are brought together, the denser the energy levels would pack in a certain energy region. For a crystal with a normal size, there would be more than 10^{22} atoms. Therefore, we could see that the energy level would broaden into a "band" in which the energy levels of electrons are almost continuous. In a single atom, there are many different discrete energy levels; therefore, in a crystal, there would be many different bands. Some of the bands might overlap with each other, but in some cases there might be an energy gap between two adjacent bands.

For a material in a ground state, the electrons should locate at the lowest available energy as possible. In other words, the electrons should fill the bands starting from the lowest energy and gradually fill up the bands until the last one. The energy levels with electrons occupied are called valence bands and the energy levels without electrons are

Organic, Inorganic, and Hybrid Solar Cells. By C.-F. Lin, W.-F. Su, C.-I Wu, and I-C. Cheng

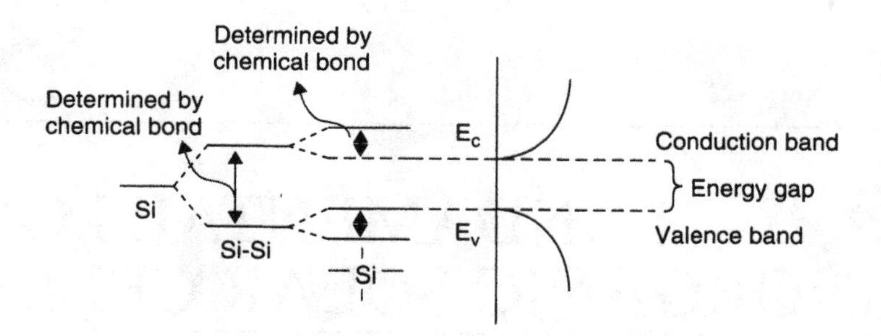

Figure 3.1. Chemical bonds and energy bands.

called conduction bands. In a material, if the electrons with the highest energy are at the middle of a band or at the top of a band that overlaps with an adjacent band of higher energy, this material would behave as a conductor. In such a case, the electrons in the upmost level can easily transport in the adjacent unoccupied level because there is no energy barrier. On the other hand, if the highest-energy electrons of a material are at the top of a band in which there is an energy gap to the next band, we call this material an insulator or a semiconductor if the energy gap is relatively small. In these materials, the electrons are normally not able to conduct current because all the energy levels in the valence band are occupied and the electron with the highest energy would need a certain amount of extra energy to overcome the energy gap and jump to the conduction band to transport.

3.2 ENERGY BANDS FROM A QUANTUM MECHANICS POINT OF VIEW

So far, we have only explained the formation of energy bands in a material in an intuitive way. To rigorously understand the origins of energy bands, we need to look at the Schrödinger equation as well as the periodic potential and its effect on electrons.

The Schrödinger equation of electrons in a periodic potential (Figure 3.2) can be written as follows:

$$\frac{-\hbar^2}{2m}\frac{d^2}{dx^2}\psi + U(x)\psi = \in \psi$$

where $U(x + na) = U(x)$, or $H\psi = \in\psi$ with $H(x + na) = H(x)$.

Before we deal with the Schrödinger equation with a periodic potential, we need to introduce the Bloch theory, which can be described as follows. The wave functions of free electrons are

$$\psi(x) \sim e^{ikx} \tag{3-1}$$

In a periodic potential with $U(x + na) = U(x)$, the wave function of the electrons must have the following properties:

Figure 3.2. Electrons in a periodic potential.

$$\psi_{\vec{k}}(x) = u_{\vec{k}}(x)e^{ikx} \quad \text{with} \quad u_{\vec{k}}(x+na) = u_{\vec{k}}(x) \tag{3-2}$$

where $u_{\to k}(x)$ has the same periodicity as that of $U(x)$. We can look at this theory in this way. For an electron in a periodic potential, the original free electron wave function e_{ikx} would be modulated by a periodic function that has the same periodicity as the potential energy. To prove the Bloch theory would take some rigorous mathematics. We do not plan to prove the theory here, but if the reader would like to know the proof, it could be find in most solid state physics textbooks.

We will focus on how to solve the Schrödinger equation in the periodic potential. Let us look at a famous example, the Kronig–Penney model, which is probably the only doable problem for a periodic potential without too much numerical programming. Here is how the Kronig–Penney model works. Assume a potential of the form shown in Figure 3.3. The potential energy is 0 between $x = 0$ and $x = a$, and is U_0 between $x = -b$ and $x = 0$. The periodicity is $a + b$. From the Bloch theory, we now are trying to solve the equations

$$\begin{aligned} &\frac{d^2\psi}{dx^2} + \frac{2m}{\hbar^2}(\in - U(x))\psi = 0 \\ &\psi(x) = e^{ikx}u\ (x) \end{aligned} \tag{3-3}$$

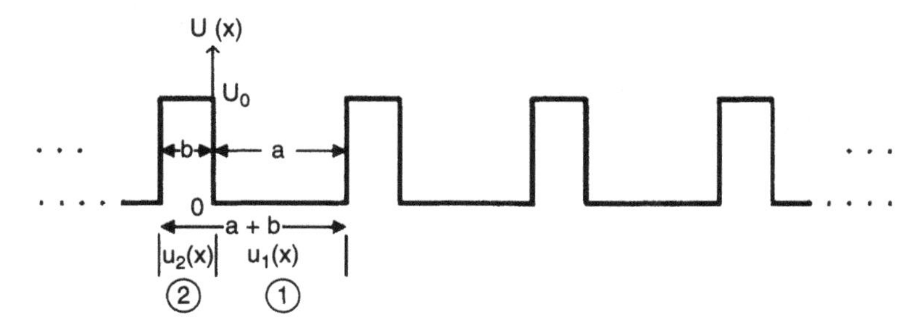

Figure 3.3. The Kronig–Penney potential.

We can rewrite the equations in both region 1 and region 2 as follows:

$$\text{Region 1}: \quad \frac{d^2u_1}{dx^2} + 2ik\frac{du_1}{dx} = (k^2 - \alpha^2)u_1(x) = 0 \quad \text{with} \quad \alpha = \left(\frac{2m \in}{\hbar^2}\right)^{1/2}$$
$$\text{Region 2}: \quad \frac{d^2u_2}{dx^2} + 2ik\frac{du_2}{dx} = (k^2 - \beta^2)u_2(x) = 0 \quad \text{with} \quad \beta = \left[\frac{2m(\in - U_0)}{\hbar^2}\right]^{1/2} \tag{3-4}$$

The general solutions of these differential equations are

$$\text{Region 1}: \quad u_1 = Ae^{i(\alpha-k)x} + Be^{-i(\alpha+k)x}$$
$$\text{Region 2}: \quad u_2 = Ce^{i(\beta-k)x} + De^{-i(\beta+k)x} \tag{3-5}$$

We now can match the boundary conditions. Since wave function is well-behaved, the boundary conditions are continuity u and du/dx:

$$u_1(0) = u_2(0)$$

$$u_1(a) = u_2(-b)$$

$$\left.\frac{du_1}{dx}\right|_{x=0} = \left.\frac{du_2}{dx}\right|_{x=0}$$

$$\left.\frac{du_1}{dx}\right|_{x=a} = \left.\frac{du_2}{dx}\right|_{x=-b}$$

You can then write four linear equations:

$$\begin{aligned} &A + B = C + D \\ &Ae^{i(\alpha-k)a} + Be^{-i(\alpha+k)a} = Ce^{i(\beta-k)b} + De^{-i(\beta+k)b} \\ &i(\alpha-k)A - i(\alpha+k)B = i(\beta-k)C - i(\beta+k)D \\ &\ldots\ A \quad \ldots\ B = \ldots\ C \quad \ldots\ D \end{aligned} \tag{3-6}$$

In these four linear equations, there would be no solutions except $A = B = C = D = 0$ unless the following determinant is zero:

$$\Rightarrow \begin{vmatrix} 1 & 1 & -1 & -1 \\ e^{i(\alpha-k)a} & e^{-i(\alpha+k)a} & -e^{-i(\beta-k)b} & -e^{i(\beta+k)b} \\ (\alpha-k) & -(\alpha+k) & -(\beta-k) & (\beta+k) \\ (\alpha-k)e^{i(\alpha-k)a} & -(\alpha+k)e^{-i(\alpha+k)a} & -(\beta-k)e^{-i(\beta-k)b} & (\beta+k)e^{i(\beta+k)b} \end{vmatrix} = 0 \tag{3-7}$$

From the determinant (3.7) above, we can then get

$$-\frac{\alpha^2+\beta^2}{2\alpha\beta}\sin(\beta b)\sin(\alpha a)+\cos(\beta b)\cos(\alpha a)=\cos k(a+b) \qquad (3\text{-}8)$$

If $0 < \in < U_0$, let $\beta = i\gamma$, so

$$\Rightarrow \frac{\gamma^2-\alpha^2}{2\alpha\gamma}\sinh(\gamma b)\sin(\alpha a)+\cosh(\gamma b)\cos(\alpha a)=\cos k(a+b) \qquad (3\text{-}9)$$

In both cases, we have the form

$$k_1\sin(\alpha a)+k_2\cos(\alpha a)=\cos k(a+b) \qquad (3\text{-}10)$$

The term on the right-hand side of the Eq. (3.10) is always between –1 and 1, whereas the terms on the left-hand side of the equation could be outside [–1, +1]. This means that there exist some values of $\in$ for which this happens, which are not allowed! In other words, there are forbidden values of energy $\in$.

The left-hand side of the equation can be rewritten as:

$$f(\in) \triangleq \left[1+\frac{(\alpha^2-\beta^2)^2}{4\alpha^2\beta^2}\sin^2\beta b\right]^{1/2}\cos(\alpha a-\delta)=\cos k(a+b) \qquad (3\text{-}11)$$

with

$$\tan\delta = -\frac{(\alpha^2+\beta^2)^2}{2\alpha\beta}\tan\beta b$$

What are we interested in these equations? We would like to solve the allowed energies for any given k. Let us plot just the $f(\in)$ versus energy $\in$ in Figure 3.4. We can start with $\in$ very small ($\in \to 0$), where $f(\in) > 1$, then follow the curve and show that it oscillates. It should be noted that as $\in \to \infty$, $f(\in)$ goes to a pure sinusoidal with amplitude 1.

Whenever the value of $\in$ leads to $-1 \le f(\in) \le 1$, we can find a corresponding k to satisfy the equation. In these cases, the solution exists or in other words, there is an allowed state. On the other hand, whenever $f(\in) < -1$ or $f(\in) > 1$, there is no solution, that is, there are no allowed states corresponding to this energy. These results imply that there are some energies that do correspond to any allowed states, which can be defined as the energy gaps. The energy gaps are separated by regions of allowed energies, which can be identified as the "band."

It is very important to know that, for a given energy, there would be more than one k corresponding to this energy. For example, when $f(\in) = \pm 1$, $\cos k(a + b)$ also equals ± 1. In this case, the solution is $k = n\pi/(a + b)$ and $n = 0, \pm 1, \pm 2, \ldots$. Now suppose we want to plot $\in$ versus k. First note that the choice of k is arbitrary within factors of

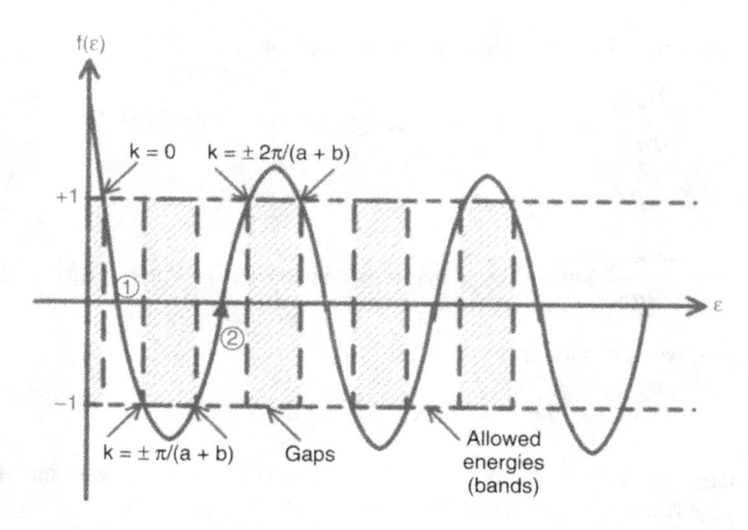

Figure 3.4. Graphic solution of electrons in the Kronig–Penney potential.

$n[2\pi/(a+b)]$ with $n = 0, \pm1, \pm2, \ldots$, and so on. The graph solution of energy versus k is shown in Figure 3.4. First, draw region 1 with smaller k, then increase k and draw region 2, and so on. The dotted line is the energy versus k for free electrons, which has the relation of $\in = (\hbar^2k^2/2m)$ and is a parabolic line. It should be noted that when energy is large, $\in = (\hbar^2k^2/2m)$, which is close to the case of a free electron. This is not surprising because when the energy of electron is very large, the binding energy is small and can be neglected and the electrons look like free electrons.

From this simple model, we can clearly see how the gaps and bands are formed in a mathematical way.

We can also replot Figure 3.5 into a so-called reduced-zone scheme. Note that the k values are arbitrary within a phase factor $\pm(2\pi/d)/n$, $n = 0, \pm1$, because all we need to have a solution is that $-1 < \cos kd < +1$ and $\cos kd$ does not care whether k is shifted by $\pm(2\pi/d)/n$. Therefore, we can fold the curve in region 2 and beyond into region 1 so that we can only focus on the k values between $\pm(2\pi/d)$, as shown in Figure 3.6. This region is called the first Brillouin zone, or B.Z.

3.3 THE ENERGY BAND IN SEMICONDUCTORS

For a typical semiconductor material, there are valence bands and conduction bands. In between the conduction band and valence band, there is a forbidden region where there are few allowed states for carriers. The energy difference between the conduction band and the valence band is the energy band gap. Without external doping, almost all the electrons are located in the valence band. The carrier distribution would follow the Fermi–Dirac distribution function:

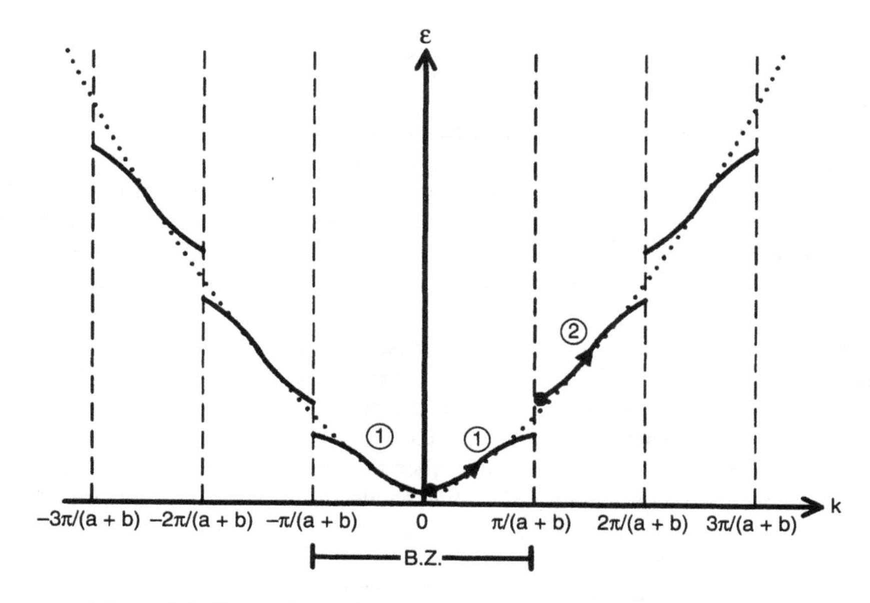

Figure 3.5. Energy band of electrons in the Kronig–Penney potential.

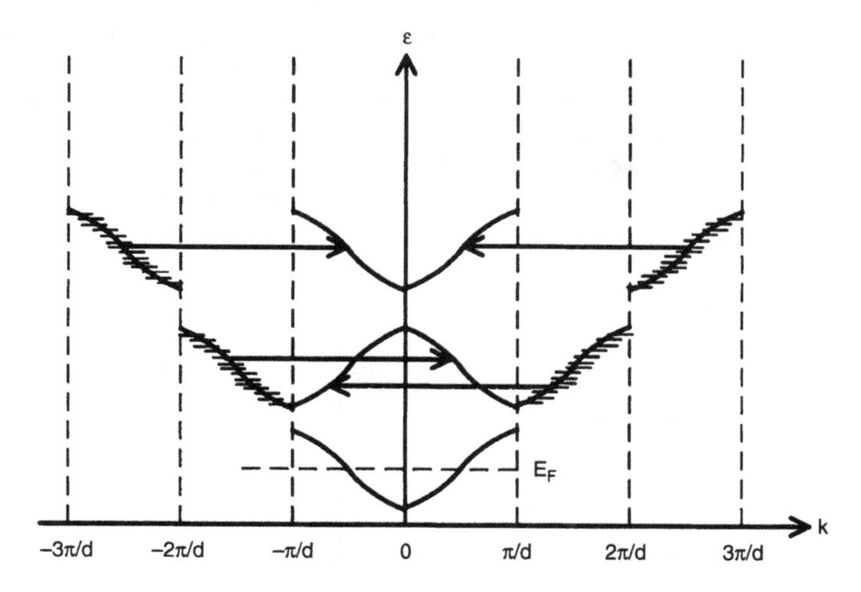

Figure 3.6. Reduced-zone scheme of an energy-band diagram.

$$f(E) = \frac{1}{1 + e^{[E - E_F(T)]/kT}} \tag{3-12}$$

where $f(E)$ is the probability of finding electrons with energy of E, k is the Boltzmann constant, T is the temperature, and E_F is the Fermi energy.

From the Fermi–Dirac distribution function, we can get some simple idea of how the electrons distribute at different energy levels. Let us first look at the Fermi–Dirac function when T is close to zero. If the energy of the electron is larger than E_F, the exponential term in the denominator will approach infinity since kT is close to zero. Therefore, the value of $f(E)$ is approximately zero, or the probability is zero. On the other hand, if the energy of the electron is smaller than E_F, the exponential term in the denominator will be zero since $(E - E_F)/kT$ is negative infinity. As a result, the value of the Fermi–Dirac function becomes one or the probability is 100%. In other words, if the energy of the electron is smaller than the Fermi energy, the probability of finding electron at that level is 100%. However, when the energy of the electron is larger than the Fermi energy, the probability of finding an electron at such levels is null. The Fermi–Dirac function at zero temperature is a step function, as shown in the Figure 3.7.

Even if the temperature of the system is not absolute zero, as long as the energy scale of E_F is much larger than that of kT, the statements above is still valid. Typically, the Fermi energy is about a few electron volts (eV). At room temperature, the value of kT is about 25 meV, which is about 100 times less than that of the Fermi level. Therefore, we can say that at the normal device-operation temperature we can always view the Fermi–Dirac distribution function as a step function.

The Fermi energy, or the Fermi level, which can be defined as the highest energy level of the electrons in a system at absolute zero, is dependent on the electron concentrations. For most of the semiconductors, the Fermi level lies between the conduction band and the valence band. This level is close to the middle of the band gap if the semiconductors do not have any impurity dopant. At temperatures above 0 K, some electrons in the valence band will gain certain thermal energy and move to the conduction

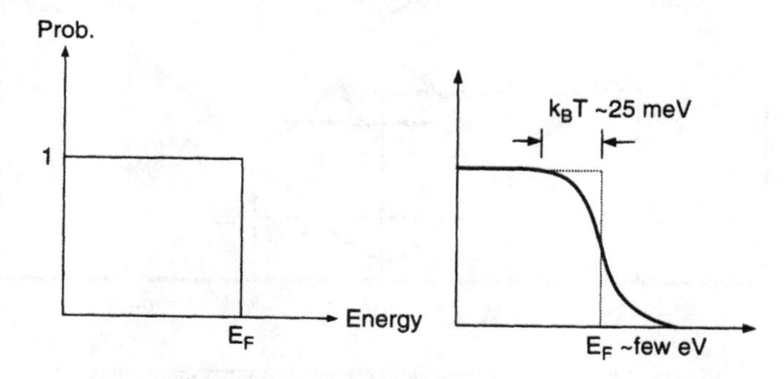

Figure 3.7. The Fermi–Dirac distribution function at absolute zero (left-hand side) and room temperature (right-hand side).

band. Those electrons in the conduction band will be free to move in the semiconductors. On the other hand, the valence band will lack electrons. In the physical space of the semiconductors, those locations that lack electrons can be filled by neighboring electrons. As they are filled, the empty locations apparently shift to neighboring locations. Thus, they are called holes, which are actually bonds that lack electrons. Because positive charges of nuclei remain the same, a lack of electrons behaves like a positive charge. Therefore, holes are regarded as positive charges. They are able to move in the semiconductors and conduct current. For a semiconductor without any dopant, called an intrinsic semiconductor, the number of electrons in the conduction band is the same as the number of holes in the valence band.

The semiconductors can be doped with other types of atoms. Then the number of electrons and holes could be different. For example, if a Si semiconductor is doped with phosphorus (P) atoms, there will be more electrons than holes. The P atom has one more electron in the most outside orbit than Si. As the Si atom is replaced with the P atom, there will be one more electron left unbonded. Consequently, this electron is free to move in the doped semiconductor. This is called an N-type semiconductor. On the other hand, if the Si is doped with boron, which has only three electrons in the most outside orbit, there will be more unoccupied bonding than with regular Si. Then there will be more holes than electrons. Such a semiconductor is called P-type.

For N-type semiconductors, the electron concentration increases, so the Fermi level will move up in the energy band diagram, close to the conduction band of the materials. On the other hand, for P-type semiconductors, the electron concentration decreases or the hole concentration increases, so the Fermi level will shift down in the forbidden band gap, close the valence band. Therefore, for N-type semiconductors, the Fermi level would be closer to the conduction band or at the top half of the band gap, whereas for P-type semiconductors, the Fermi level would be closer to the valence band or at the bottom half of the band gap.

3.4 P–N JUNCTION

One of the simplest structures of inorganic solar cells is a semiconductor diode that consists of p-type and n-type junctions. Understanding the physics of the P–N junction is an important step toward understanding the solar cells, because the traditional inorganic solar cell is just a P–N junctions operated under luminescence. The P–N junction is also a fundamental component of microelectronic devices. In this section, we will review the basic principles of P–N junction diodes. The following concepts will be discussed:

1. The depletion region in the P–N junction and how it forms
2. The built-in voltage and where it comes from
3. The energy band diagram for a P–N junction at equilibrium
4. The carrier distribution, the band diagram, and the current formation of the P–N junction under forward bias

5. The carrier distribution, the band diagram, and the current formation of the P–N junction under reverse bias
6. Breakdown of P–N junction diodes

When we put N-type and P-type materials in contact, a P–N junction is formed at the interface of the two materials. In the N-type material side, the electron concentration is much larger than that of the hole; however, in the P-type material side, the hole concentration would be much greater. The carrier imbalance in the two sides of the junction would induce charge redistribution through carrier diffusion.

Diffusion would drive majority carriers across the junction, in response to the concentration difference (gradient), leaving ionized dopant ions in the vicinity of the junction. A schematic diagram is shown in Figure 3.8. If the particles involved in the diffusion process are neutral, such as in a classical ideal gas, the redistribution of the particles would not stop until there were equal concentrations on both sides of the junction. However, the behavior of diffusion involved with charged particles would be drastically different. In the case of carrier redistribution of a P–N junction, once the diffusion process starts, an electric field builds up (due to the charged ions) in a direction opposing diffusion. The more the charged carriers diffuse to the other side of the junction, the larger the electric field builds up. There would be two forces driving the movement of charged carriers: the diffusion due to the concentration gradient would drive the carrier in one direction, whereas the build-up of an electric field would force the carrier shift in the opposite directions. At some point these two competing processes would balance each other and the equilibrium state would be reached.

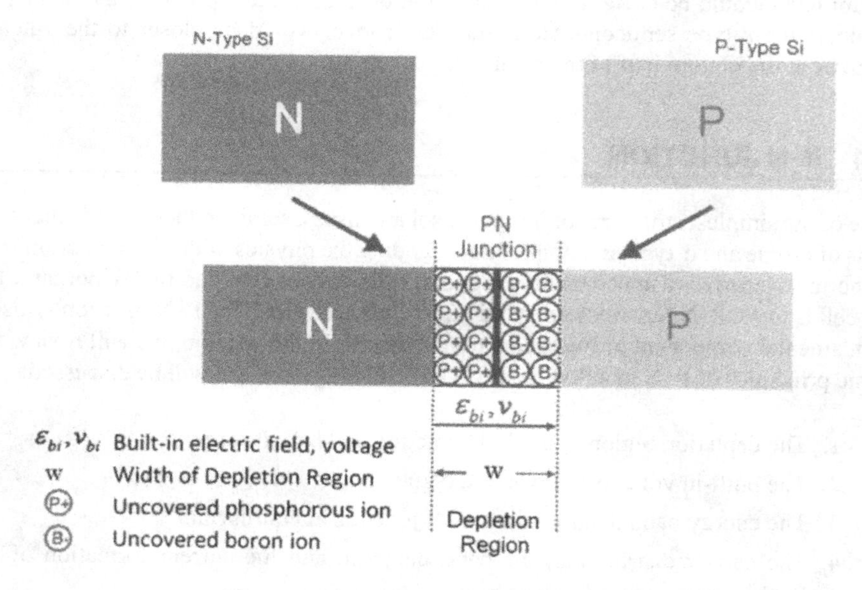

Figure 3.8. Schematics of a P–N junction.

At equilibrium, there is a *depletion region* near the junction. The depletion region means that there are almost no mobile carriers close to the junction, or depletion of mobile charges. It is intuitive to understand the concept of the depletion region in this way. Since there is an electric field pointing in a certain direction at the vicinity of the junction, the mobile or nearly free carriers in that region would be swept to the either side of the junction, depending on the polarity of the charges. In the meantime, since there are net charges and an electric field in the depletion region, a built-in electric voltage, or built-in potential, would be formed in this region.

3.5 ENERGY BAND DIAGRAM OF THE P–N JUNCTION

The best way to understand the electrical behaviors, such as carrier distributions, energy levels of electrons and holes, and the current formation, of the P–N junctions is to understand the band diagram of the device. We can construct an energy band diagram for the P–N junction. We start with the energy band diagram for N-type and P-type materials.

When P-type and N-type semiconductors are in contact, the key to combining them into the band diagram is to understand that the Fermi levels must align at equilibrium. As long as the Fermi levels are not aligned, which means that the electrons have higher energy in one side than those in the other side, carriers will flow across the junction because it is energetically favorable for them to do so. A simple analogy is the water levels of two buckets. If buckets with different levels of water are connected with a hose, the water level in the two buckets would eventually be at the same level, unless the water in one bucket is all gone.

To construct the band diagram for a P–N junction, as shown in Figure 3.9, we can first put the top two diagrams together, then slide them vertically until the Fermi levels

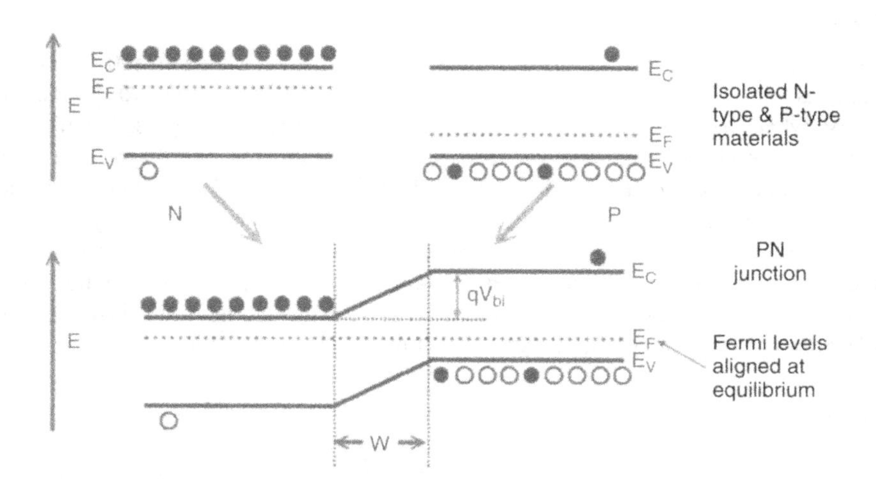

Figure 3.9. The band diagram of a P–N junction diode.

line up. Since the energies of the bands, either conduction band or valence band, are different in the two sides after the P-type and N-type semiconductors are brought into contact, there would be a potential gradient in the junction region, which is the depletion region. The potential difference would result in an electric field in the depletion region. The value of the potential difference in the two sides is defined as the built-in potential. Outside of the depletion region, there is negligible electric field since there is no net charge and the materials remain neutral.

It is noteworthy that in the aforementioned discussion we define the Fermi level as the highest energy level of the electrons. Strictly speaking, this is only true at absolute zero temperature. If the temperature of the system is not in 0 K, the Fermi–Dirac distribution function would not be a step function and there would be a tail in the distribution function extending to infinity in energy. That means there is a possibility of finding electrons at very high energy and we cannot define the Fermi level in the previous way. However, at room temperature or even at higher temperature at which the device might operate, the distribution function of the electrons are still very close to a step function with error of less than a few tens of milli-electron-volts. Therefore, we can almost always approximate the Fermi–Dirac distribution function as a step function and the definition of the Fermi level described above will mostly be valid.

3.6 CARRIER TRANSPORT IN A P–N JUNCTION

As mentioned in the previous section, there are two driving forces to the carriers in the depletion regions of the P–N junction: diffusion and drift. In this section, we briefly discuss these two driving forces in detail.

3.6.1 Diffusion

In extrinsic doped semiconductors, N-type materials have a high concentration of free electrons and P-type materials have a concentration of holes. Similar to the ideal gas theory, when the two materials are placed in contact, diffusion begins. Electrons diffuse from the N-type side (higher electron concentration) into the P-type side (lower electron concentration). Holes diffuse from the P-type side (higher hole concentration) into the N-type side (lower hole concentration). If there were no charge particles involved, this process would continue until a uniform concentration of electrons and holes were achieved throughout the entire sample, at which point a concentration gradient would cease to exist and diffusion would stop. An example of this is the ideal gas diffusion with oxygen gas on one side and nitrogen gas on the other side.

3.6.2 Drift

However, diffusion is not the only force at work in a system with charged particles involved. We should notice that, as diffusion takes place, dopant ions are ionized and left near the junction. In other words, when an electron diffuses out of the N-type side, it leaves behind a positively charged phosphorous dopant ion; when a hole diffuses out of the P-type side, it leaves behind a negatively charged boron ion. It should be noted that these ions are bonded into the silicon lattice and are not mobile. Only the free electrons and holes can move. Furthermore, the electrons and holes would recombine as they traverse the junction, since that would be the lowest energy configuration for the carriers.

This region of uncovered, fixed charge near the junction is called the depletion region because it is depleted of mobile charge carriers (i.e., free electrons and holes). This uncovered charge gives rise to an electric field in the depletion region and, as diffusion continues, more charge is uncovered (the depletion region grows wider) and the electric field increases. This is called the *built-in electric field,* giving rise to a *built-in voltage*. Electrons and holes are charged particles, hence their motion will be influenced by this electric field. Motion of a charged particle due to an electric field is normally called *drift.*

The electric field is oriented so as to oppose diffusion, that is, diffusion is driving electrons from the N-type side to the P-type side, whereas the electric field drives electrons from the P-type side to the N-type side. Obviously, there will come a point at which these two forces, drift and diffusion, are equal but opposite, and equilibrium will be established under such cases. The depletion region and built-in voltage would be formed and be stable under equilibrium. At this point, the Fermi levels will be aligned on the band diagram, as shown in the previous section.

When P-type materials and N-type materials make contact, electrons diffuse from the N-side to the P-side. Holes diffuse from the P-type side to the N-type side. They recombine in the depletion region. As the carriers start diffusing, the band diagrams slide vertically with respect to each other. The potential difference from the energy shift would induce drift forces for carriers to compete with diffusion. The potential of the P-side is being raised with respect to the N-side, as the ionized charge gives rise to a built-in voltage. The potential "hill" or "barrier" (built-in voltage) across the depletion region gets bigger as the bands slide, making it more difficult for carriers to diffuse across the junction. When the Fermi levels are aligned, it is no longer energetically favorable for carriers to move from one side to the other. The barrier is sufficiently high so as to impede further diffusion. Drift and diffusion have balanced each other, and there is no net movement of carriers across the junction—equilibrium has been established.

The equilibrium is established very rapidly after contact; it is instantaneous for all practical purposes. It is important to note that this is a *dynamic* equilibrium (as opposed to a static equilibrium) because there are carriers crossing the junction, but there is no *net* movement of carriers. This tiny population of carriers arises from thermal generation of minority carriers inside the depletion region or in the quasineutral regions (described later), where they can drift across the junction due to the built-in voltage, V_{bi}. It also arises from the fact that, statistically, some majority carriers in the far "tail" of the Fermi distribution will have enough energy to overcome the barrier and diffuse across the junction. The key point is that, at equilibrium, drift counterbalances diffusion. At any period in time, the net flow of electrons across the junction is zero, and the net flow of holes across the junction is zero.

The N-type and P-type *bulk regions,* away from the junction, are still in neutral because the electrons or holes leave the original dopant atoms. However, in the depletion region on the N-side, there are ionized, positively-charged N-type dopants, such as phosphorus ions. The free electrons (i.e., the fifth valence electrons for N-type dopant atoms) diffuse out of this region and into the P-side.

Similarly, in the depletion region on the P-side, there are ionized and negatively charged boron ions. One way to understand this, treating holes as independent positively charged particles, is to think that the holes diffuse to the N-side, leaving behind the P-type dopants, such as boron, with a net negative charge. An equivalent way of understanding this is to observe that the electrons that diffuse into the P-side from the N-side now fill the holes in the boron-doped bonding. The recombination process therefore happens in the depletion region. The presence of this electron gives the boron atom a net negative charge.

3.7 P–N JUNCTION DIODES

The built-in voltage (V_{bi}) in the depletion region can be seen as a hill or energy barrier for the carriers on the band diagram. As mentioned in the previous section, the built-in voltage or the built-in electric fields exists only across the depletion region, as indicated by the sloped bands, due to the net charges that exist there. In an ideal case, there is no electric field or potential difference in the bulk region of either N-type or P-type materials outside of the depletion regions since the mobile carriers, electrons and holes, do not diffuse away from the bulk so that there are no net charges.

The energy band after N-type and P-type materials make contact shifts exactly the difference of the N-type and P-type Fermi levels with respect to each other since that is how much the band should move to align the Fermi level at equilibrium. The magnitude of the built-in voltage (i.e., the height of the hill) is therefore equal to the difference between the two Fermi levels before contact. In N-type materials, the more the dopant concentration, the closer the Fermi level will be to the conduction band. In P-type semiconductors, the Fermi level would move closer to the valence band if the doping concentration increases. As a consequence, increasing the dopant concentration in one or both sides of the junction will result in a larger built-in voltage.

An ideal diode equation is as follows:

$$I = I_0\left[\exp\left(\frac{qV_A}{k_BT}\right)-1\right] \qquad (3\text{-}13)$$

where V_A is the applied voltage.

In Equation (3-13), I is the reverse saturation current and can be written as

$$I_0 = qAn_i^2\left[\frac{D_n}{L_nN_A}+\frac{D_p}{L_pN_D}\right] \qquad (3\text{-}14)$$

where
D_n = electron diffusion coefficient
D_p = hole diffusion coefficient
L_n = electron minority carrier diffusion length
L_p = hole minority carrier diffusion length

The preceding current equations can be combined with the following assumptions to derive the *ideal diode equation*. We will show the derivation of the ideal diode equation in a later chapter.

At the forward bias, the intersection of I and V occurs near $V_A = V_{bi}$ and the barrier height is reduced to zero. Increasing the voltage even a small amount beyond this point would result in a very sharp increase in current. The P–N junction is sometimes referred to as a *rectifying contact*, as opposed to an ohmic contact, because the forward-bias voltage drop across the device is approximately limited to the built-in voltage, as shown by the sharply increasing portion of the I–V curve shown in Figure 3.10.

3.8 SOLAR CELL DIODES

The ideal diode current–voltage characteristics described above is at equilibrium and under no external excitation. If the diode is under illumination, there will be an additional current source added to the diode circuit and the I–V characteristics would be modified accordingly. As discussed in Chapter 2, when the photon energy is larger than the bandgap of a semiconductor, the photon can be absorbed and an electron in the

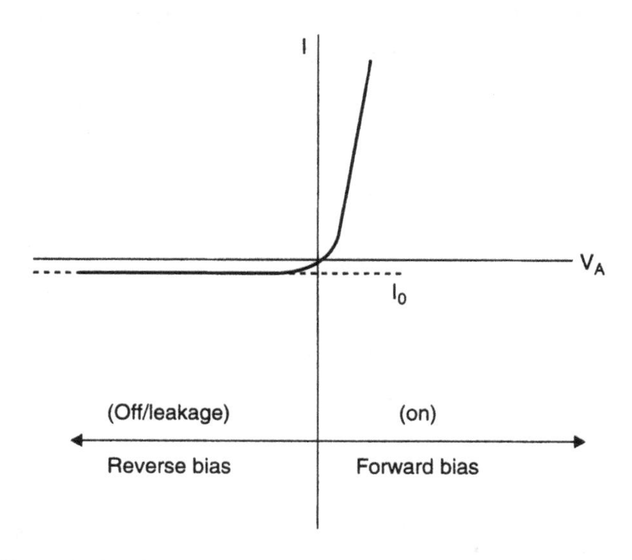

Figure 3.10. The current–voltage characteristics of an ideal diode.

valence band will transit to the conduction band. Thus, there will be one more electron in the conduction band and the valence band will lack an electron, equivalent to a hole. In other words, an electron–hole pair is generated. The generated electron and hole are free to move in the semiconductor.

With the above concept in mind, let us now consider an ideal diode illuminated with light and connected with an external resistor, R, as shown in Figure 3.11 (a). The direction of current is defined from P-type side to N-type side as positive. If there is no external load and electrical bias, such as a short circuit, then the only current source in the circuit is due to the photo-generated current, I_{ph}, as shown in Figure 3.11 (b). The photo-generated current, I_{ph}, only depends on the intensity of the light illuminating the P–N-junction diode and would not be affected by the load and bias in the circuit. There are a few factors that would determine the photo-generated current, such as the absorption rate of the materials, the recombination rate of the photo-generated carriers, the depletion width of the P–N junction, and the defect centers in the materials. In general, the higher the light intensity is, the larger the photo-generated current in the device. If the incident photon flux is F, the photo-generated current can be written as $I_{ph} = KF$, where K is a proportionality constant depending on the factors mentioned above. If there are no external electrical bias and load in the circuit, this photo-generated current is the short-circuit current with a minus sign since we define the positive current direction as from the P-type side to the N-type side. Therefore, we can have $I_{sc} = -I_{ph}$.

Now consider that there is an external load, R, in the circuit as shown in Figure 3.12 (c). There would be a voltage drop on the two sides of the load. As a result, the voltage across the diode would not be zero as in the case of the short-circuit condition. The voltage across the diode would then induce the diffusion current and drift current as in the normal diode that we described in the previous sections. There would be an ideal diode current in addition to the photo-generated current in the circuit. In the last section, we showed that the ideal diode current is

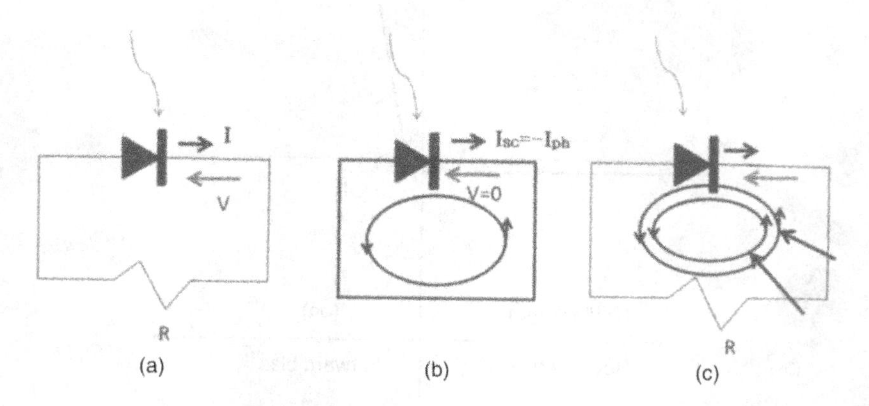

Figure 3.11. (a) The ideal diode under illumination and connected to an external resistor to form a circuit. (b) The current and voltage of a solar cell in short-circuit condition (without external resistance). (c) The current and voltage of a solar cell with an external load R.

$$I_d = I_0 \left[\exp\left(\frac{qV_A}{k_B T} \right) - 1 \right] \tag{3-15}$$

Therefore, we can determine that the total current for a solar cells with an external load is

$$I = -I_{ph} + I_0 \left[\exp\left(\frac{qV_A}{k_B T} \right) - 1 \right] \tag{3-16}$$

The typical *I–V* characteristics of a conventional solar cell, in the ideal case, is shown in Figure 3.12. The intercept of the *I–V* characteristics with the current axis is the short-circuit current, because at that point the voltage across the diode is zero, which is the case we described in Figure 3.11 (b). On the other hand, the intercept the *I–V* curve with the voltage axis is defined as the open-circuit voltage, V_{oc}, since the current is zero at that point, as if the circuit is open.

In the practical case, there might be two additional resistances associated with the ideal solar cells. The series resistance, R_s, is caused by the parasitic resistance from the circuit; for example, the contact resistance to the electrodes. The other one is that shunt resistance, R_{sh}. There is some fraction of photo-generated carrier that might not flow through the external load. Instead, it could flow through the grain boundary, the edge of the surface, or recombine during transport. These effects can be represented by a parallel resistance in the circuit and this parallel resistance is called shunt resistance. The equivalent circuit is shown in Figure 3-13.

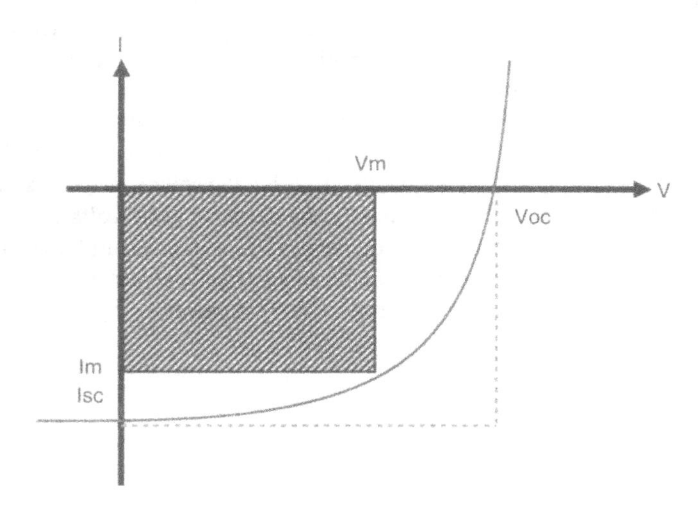

Figure 3.12. *I–V* characteristics of a conventional solar cell.

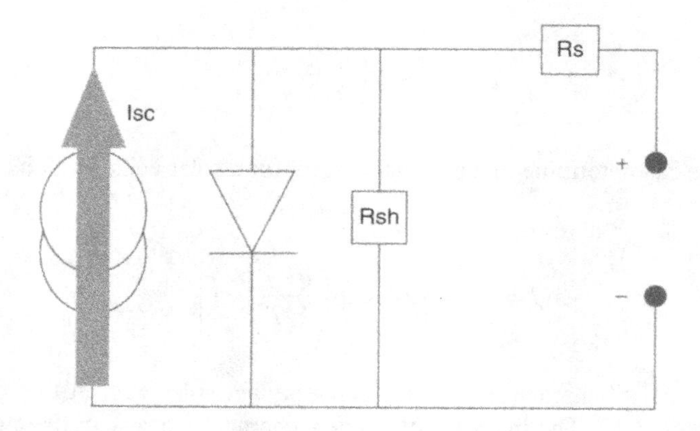

Figure 3.13. The equivalent circuit of a solar cell with a shunt resistance and a series resistance.

3.9 INTERACTION OF LIGHT AND MATERIALS

To realize an ideal photo-converter and better power conversion efficiency, certain requirements should be fulfilled when choosing materials for constructing solar cells or photovoltaic cells (PVCs). First of all, the band gap of materials plays an important role in the performance of PVCs. A suitable band gap for materials is needed to enable valid exciton formation and charge separation.

For metals, the highest occupied states lie in the conduction band and lead to no energy gap from the lowest unoccupied states. Therefore, carriers in excited states relax rapidly (on the order of femtoseconds) through the continuum of energy levels within the band, namely intraband transition, and charge separation cannot be realized as a result. Semiconductors and insulators both possess energy gaps. Nevertheless, insulators are not suitable for photovoltaics because their band gaps are so large, beyond the photon energies of visible light that is available from solar illumination. Besides, the conductivity of insulating materials is too low to facilitate effective charge separation. The situation is quite different when it comes to semiconductors. Since semiconductors have band gaps in the range of 0.5 to 3 eV, which is within the solar spectrum, and could provide reasonable conductivity, they are great candidates for photovoltaics.

The photocurrent of an ideal solar cell, which has perfect absorption so that all incident photons with energies larger than the band gap can be absorbed and used to pump one electron per photon to the excited states, can be represented by the following equation:

$$J_{sc} = q \int_{E_g}^{x} S(E)dE \tag{3-17}$$

where J_{sc} is the short-circuit current, q is the elementary charge, E_g is the band gap of materials, and $S(E)$, which is a function of photon energy, is spectral photon flux of light sources. This equation states the rule of thumb that photocurrent is actually a function of E_g and the spectrum of incident light. For a fixed-light spectrum, higher short-circuit current can be achieved with materials of lower band gaps. This result is obvious since lower band gap means that a wider range of light can be absorbed to pump electrons to excited states and to generate more conducting carriers. However, there is a trade-off between output current and output voltage in solar cells due to the effect that band gap has on the open-circuit voltage, V_{oc}, of solar cells, which is related to and limited to less than E_g according to the following equation:

$$V_{oc,max} = E_g(1 - T_{\text{ambient}}/T_{\text{sun}}) \quad (3\text{-}18)$$

where $V_{oc,max}$ is the theoretical maximal V_{oc}, T_{ambient} is ambient temperature, and T_{sun} is the temperature of our lovely sun. In an ideal solar cell, Voc could not exceed $V_{oc,max}$. If materials with small band gap are used, low output voltage will pull down power conversion efficiency (PCE), although the photocurrent is high. Likewise, small photocurrent in large band gap materials could not achieve high PCE even with higher output voltage. For a certain spectrum there is an optimal value of band gap for maximal PCE, and this is called the Shockley–Queisser efficiency limit after William Shockley and Hans Queisser, who published their work in 1961.

Figure 3.14 illustrates the Shockley–Queisser efficiency limit as a function of band gap under air mass (AM) 1.5 solar illumination. Requirements for band gap restrict the selection for materials. For III–V compounds gallium arsenide (GaAs) and in-

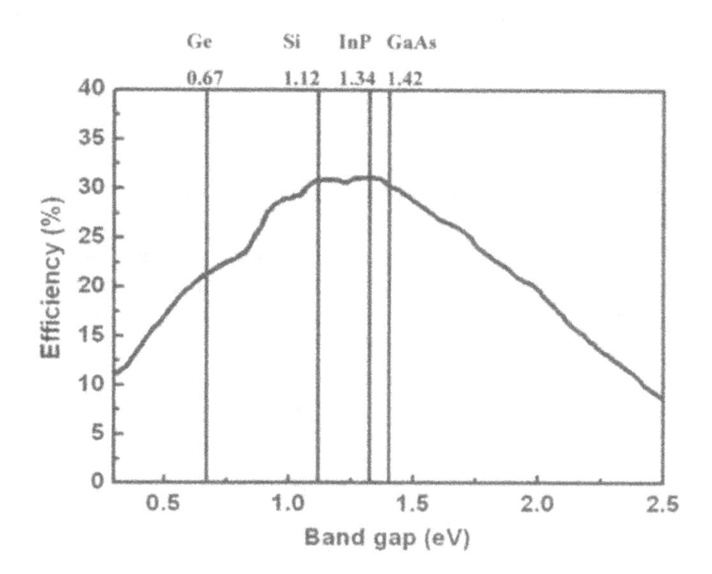

Figure 3.14. Shockley–Queisser efficiency limit as a function of band gap under AM 1.5 solar spectrum. Band gaps of four common semiconductors are specified as well.

dium phosphide (InP) have band gaps close to peak efficiency and have been developed for high-performance photovoltaics. Silicon is also favorable due to its availability on Earth and a mature silicon-based industry. Cadmium telluride (CdTe, $E_g = 1.44$ eV) and copper indium gallium diselenide ($CuIn_xGa_{(1-x)}Se_2$, Eg = 1.0 ~ 1.7eV) are used to fabricate thin-film PVCs.

The physical reason that the open-circuit voltage is limited to be less than E_g is due to the fast scattering process of the carriers with phonons in semiconductors. When electrons are excited to an energy level higher than the conduction band minimum, they quickly release their energy to the crystal by electron–phonon scattering until their energy is near the conduction band minimum. This process is very fast, in a time scale of picoseconds. Holes behave similarly. Therefore, before they diffuse to the P–N junction region, they have already lost some of their energy. As they move across the depletion region, the electrons only have energy near the conduction-band minimum and holes only have the energy near the valence-band maximum. Then the energy difference between electrons and holes that are extracted to the external circuit cannot be more than the band gap.

Light absorption of materials is another key factor in deciding the PCE of photovoltaics. Although photons with energy larger than the band gap of PV materials could be absorbed, not all the light go into this process. Although tens of microns of active layer could be used to absorb incident photons completely, thicker active layers also mean higher cost of unit cells, and the quality issues of active layer need to be dealt with more seriously.

The absorption length of PV materials is an index that can be used to address the absorbance of materials. Furthermore, absorption length is decided by the absorption coefficient α, with assumption that there is no reflection and scattering, as in the following equation:

$$I = I_0 e^{-\alpha x} \tag{3-19}$$

where I is light intensity after traveling a depth x in film, and I_0 is initial intensity of incident light. To effectively absorb incident light and elevate the PCE, α needs to be large in the chosen materials.

In Section 3.2, we describe how the band diagrams (energy vs. k curves), are calculated. From the band diagrams, the semiconductors can be categorized into two different kinds: one with direct band gap, in which the conduction band minimum and the valence band maximum are located at the same k point; and another one with indirect band gap, in which the conduction band minimum and the valence band maximum are at different k points. For direct band gap materials, absorption coefficient α can be typically calculated by the Fermi's Golden Rule and be approximated as

$$\alpha(E) \propto \alpha_0 (E - E_g)^{1/2} \tag{3-20}$$

where α_0 is a material dependent constant and E_g is band gap. On the other hand, α for indirect band gap materials can be expressed as

$$\alpha(E) \propto (E - E_g)^2 \tag{3-21}$$

By comparing Eq. (3-20) and Eq. (3-21), one can infer that α in indirect band gap semiconductors is smaller and grows more slowly than that in direct band gap semiconductors near the band gap region. From physical point of view, extra phonon absorption or emission is needed to fulfill momentum conservation when electrons are pumped up to the conduction band edge by photon absorption in indirect band gap materials, as shown in Figure 3.15. That requirement makes light absorption less likely to happen. This fact leads to an important conclusion: direct band gap semiconductors, such as GaAs, InP, CdTe, and $CuIn_xGa_{(1-x)}Se_2$, have better absorbance of light than indirect band gap semiconductors such as silicon or germanium. For this reason, silicon solar cells are typically thicker—about hundreds of micrometers—due to its inferior light absorbance. On the other hand, thin-film solar cells—less than 1 micrometer usually—could be fabricated from direct band gap semiconductors, as mentioned above.

In amorphous material, momentum conservation is always fulfilled in an optical transition because there is no long-range order and the band gap is virtually direct. Therefore, amorphous silicon could have higher absorption coefficient than that of its crystalline form.

3.10 SOLAR CELL MATERIALS

The development of the solar cells can be divided into three generations. The solar cells of the first generation are made of single crystals, such as silicon and gallium arsenide. The second-generation solar cells normally refer to those fabricated using thin film. The third-generation solar cells are those developed in recent years, such as polymer solar cells or dye-sensitized solar cells. In this section, we will briefly describe the properties of these solar cell materials.

3.10.1 Crystalline Silicon

Single-crystal silicon solar cells are one of the most effective solar cell technologies. They can achieve more than 20% efficiency, whereas other types of less expensive cells, such as thin film and polycrystalline, are only capable of achieving about 10%

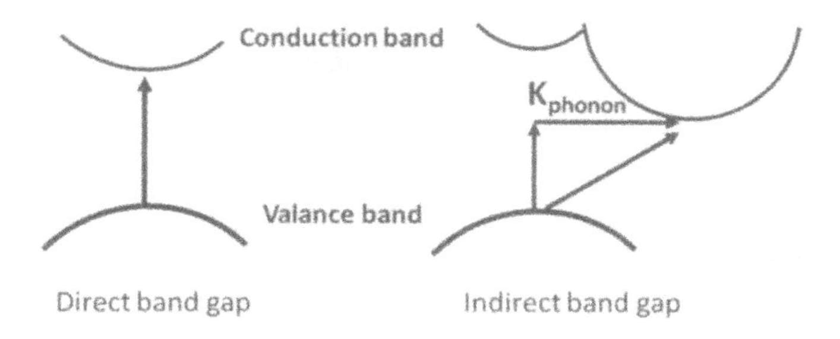

Figure 3.15. Electron excitation in direct and indirect band structures. Momentum conservation requires a phonon momentum K_{phonon} to allow photon absorption in indirect band gap materials.

efficiency. Thus, the single-crystal silicon-based solar panel is a great choice for urban area where space is limited.

Silicon can tolerate heavier doping of impurities and higher operation temperature. The silicon substrate can easily grow a layer of silicon dioxide in a thermal oxidation process. Silicon dioxide is a stable dielectric and can serve as a doping mask since most dopants, except phosphorus, diffuse more slowly in silicon dioxide than in silicon. Silicon has an indirect band gap of 1.1 eV, that is, the top of the valence band and the bottom of the conduction band are at different *k* points. The direct band gap of silicon is 2.2 eV, which is important for optical transitions since the absorption coefficient for photon above this energy is very strong. The crystal structure of silicon is like diamond: a face-centered-cubic with two atoms in a basis. Due to the symmetry of the fcc lattice, there are six degenerate X-points, which means six conduction-band edge valleys. Even though there is strong anisotropy in each valley, there is not as much anisotropy in the transport of electrons, involving the electrons moving in all of the valleys. The electron transport in silicon is poor because of the six-fold degeneracy of the conduction-band edge. The very large density of the states near the band edge leads to a high scattering rate. We can see that the electronic properties of silicon are far from ideal. It is other material properties, such as robustness, abundance in nature, ease of processing, and stability of silicon oxide, that make silicon the most important semiconductor for the electronics industry.

Silicon is the second-most element in the Earth's crust, about 27% by mass. It rarely occurs as the pure, free element in nature, but is usually found in the forms of silicon oxides. Silicon dioxide is the main component of quartz sands, and it can react with carbon to form silicon and carbon monoxide or carbon dioxide at high temperature. Silicon generated after this process is polycrystalline, called metallurgical-grade silicon (MGS). The chemical reaction from silicon dioxide to MGS can be expressed as

$$SiO_3 + 2C \rightarrow Si + 2CO \tag{3-22}$$

MGS needs to be purified in several steps for use in semiconductor-device manufacturing. First, MGS is ground into powder. Then the powdered silicon reacts with hydrochloride vapor to form trichlorosilane vapor. After going through filters, condensers, and purifiers, trichlorosilane vapor is turned into ultrahigh-purity liquid trichlorosilane. The chemical reaction during this step can be expressed as

$$Si + 3HCl \rightarrow SiHCl_3 + H_2 \tag{3-23}$$

Ultrahigh-purity liquid trichlorosilane is usually used as a silicon source. It can react with hydrogen to form high-purity polycrystalline silicon, which is called electronic-grade silicon (EGS). The chemical reaction is

$$SiHCl_3 + H_2 \rightarrow Si + 3HCl \tag{3-24}$$

The Czochralski (CZ) method is one of the most common methods to generate single-crystal silicon from EGS. In the Czochralski method, high-purity EGS is melted in a slowly rotating crucible, usually made of quartz. A rod-mounted seed of single-crystal silicon is dipped into the molten silicon. The temperature of the seed crystal is precisely controlled just a little lower than the melting point of silicon. After thermal equilibrium is reached, the seed crystal is slowly pulled upward and rotated simultaneously. Then a single-crystal silicon ingot is created by precisely controlling the temperature gradients, rate of pulling, and speed of rotation. Although the crucible may introduce contaminations of oxygen and carbon, the Czochralski method is more popular than the floating-zone method due to its low cost and capability of producing large size crystals. Some important parameters of silicon are listed in Table 3.1.

3.10.2 GaAs

In the early stage of the development of inorganic solar cells, the III–V compound semiconductors attracted lots of research because of their potential of high power-conversion efficiency and high thermal stability, which are particularly essential for applications in space power supply. Among several III–V compounds, gallium arsenide (GaAs) is one of the most promising choices owing to its excellent electronic and optical properties, such as direct band gap and large absorption coefficient. At present, the incorporation of GaAs is still the best candidate to produce highly efficient multijunction solar cells. In this section, the basic material properties and the history of the GaAs-based solar cell are briefly reviewed. Then the advantages of GaAs solar cells over the silicon-based devices as well as the limitations of the GaAs solar cells are dis-

TABLE 3.1. Properties of silicon

Characteristics	Values
Atomic number	14
Standard atomic weight	28.0855
Electron configuration	$3s^2\ 3p^2$
Melting point	1687 K
Boiling point	3538 K
Density of solid (RT)	2.33 g cm^{-3}
Molar volume	12.06 cm^3
Molar heat capacity	19.789 J mol^{-1} K^{-1}
Mohs hardness	6.5
Velocity of sound (RT)	8433 m sec^{-1}
Reflectivity	28%
Thermal conductivity	150 W m^{-1} K^{-1}
Coefficient of linear thermal expansion	2.6×10^{-6} K^{-1}
Crystal structure	Diamond cubic
Bond length in single-crystal silicon	235.2 pm

cussed. The final part includes the application and types of GaAs solar cells and the manufacturing process of GaAs.

GaAs is one of the typical III–V compounds with a zinc-blended structure. The lattice constant of GaAs is about 5.65 Å, which matches well with the lattice constant of AlAs. As a result, GaAs and AlAs are ideal material combinations to achieve optimal band gap or other parameters in heterostructure semiconductor devices. As for the optical properties, GaAs has a direct band gap of 1.42 eV, which is very close to the maximum of the solar spectrum, and its absorption coefficient in the frequency range of sunlight is also quite high, which makes GaAs a perfect material for applications in solar cells. The direct band gap of GaAs also expands its application to LEDs in visible red light. In addition, GaAs has extremely high electron mobility, which is required in high-frequency transistors, and relatively good thermal stability and resistance to radiation, which further ensures its superiority in practical applications over other semiconductor materials.

To provide a light and reusable energy source for space exploration, solar cells were intensively developed and the manufacturing techniques were greatly improved after the 1950s. The first solar cell used in spacecraft was made of single-crystal silicon and had efficiency of about 10%. However, after the demonstration of highly efficient GaAs heterostructure solar cells in 1970, the GaAs-based solar cells got more and more attention and the efficiency was increased to 19%. Owing to its high thermal stability and resistance to radiation, GaAs is more suitable than silicon in applications in outer space. As a result, both the physical investigation and the processing technology of GaAs-based solar cells made great progress in this period because of support from the USA and USSR. Nowadays, the multijunction solar cells with GaAs still exhibit the highest efficiency—above 40%—and are one of the best choices for energy sources in outer space.

Despite silicon's abundance and mature processing technology, GaAs has several superior electronic, optical, and physical properties for applications in photonic devices, such as LEDs, transistors, and solar cells. The reasons are briefly discussed in the following.

First, GaAs has a direct band gap, whereas silicon has an indirect band gap. The direct band gap allows the material to emit and absorb photons more efficiently because it does not require the participation of phonons to complete the electron transition, as illustrated in Figure 3.16. Besides, the band gap of GaAs is 1.42 eV, which is much more efficient than silicon (1.12 eV) in absorbing photons in the range of solar radiation.

Since GaAs has an outstanding absorption coefficient in a wide range of solar spectrum, it only needs a very thin film (several microns) to absorb sun radiation, whereas a traditional silicon-based solar cell requires a thickness of more than 100 microns. A thinner film can not only save the materials but also provide the possibility of the flexible devices.

Another important advantage of GaAs for application in solar cells is its great stability in high-temperature situations and high resistance to radiation damage. This property indicates that the performance of the GaAs-based solar cells will not be af-

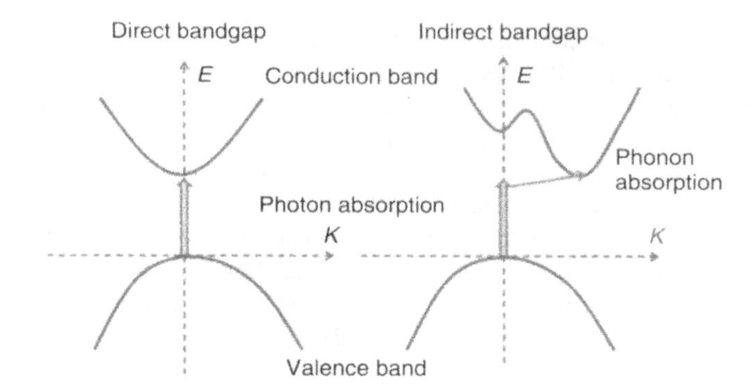

Figure 3.16. Scheme of direct bandgap and indirect bandgap. The indirect bandgap requires the participation of phonons to perform a transition, whereas the direct bandgap does not.

fected enormously as the temperature increases to a certain extent, such as 100 °C. On the other hand, the power conversion efficiency degradation under thermal stress in silicon-based solar cells is quite severe, which hinders the devices from operating effectively in outer space or under other high-temperature conditions.

The main limitation of GaAs for the practical applications is its extremely high cost and relatively complex and difficult manufacturing. First, both gallium and arsenic are not as abundant as silicon, thus the cost of materials is quite high. Since GaAs is a compound instead of a pure material, the fabrication of GaAs-based devices requires strict controls of the material composition and crystallization. In reality, the yield and quality of GaAs processing are not ideal, and the fabrication of large-area GaAs films is too difficult to be realized.

Due to its excellent material properties but the high cost of fabrication, the GaAs-based techniques were only incorporated in the devices for space exploration in the past. With the recent development of new technology and the use of renewable green energy, GaAs is now used in highly efficient solar cells to achieve lowest cost per unit energy. Multijunction solar cells, formed by stacking several single-junction cells, is one of the effective methods to achieve high power-conversion efficiency. With several single junctions of different materials and band gaps, the absorption of multijunction cells can cover a wider wavelength range and work more effectively. However, the fabrication of multijunction solar cell is not so easy and straightforward. There are many issues that need to be taken into consideration, for instance, the optical loss and reflection, the current match, and the stability of mechanically stacking multiple thin layers. Today, there are still numerous attempts being made to investigate and resolve these problems to improve the efficiency of multijunction solar cells.

A concentrator photovoltaic (CPV) is a system containing some optical elements, such as a Fresnel lens, to condense sunlight from a larger area to a smaller spot on photovoltaic devices. With this technique, it does not require a huge amount of solar cells to cover the region of sunlight radiation, greatly saving the cost of devices per unit area. Nonetheless, as the condensation efficiency is strongly dependent on the direction of the light source, the CPV systems need the capability of tracking the sunlight to reach the expected power conversion efficiency, which further increases the complexity of processing and limits the applications.

The manufacturing of GaAs-based solar devices includes the growth of single-crystal GaAs with various dopants for each thin layer. The fabrication of single-crystalline GaAs is often performed by molecular-beam epitaxy (MBE) or metal–organic chemical-vapor deposition (MOCVD). Nowadays, lowering the processing cost and improving the quality of the crystalline GaAs are the critical issues for extending the practical applications of GaAs-based solar cells.

3.10.3 Thin-Film Silicon

Since the 1960s, amorphous silicon (a-Si) has been widely studied and used in the semiconductor industry. In the 1970s, amorphous silicon was reported to be suitable for solar cells and started to be used, and it was put into commercial production in 1980s. Amorphous silicon is one of the best-developed materials among thin-film solar cell materials and it is relatively cheap because of the abundance of silicon.

The lattice structures of amorphous silicon are different from ordered crystalline silicon (c-Si), as shown in Figure 3.17. Every silicon atom in amorphous silicon still bonds to the other four atoms but there is no regular and periodic arrangement. In other words, it does not have a long-range order but has a local order. Compared with single-crystalline silicon, amorphous silicon has many advantages. For example, the absorption coefficient of visible light is about one order of magnitude higher than crystalline silicon; that is, the thickness needed to absorb sunlight is less than 1 micrometer for amorphous silicon, which is much thinner than that of crystalline silicon used in solar cells. Moreover, amorphous silicon has a relatively low processing temperature owing to having no requirement of high crystallization. Therefore, it is possible to grow it on a variety of substrates, like plastic, glass, and metal. However, there are also some dis-

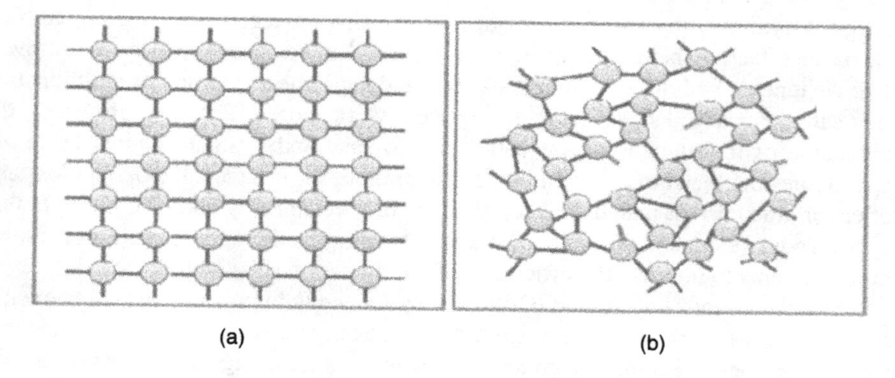

Figure 3.17. Single-crystal silicon (a) and amorphous silicon (b).

advantages of amorphous silicon, including lower efficiency for transferring energy and severer decay under sunlight.

Although the majority of silicon atoms in amorphous silicon still bond to the four nearest neighboring atoms as in single-crystalline materials, there are still some atoms bonding to only three neighboring atoms and leaving one unbonded valence orbital. This defect is called dangling bond and is shown in Figure 3.18. The dangling bonds will give rise to extra states in the energy band gap. As a result, the atomic structure in amorphous silicon is no longer long-range compared with single-crystalline silicon; that is, the distances and orientations in silicon atoms are various. This disordered arrangement causes more defects in states in band gap and alters the band structure of silicon. The distribution of energy states will be broadened into a tail at the top of valence band and the bottom of the conduction band as compared to crystalline silicon. Figure 3.19 shows the schematic representation of the band-tail effect on the density of states, known as the Urbach tail. In addition, the single-crystalline silicon has an indirect energy band gap, and the absorption of photons in visible light is limited by momentum conservation. However, the absence of an ordered arrangement indicates that the E–k relationship is not well defined in amorphous materials. In other words, the absorption of photons does not require the simultaneous participation of phonons. Therefore, the absorption coefficient of amorphous silicon in visible light is higher than that of crystalline silicon. Moreover, the energy band gap can be adjusted by adding alloys and hydrogen.

In order to fabricate a solar cell device, we need doping to form the n-type or p-type amorphous silicon. Hydrogenated amorphous silicon has many advantages over unpassivated amorphous silicon. First, the defect density in amorphous silicon is so high that it is hard to be doped effectively, because the extra electrons or holes provided by the impurity doping will be captured by defects of dangling bonds. Fortunately, in hydrogenated amorphous silicon, the defect density and the number of dangling

Figure 3.18. Some silicon atoms are bonding to only three other atoms and leaving a valance orbital in a-Si material. This is known as a "dangling bond."

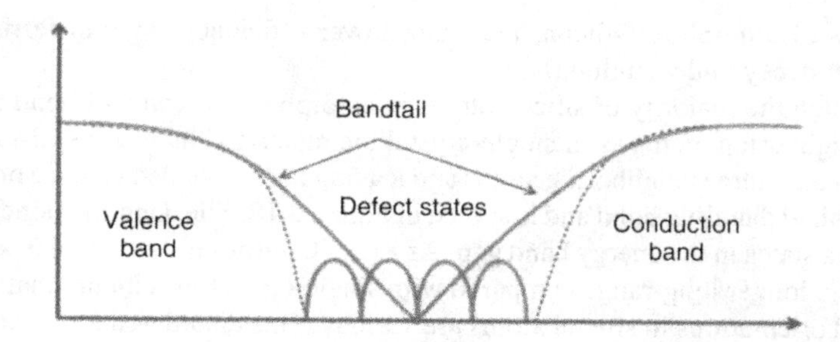

Figure 3.19. Schematic representation of the density of states in a-Si material.

bonds are much smaller, which helps realize the doping of both n-type and p-type materials. However, the doping efficiency is still not high enough due to the remaining defects. As in the case of single-crystalline silicon, the n-type doping and p-type doping in amorphous silicon are realized by the incorporation of group V materials, such as phosphorus, and group IV materials, such as boron, respectively.

The basic device structure is a p–i–n junction instead of a p–n junction in amorphous silicon solar cells. The intrinsic region is used as the absorption layer and the p-type region is used as the window. In the p–i–n structures, the n-type area and the p-type area will provide a built-in voltage, which can help the photo-induced electrons and holes move in opposite directions to the cathode and the anode, respectively.

The passivation of dangling bonds with hydrogen in amorphous silicon will decrease the density of defects and increase the absorption of photons. The energy band gap is also altered in hydrogenated amorphous silicon. However, there is a problem in this material compared with other solar cell materials. The density of defects in hydrogenated amorphous silicon will increase as the time of light exposure increases, which largely drops both the dark current and the photon current and decreases the efficiency of energy conversion. Nevertheless, after annealing at high temperature, about a few hundred degrees, the devices will be restored to perform efficiency as before. It is reported that the bonds of Si-H will be broken under sunlight to increase the density of dangling bonds and cause hydrogen to diffuse. This will produce extra recombination centers and reduce the device's efficiency. The reduction in efficiency is dependent on the initial atomic bonding, the power of sunlight, and the environment temperature. This light-induced degradation is the major issue for hydrogenated amorphous silicon in applications of solar cells.

3.10.4 Cu-In-Ga-Se (CIGS)

Recent interest in the development of thin-film photovoltaics stems from the promise of achieving low-cost, high-power conversion efficiency and good thermal stability of solar cells. For these purposes, a novel material with the properties of feasibility in modern technology, narrow band gap, and low light recession is needed. As such, an

I–III–VI compound semiconductor material, composed of copper, indium, and selenium, was synthesized at Bell Labs to match these properties in 1974. With strong light absorption, the first copper–indium–selenide solar cell possesses significantly high power conversion efficiency, over 6%. Later, in 1982, the Boeing company produced a copper–indium–selenide solar cell with higher power conversion efficiency of 10% by introducing cadmium sulfide (CdS) as a buffer layer. After that, more and more reports proposed gallium as an alternative to indium since there is a shortage of indium deposits and the use of gallium increased the open-circuit voltage due to its relatively large optical band gap to achieve a tremendous power conversion efficiency of 19.9%. Therefore, growing research in copper–indium–gallium–selenide will provide high power conversion efficiency photovoltaic solar cells in the future.

Basically, copper–indium–gallium–selenide is an I–III–VI compound semiconductor material often abbreviated as CIGS. With a chemical formula of $CuIn_xGa_{(1-x)}Se_2$, this material is actually a solid solution that consists of copper indium selenide (CIS) and copper gallium selenide. By varying the value of x from 0 (pure copper gallium selenide) to 1 (pure copper indium selenide) the electrical properties and mechanical properties of CIGS can be tuned easily. For instance, the optical band gap of CIGS varies continuously from 1.68 eV to about 0.98 eV with value of x ranging from 0 to 1. In this way, the composition of CIS and copper gallium selenide makes CIGS flexible, which will lead to applications in photovoltaic cells.

Another advantage of using CIGS in photovoltaic cells originates in the intrinsic property of CIGS, a direct band gap material. Since it is a direct band gap material, CIGS possesses strong light absorption, so only a few micrometers of CIGS is needed to absorb sufficient sunlight. In addition, CIGS has a relatively high absorption coefficient of more than 10^5 per centimeter for 1.5 eV or higher energy photons. This is quite a large value that is about 100 times the absorption coefficient of single-crystal silicon. Accordingly, compared with conventional silicon solar cells, a much thinner layer of CIGS is required for the same quantity of absorption, which leads to less demand for raw materials.

Structurally, CIGS is a tetrahedrally bonded semiconductor with density of 5.7 g/cm^3 and a chalcopyrite crystal structure as shown in Figure 3.20. However, heat can deform the crystal structure of CIGS into the zincblende form when the transition temperature increases from 805 °C for $x = 1$ to 1045 °C for $x = 0$. Owing to the blend of CIS and copper–gallium–selenide, the lattice constants of CIGS are no longer constant. Both a and c depend on the value of x with slight variance; a increases from 0.56 nm for $x = 0$ to 0.58 nm for $x = 1$, whereas c increases from 1.10 nm for $x = 0$ to 1.15 nm for $x = 1$.

Unlike the conventional silicon photovoltaics form by single-crystalline homojunction, the structure of CIGS photovoltaics is a more complex heterojunction system in the form of polycrystalline α-phase with the chalcopyrite crystal structure. In an indium-rich CIGS, there is a copper deficiency so that the surface of the CIGS film forms an ordered defect compound. With an overall copper deficiency in CIGS structure, the majority carrier concentration—the hole concentration—can be enhanced by increasing the number of copper vacancies as the copper vacancies act as electron acceptors.

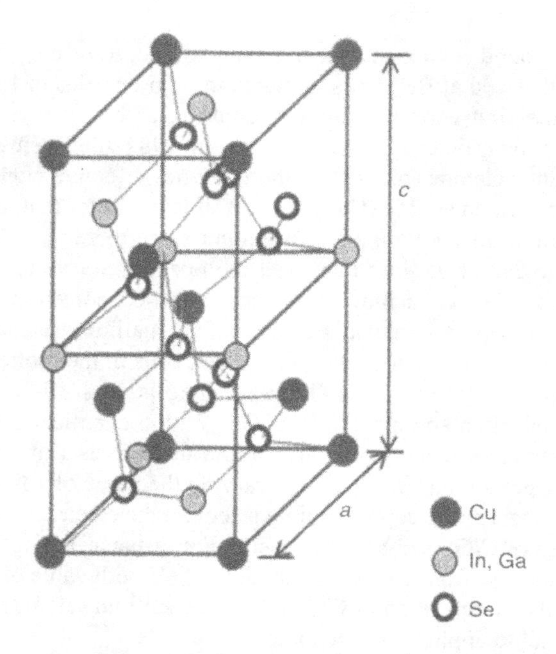

Figure 3.20. CIGS unit cell.

A p–n homojunction can also be formed in CIGS films at the interfaces between the ordered defect compound and the α phase by treating the ordered defect compound as n-type. Even though CIGS photovoltaic effect decreases when the ratio of copper to group III elements in CIGS is larger than 0.95, the CIGS photovoltaic still draws lots of attention based on its thermal stability. This I–III–VI compound semiconductor material was reported to have no degradation and merely lost several percentages of power conversion efficiency in light recession over 10 years.

To be considered for use in the important next-generation solar cells, CIGS can be deposited in several different ways, either directly or by crystallizing precursor films. The most common method is vacuum-based process combining coevaporation or cosputter of copper, indium, and gallium, following by annealing the resulting film in the surrounding of selenium vapor to form the final CIGS compound semiconductor. On the other hand, there are still some nonvacuum-based processes that can provide CIGS structure. One can first subject the the substrate to deposition of nanoparticles of the precursor materials and sinter the resulting structure in situ to get a CIGS absorber. Other applicable ways of CIGS fabrication are electroplating, nano-imprint lithography, and solution processes. All of them are lower-cost alternatives to expensive vacuum processes and provide increased throughput uniformity over large area, making them commercially viable.

Moreover, it is not necessary to use conductive glass as a substrate for CIGS photovoltaics. Even a common building material such as sodium glass can be used as the substrate for CIGS photovoltaics due to their superior optical and electrical properties. Also, replacement of the glass substrate by novel flexible substrates such as polyimide or metal foils allows CIGS photovoltaics to be light weight and flexible, which leads to their adaptability to roll-to-roll fabrication. Based on these superiorities, CIS and CIGS have been recognized and partially realized as next-generation photovoltaic materials due to their attractive features that include being able to be used on flexible substrates, generally lower processing cost, easy fabrication, high stability, and high power-conversion efficiency.

3.10.5 Polymer Solar Cell Materials

Solution-processable solar cells show great possibility for providing a low-cost route for producing flexible, light-weight, and high-throughput solar energy conversion devices. Polymer-based photovoltaics have attracted considerable attention due to their potential of a low-cost source to harvest solar radiation energy. The conjugated polymers have led to the development of organic solar cells, with the ability for roll-to-roll manufacturing onto many kinds of substrates such as plastic and cloth. The optimization of device structures and the introduction of more stable conjugated polymers are critical in order to achieve better performance in organic solar cells. The typical device is made by spin coating the conjugated polymers (electron donor), which absorb solar radiation, onto the substrate. The blend of conjugated polymers and electron acceptors such as [6,6]-phenyl-C61 butyric acid methyl ester (PCBM) has been most intensively studied as an active layer in recent year due to its great power conversion efficiency of 5–6%.

In order to increase the charge extraction rate in the active layer, it is also important to consider the orientation of the polymer component in the organic blend. Many conjugated polymers are semicrystalline polythiophenes, in which high charge mobility occurs along the backbone chain of the polymer parallel to the π–πstacking direction of the thiophene rings, as shown in Figure 3.21. In polymer/fullerene bulk heterojunction (BHJ) solar cells, charge must be transported between two electrodes, perpendicular to the substrate in Figure 3.22. Therefore, theπ–πstacking structure should be largely perpendicular to the substrate to form a high-mobility pathway for charge carriers.

The performance of BHJ solar cells depends on the efficiency of photogenerated excitons to reach the heterojunction between the donor materials and acceptor materials. The ability of photogenerated free carriers to escape to the electrodes is also a critical point. The exciton diffusion length of conjugated polymers such like poly(3-hexylthiophene) (P3HT) has been reported to be in the range of 10 nm. Based on this diffusion length, the thickness of an active layer of about 200 nm would be ideal. The efficient pathways require the donor and acceptor to be distributed within a maximum distance of 10 nm because of the limited diffusion distance of photogenerated excitons. Also, the length of the pathway for electron and hole transporting to the anode and cathode is about 100–500 nm. However, the phase-separated morphology is difficult to control and increasing the efficiency of the devices is difficult. The coulombically

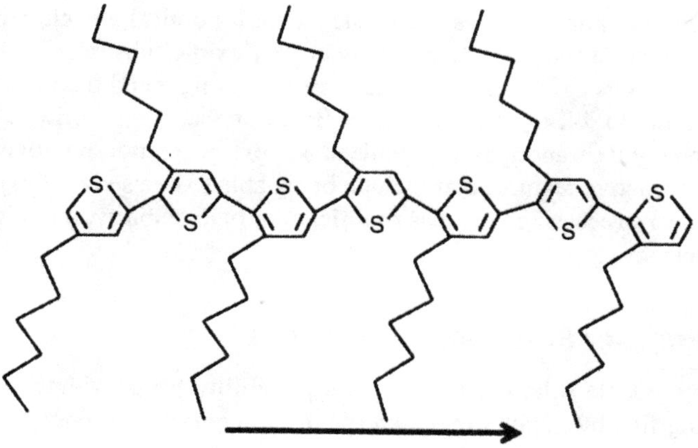

Figure 3.21. The π–π stacking direction of P3HT conjugated polymer.

bounded electron–hole pair is formed upon light radiation, which governs the performance of device by the nanoscale morphology of the active layer. The interaction energy of electrons and holes varies inversely with the dielectric constant of the surrounding medium. Particles in a lower dielectric material such as polymer matrix have a larger binding energy than particles in an inorganic semiconductor, which has a higher dielectric constant. Hence, the lifetime of the electron–hole pair is short and the exciton diffusion length in the polymer-based active layer is limited. Some groups synthesize polymers with higher dielectric constants or accurately control the morphology of the thin film to overcome the high binding energy issue.

On the other hand, it has been established that the value of the open-circuit voltage (V_{oc}) in an organic solar cell is related to the energy difference between the lowest unoccupied molecular orbital (LUMO) of the acceptor and the highest occupied molecular orbital (HOMO) of the donor. For the P3HT:PCBM based solar cells, the Voc is relatively small, around 0.6 V, due to the limitation of energy levels. Controlling the HOMO level of the donor and/or the LUMO level of the acceptor is an efficient way to enhance the V_{oc} of solar cells. Several research groups have put a lot of effort into the

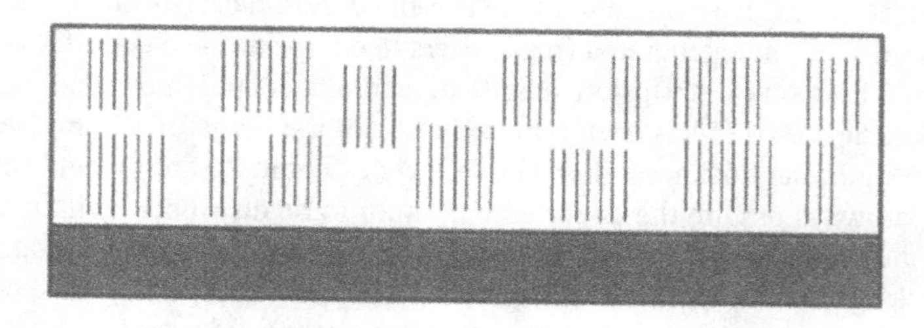

Figure 3.22. Polymer lamellae orientated perpendicular to the substrate.

synthesis of conjugated polymers with deep HOMO levels. By modifying the chemical structure of polymers, the energy levels and band gap of donor materials can be easily changed. For instance, several deep-HOMO conjugated polymers like BisDMO-PFDTBT, PSiFDBT, and PCDTBT have been applied in polymer-based photovoltaics, achieving high PCEs of about 5–6% due to the high V_{oc} of about 1 V. Typically, varying the length of the alkyl side chain and chemical structure of the polymer are effective methods to synthesize polymers with larger HOMO values. The energy level difference of polymers between the HOMO and LUMO decreases as the length of conjugation increases. However, the band gap decreases when the number of monomer units exceeds a saturated value.

To achieve the maximum potential, an ideal conjugated polymer for photovoltaics should not only have a deeper HOMO level but also low band gap. Consequently, side-chain-conjugated polythiophenes are used to enhance the absorption of solar irradiation. Since the photon flux of sunlight has a peak at about 680 nm, several low band gap polymers have been synthesized and used in organic solar cells. Commonly, polythiophenes with conjugated side chains present a broader response range to the solar irradiation spectrum, leading to an obvious improvement in conversion efficiency of devices. For example, the polymer solar cells using PCPDTBT and PSBTBT, two low band gap materials, showed a higher short-circuit current (J_{sc}) because of their broad absorption.

The polymer solubility and carrier mobility are the two critical factors when considering polymer synthesis to achieve great performance in devices. The solubility issue can directly affect many other important polymer physical parameters such as surface morphology, phase behavior, and crystallinity, which strongly govern the device performance of solar cells. The solubility of one polymer is determined by several factors of structure, including the chain length of the aliphatic groups, the degree of polymerization, the polarity of the attached substituents, polymer regularity, and intermolecular interactions. Generally speaking, polymers with long side chains on backbones usually have good solubility. Without affecting the energy levels of polymer materials, the attached side chains can be saturated alkyl groups. Furthermore, branched alkyl chains are more critical for inducing solubility than their straight-chain counterparts. Also, it is known that strong interchain π–π stacking interactions are the major reason that most polyaromatic conjugated polymers are insoluble. It is necessary to improve the solubility by introducing aliphatic side chains that are covalently attached to the conjugated main chain of polymers. Moreover, interchain hopping of carriers requires a favorable overlapping of the electron wave function of neighboring conjugated units on the main chains. The use of bulky side chains, increasing the content of insulating alkyl chains relative to the hole–conductor portion in the polymer, may lead to deterioration in the charge mobility function. However, a short side chain will result in poor solubility and processibility. Therefore, choosing an appropriate size for the side chain is very important in conjugated polymer synthesis.

In conclusion, the ideal materials for donor–acceptor pairs should exhibit (1) appropriate HOMO and LUMO levels to ensure a large V_{oc} and a small energy offset for exciton dissociation, (2) formation of an interpenetrating network with the ideal morphology for creating two distinct bicontinuous pathways for photogenerated charge carrier transport, (3) adequate solubility of donor polymers to guarantee solution processability as well as miscibility with acceptor materials, (4) low optical band gap for a strong and broad absorption range in the solar irradiation spectrum to capture more solar energy, and (5) high hole mobility for speeding up charge transport, which results in a thicker active layer required for increased light harvesting to suppress charge recombination and series resistance.

REFERENCES

1. A. Luque (2011), "Will we Exceed 50% Efficiency in Photovoltaics?" *J. Appl. Phys., 110,* 031310.
2. T. Markvart, L. Castañer, and V. M. Andreev (2005), "High-efficiency Concentrator Silicon Solar Cells," *Solar Cells Materials, Manufacture and Operation,* Elsevier, pp. 353–369.
3. Y. Shimizu and Y. Okada (2004), "Growth of High-Quality GaAs/Si Films for Use in Solar Cell Applications," *Journal of Crystal Growth, 265,* 99–106.
4. J. Nelson (2003), "GaAs Solar Cell Design." *The Physics of Solar Cells,* Imperial College Press, pp. 204–210.
5. A. Romeo, R. Gysel, S. Buzzi, D. Abou-Ras, D. L. Bätzner, D. Rudmann, H. Zogg, and A. N. Tiwari, "Properties of CIGS Solar Cells Developed with Evaporated II-VI Buffer Layers," in *Technical Digest of the International PVSEC-14,* Bangkok, Thailand (2004).
6. E. Verploegen, R. Mondal, C. J. Bettinger, S. Sok, M. F. Toney, and Z. Bao (2010), "Effects of Thermal Annealing Upon the Morphology of Polymer–Fullerene Blends," *Adv. Funct. Mater., 20,* 3519.
7. H. Xiao (2001), *Introduction to Semiconductor Manufacturing,* Prentice-Hall.
8. H. Okaniwa, K. Nakatani, M. Yano, M. Asano, and K. Suzuki (1982), "Proceedings of 3rd Photovoltaic Science and Engineering Conference," *Japanese Journal of Applied Physics, 21,* 239–244.
9. D. Staebler and C. Wronski (1977), "Reversible Conductivity Changes in Discharge-Produced Amorphous Si," *Appl. Phys. Lett., 31,* 292–294.
10. L. Yang and L. Chen (1991), "Proceedings of the Fourteenth International Conference on Amorphous Semiconductors-Science and Technology," *Journal of Non-Crystalline Solids, 137–138,* 1189–1192.
11. A. Klaver (2008), R.A.C.M.M. van Swaaij, "Modeling of Light-Induced Degradation of Amorphous Silicon Solar Cells." *Solar Energy Materials and Solar Cells, 92,* 50–60.
12. X. Deng and E.A. Schiff (2003), "Amorphous Silicon-Based Solar Cells," in *Handbook of Photovoltaic Science and Engineering,* Edited by A. Luaue and S. Hegedus, Wiley, pp. 530–534.
13. A. Banerjee, G. DeMaggio, K. Lord, B. Yan, F. Liu, X. Xu, K. Beernink, G. Pietka, C. Worrel, B. Dotter, J. Yang, and S. Guha (2009), "Advances in Cell and Module Efficiency of a Si-Based Triple-Function Solar Cell Made Using Roll-to-Roll Deposition," in *Proceedings of 34th IEEE Photovoltaic Specialists Conference,* pp. 116–119.

14. W. N. Shafarman, R. Klenk, and B. E. McCandless (1996), "Characterization of $Cu(InGa)Se_2$ Solar Cells with High Ga Content," in *Proceedings of 25th IEEE Photovoltaic Specialist Conference,* pp. 763–768.
15. V. K. Kapur, A. Bansal, P. Le, and O. I. Asensio (2003), "Non-vacuum Processing of $CuIn_{1-x}Ga_xSe_2$ Solar Cells on Rigid and Flexible Substrates Using Nanoparticle Precursor Inks," *Thin Solid Films,* 431–432, pp. 53–57.
16. S. Yoon, T. Yoon, K.-S. Lee, S. Yoon, J. M. Ha, S. Choe (2009), "Nanoparticle-based Approach for the Formation of CIS Solar Cells," *Solar Energy Materials and Solar Cells, 93,* 783–788.
17. E.-J. Bae, J.-M. Cho, J.-D. Suh, C.-W. Ham, K.-B. Song (2010), "Volatile Electrical Switching and Static Random Access Memory Effect in a Functional Polyimide Containing Oxadiazole Moieties," in *Proceedings of 35th IEEE Photovoltaic Specialists Conference,* 3391–3393.
18. D. L. Schulz, C. J. Curtis, R. A. Flitton, H. Wiesner, J. Keane, R. J. Matson, K. M. Jones, P. A. Parilla, R. Noufi, and D. S. Ginley (1998), "Cu-In-Ga-Se Nanoparticle Colloids as Spray Deposition Precursors for $Cu(In,Ga)Se_2$ Solar Cell Materials," *Journal of Electronic Materials, 27,* 433–437.
19. S.-H. Wei, S. G. Zhang, and A. Zunger (1999), "Effects of Na on the Electrical and Structural Properties of $CuInSe_2$," *Journal of Applied Physics, 85,* 7214–7218.
20. L. Kronik, D. Cahen, and H. W. Schock (1998), "Effects of Sodium on Polycrustalline $Cu(In,Ga)Se_2$ and Its Solar Cell Performance," *Advanced Materials, 10,* 31–36.
21. D. Rudmannm A. F. da Cunha, M. Kaelin, F. Kurdesau, H. Zogg, A. N. Tiwari, and G. Bilger (2004), "Efficiency Enhancement of $Cu(In,Ga)Se_2$ Solar Cells Due to Post-Deposition Na Incorporation," *Applied Physics Letters, 84,* 1129–1131.
22. T. Nakada, D. Iga, H. Ohbo, and A. Kunioka (1997), "Effects of Sodium on $Cu(In,Ga)Se_2$-Based Thin Films and Solar Cells," *Japanese Journal of Applied Physics, 36,* pp. 732–737.
23. D. Rudmann, G. Bilger, M. Kaelin, F.-J. Haug, H. Zogg, and A. N. Tiwari (2003), "Effects of NaF Coevaporation on Structural Properties of $Cu(In,Ga)Se_2$ Thin Films," *Thin Solid Films, 431–432,* 37–40.
24. S. Ishizuka, A. Yamada, P. Fons, and S. Niki (2009), "Flexible $Cu(In,Ga)Se_2$ Solar Cells Fabricated Using Alkali-Silicate Glass Thin Layers as an Alkali Source Material," *Journal of Renewable and Sustainable Energy, 1,* 013102.
25. D. Cahen and R. Noufi (1989), "Defect Chemical Explanation for the Effect of Air Anneal on $CS/CuInSe_2$ Solar Cell Perfomance," *Applied Physics Letters, 54,* 558–560.

EXERCISES

1. What is the built-in voltage of a P–N junction?
2. Calculate the depletion region width of a P–N junction diode with an abrupt junction as a function of doping concentration and built-in voltage.
3. Calculate the depletion region width of a P–N junction diode with a linear graded doping profile at the interfaces. Set the highest doping concentration at the interfaces and then linearly decrease to zero away from the interfaces.

4. Derive the ideal diode current as shown in Eq. 3-13:

$$I = I_0 \left[\exp\left(\frac{qV_A}{k_B T} \right) - 1 \right]$$

5. Explain why the open-circuit voltage is always less than bandgap for inorganic semiconductor solar cells.
6. Why does an indirect-bandgap semiconductor have a smaller absorption coefficient than a direct-bandgap semiconductor?

4

ORGANIC MATERIALS

Wei-Fang Su

Organic materials are made of organic molecules joined together either by primary chemical bonding or by secondary bonding, such as van der Waals forces, hydrogen bonding, and coordinate bonding. The molecule can be either small or large (molecular weight larger than 10,000 dalton). The large molecules are usually more flexible and durable than the small molecules. Small molecules are held together mostly by weak van der Waals forces. Large molecules are made by chemically bonding small molecules called monomers. Large molecules may also be helding together by van der Waals forces. Organic materials are derived mainly from carbon and hydrogen and have a large free volume point, so they exhibit the advantages of light weight, flexibility, low melting and ease of processing as compared with inorganic materials.

Organic materials can be designed and synthesized to meet required physical properties for specific applications such as solar cells. The synthesis of organic material can be accomplished at relatively low temperature (< 200°C) using raw materials derived from fossil fuels. Recently, raw materials made from renewable sources such as plants and biomass using bioengineering processes have become available. This subject is beyond the scope of this book. Table 4.1 shows some examples of organic molecules obtained from fossil fuels. They serve as raw materials that will undergo further chemical reactions to generate materials with desired chemical structure and composition for electrical and optical applications. The thiophene molecule can undergo polymerization to form polythiophene, which has been used in polymer solar cells. Instead of burning fossil fuel to generate energy with the release of environmentally harmful CO_2, the fossil fuel is used to synthesize organic materials with the capability of generating renewable energy. Thus, the use of organic materials for solar cells conserves material and energy. This technology is presently being explored extensively.

In this chapter, we will discuss the formation of small and large organic molecules and how their chemical structure is related to their physical properties, with the emphasis on the electrical and optical properties.

Organic, Inorganic, and Hybrid Solar Cells. By C.-F. Lin, W.-F. Su, C.-I Wu, and I-C. Cheng

TABLE 4.1. Examples of organic molecules derived from fossil fuels

Chemical structure	Common name
$H_2C{=}CH_2$	Ethylene
$HC{\equiv}CH$	Acetylene
CH_3CH_2OH	Ethanol
$H_3C{-}CH{-}CH_2$	Propylene
[benzene ring]	Benzene
[benzene ring]$-CH_3$	Toluene
[thiophene ring] S	Thiophene

4.1 BONDING AND STRUCTURE OF ORGANIC MOLECULES

Organic molecules are made by covalent bonding of carbon atoms. The electron configuration of the carbon atom is shown in Figure 4.1. The 1*s* shell is filled and the four valence electrons are in the 2*s* orbital and two different 2*p* orbitals [Fig. 4.1(a)]. The 2*s* orbital has a slightly lower energy than the three 2*p* orbitals, which have equal energies. The electron orbitals can be hybridized into three different orbitals: sp^3 orbital [Fig. 4.1(b)], sp^2-p orbital [Fig. 4.1(c)], or sp-p-p orbital [Fig. 4.1(d)]. A sigma (σ) single bond is formed by overlapping two sp^3 orbitals of carbon atoms. A pi (π) double bond results from the overlapping of two sp^2-p orbitals of carbon atoms, and a triple bond is obtained by overlapping two sp-p-p orbitals. Table 4.2 shows the comparison of C–C, C=C, and C≡C bonds. The comparison of bond length is in the decreasing order of single bond > double bond > triple bond. The freedom of rotation is in decreas-

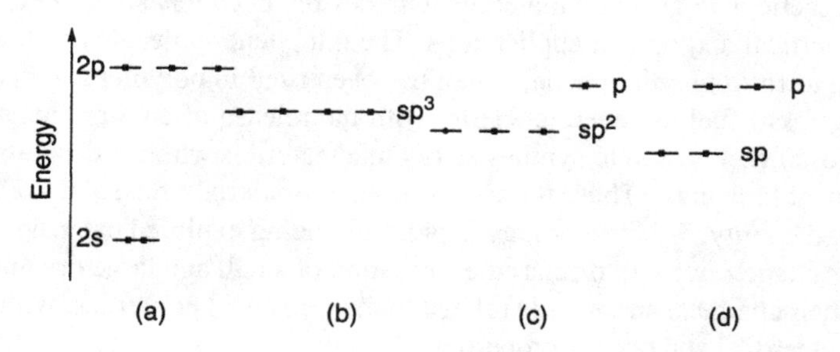

Figure 4.1. Electron configuration of the carbon atom. (a) Unhybridized orbitals, (b) hybridized sp^3 orbitals, (c) hybridized sp^2-p orbitals, (d) hybridized sp-p-p orbitals.

TABLE 4.2. Comparison of C–C, C=C, and C≡C bonds

Property	C–C	C=C	C≡C
Number of atoms attached to a carbon	4 (tetrahedral)	3 (trigonal)	2 (linear)
Rotation of bond	free	restricted	restricted
Bond angle	109.5°	120°	180°
Bond length	1.54 Å	1.34 Å	1.21 Å

ing order as well. The comparison of bond angle is in reversed order; the triple bond is rather flat with the largest bond angle of 180°. Many molecules contain electrons that are not directly involved in bonding. These are called nonbonding or *n* electrons and are mainly located in atomic orbitals of oxygen, sulfur, nitrogen, halogens, and so on.

4.2 PROPERTIES OF ORGANIC MOLECULES

Organic molecules are classified and recognized according to their chemical structures and functional groups, as shown in Table 4.3. The properties of organic molecules depend on their chemical structures. If the organic molecules are made only of carbon and hydrogen atoms, they are usually nonpolar molecules such as ethane, benzene, and so on. The polar molecules usually contain unshared electron elements such as oxygen, nitrogen, and sulfur, and have asymmetrical structures. Alcohol, amines, and acid are some examples. Functional groups have chemical properties that depend only moderately on the hydrocarbon frame structure of the molecule. For example, if a hydroxyl group, –OH, is attached to the hydrocarbon framework, the compound is called an alcohol and exhibits hydrophilic properties and hydrogen bonding. The use of functional groups in organic molecules greatly simplifies their identification, prediction of their properties, and chemical reactions.

The functional group of organic compounds not only provides specific chemical properties for the molecules but also the capability to link organic molecules together to form larger molecule called polymers. As shown in Figure 4.2, di-acids (di-carboxylic acids) and di-alcohol (diols) react to form ester-linked polymers such as polyethylene terephthalate (PET) [Eq. (4-1)]. PET is used widely in our daily life in such items as drink bottles, clothing, and flexible electronics. Equation (4-2) shows the formation of poly(3-hexyl thiophene) (P3HT), commonly used in transistors and organic solar cells. Equation (4-3) shows the formation of poly(methyl methacrylate) (PMMA), which is used extensively as a photo-resist in the fabrication of semiconductor devices.

4.3 OPTICAL PROPERTIES OF ORGANIC MATERIALS

The optical properties of organic materials are determined by both intramolecular structure and intermolecular structure. We will concentrate on the optical properties in

TABLE 4.3. Types of functional groups in organic molecules

Structure	Class	Example	Name
C–C	Alkane	$H_3C{-}CH_3$	Ethane
>C=C<	Alkene	$H_2C{=}CH_2$	Ethylene
–C≡C–	Alkyne	HC≡CH	Acetylene
(benzene ring)	Arene	(benzene ring)	Benzene
–OH	Alcohol	CH_3CH_2OH	Ethanol
–O–	Ether	$CH_3CH_2OCH_2CH_3$	Diethyl ether
>C=O	Aldehyde, ketone	$H_3C{-}C({=}O){-}CH_3$	Acetone
–C(=O)OH	Carboxylic acid	$H_3C{-}C({=}O)OH$	Acetic acid
–C(=O)OR	Ester	$H_3C{-}C({=}O)OCH_2CH_3$	Ethyl acetate
$-NH_2$	Amine	$CH_3CH_2NH_2$	Ethyl amine
–C≡N	Nitrile	$CH_3{-}C{\equiv}N$	Acetonitrile
$-C({=}O)NH_2$	Amide	$H_3C{-}C({=}O)NH_2$	Acetyl amide
–X	Alkyl halide or aryl halide	CH_3Cl	Methyl chloride
–SH	Thiol	CH_3CH_2SH	Ethane thiol

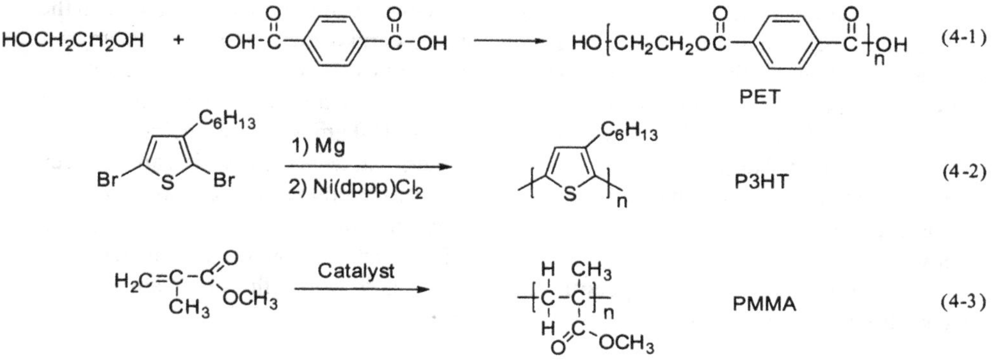

$$HOCH_2CH_2OH + HO\text{-}C(=O)\text{-}C_6H_4\text{-}C(=O)\text{-}OH \longrightarrow HO\!\left[CH_2CH_2OC(=O)\text{-}C_6H_4\text{-}C(=O)\right]_n\!OH \quad \text{PET} \tag{4-1}$$

$$\text{2,5-dibromo-3-}C_6H_{13}\text{-thiophene} \xrightarrow[\text{2) Ni(dppp)}Cl_2]{\text{1) Mg}} \text{poly(3-}C_6H_{13}\text{-thiophene)}_n \quad \text{P3HT} \tag{4-2}$$

$$H_2C{=}C(CH_3)\text{-}C(=O)OCH_3 \xrightarrow{\text{Catalyst}} \left[CH_2\text{-}C(CH_3)(C(=O)OCH_3)\right]_n \quad \text{PMMA} \tag{4-3}$$

Figure 4.2. Polymer synthesis through the chemical reactions of functional groups of organic molecules.

the ultraviolet and visible regions (UV-Vis) because they are in the wavelength of the solar spectrum. Optical properties of absorption [1] and fluorescence [2] behaviors of organic materials will be discussed below.

4.3.1 Absorption Properties

The absorption of ultraviolet and visible light promotes the electrons in the sigma (σ), pi (π), and n orbitals of organic molecules from the ground state to higher energy states. These higher energy states are called antibonding orbitals. The antibonding orbital associated with the sigma bond is called the sigma star (σ^*) orbital and that associated with the pi bond is called the pi star (π^*) orbital. As the n electrons do not form bonds, there are no antibonding orbitals associated with them.

The electronic transitions ($\rightarrow$) that are involved in the ultraviolet and visible regions are of the following types: $\sigma \rightarrow \sigma^*$, $n \rightarrow \sigma^*$, $n \rightarrow \pi^*$, and $\pi \rightarrow \pi^*$. The energy required for the $\sigma \rightarrow \sigma^*$ transition is very high; consequently, molecules containing single bonds do not show absorption in the ordinary ultraviolet region (200–400 nm). For example, propane shows λ_{max} of about 135 nm. The molecules that contain nonbonding electrons of oxygen, nitrogen, sulfur, or halogen atoms are capable of showing absorption, owing to $n \rightarrow \sigma^*$ transitions. These transitions are of lower energy than $\sigma \rightarrow \sigma^*$ transitions; consequently, molecules containing nonbonding electrons usually exhibit absorption in the ordinary ultraviolet region. Examples of $n \rightarrow \sigma^*$ transitions are shown by methyl alcohol vapor of λ_{max} 183 nm and trimethyl amine vapor of λ_{max} 227 nm. Whether an organic molecule contains a particular spectral pattern above 210 nm or not, it will generally show some absorption that increases in intensity toward shorter wavelengths in this region. This regional absorption is due in part to $n \rightarrow \sigma^*$ transitions near 200 nm if the molecule contains oxygen, nitrogen, sulfur, or halogen atoms. This absorption is generally called end absorption.

Transitions to antibonding π^* orbitals are associated only with double bonds in the molecule; these have still lower energy requirements and occur at longer wavelengths, usually well within the ordinary ultraviolet region. For example, aldehydes and ketones exhibit an absorption of low intensity around 285 nm, which is attributed to an $n \rightarrow \pi^*$ transition, and an absorption of high intensity around 180 nm, which is attributed to a $\pi \rightarrow \pi^*$ transition. The $\pi \rightarrow \pi^*$ transitions are of intermediate energy, which are between $n \rightarrow \pi^*$ and $n \rightarrow \sigma^*$ transitions. Figure 4.3 shows the general relative electronic excitation energies for these transitions. The high-energy transitions of $\sigma \rightarrow \sigma^*$ occur at shorter wavelengths and the low-energy transitions of $n \rightarrow \pi^*$ occur at longer wavelengths.

The UV-Vis absorption spectrum is recorded as wavelength versus absorbance. The absorbance A or optical density is given by

$$A = \log\frac{I_0}{I} \tag{4-4}$$

where I_0 is the intensity of incident light and I is the intensity of transmitted light. The calculation of the intensity of an absorption band involves the use of Beer–Lambert's law, as shown in the following equation:

$$\varepsilon = \frac{A}{Cl}$$

where ε is the molar extinction coefficient, C is the molar concentration, and l is the path length in centimeters. The magnitude of the molar extinction coefficient for a particular absorption is directly proportional to the probability of the particular electronic transition; the more probable a given transition, the larger the extinction coefficient. Transitions in identical functional groups in different molecules will not necessarily have exactly the same energy requirement because of different structural environ-

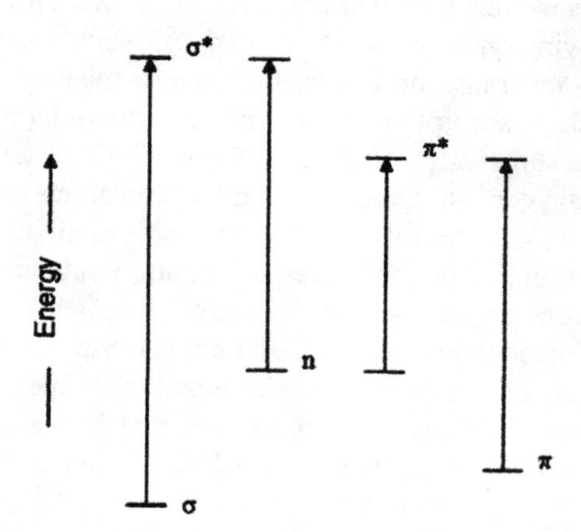

Figure 4.3. Relative electronic excitation energies for different transitions (not at scale).

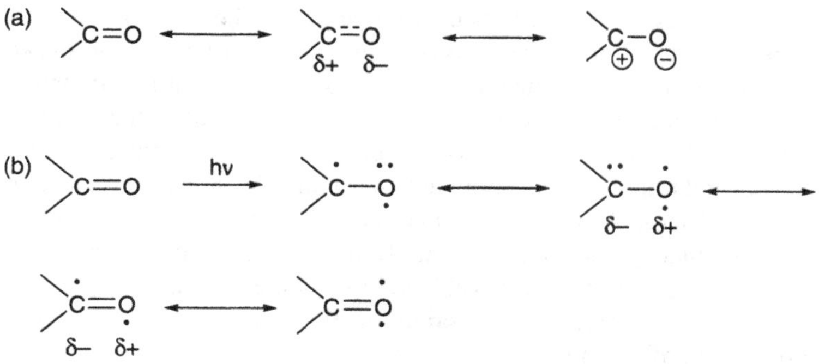

Figure 4.4. Possible ground state structures (a) and excited state structures (b) of carbonyl molecule.

ments. Electronic transitions result in a redistribution of electrons within the molecule. Both ground state and excited state contain important contributions from the polar form, as shown in Figure 4.4. The position of an absorption that involves nonbonding electrons ($n \rightarrow \pi^*$ and $n \rightarrow \sigma^*$) is particularly sensitive to the polarity of the solvent used. If the group is more polar in the ground state than in the excited state, the nonbonding electrons in the ground state are stabilized (relative to the excited state) by hydrogen bonding or electrostatic interaction with a polar solvent; the absorption is shifted to shorter wavelengths with increasing solvent polarity. Conversely, if the group is more polar in the excited state, the nonbonding electrons of the excited state are stabilized by interaction with a polar solvent; the absorption is shifted to longer wavelengths with increasing solvent polarity. Polar solvents generally shift the $n \rightarrow \pi^*$ and $n \rightarrow \sigma^*$ bands to shorter wavelengths and the $\pi \rightarrow \pi^*$ band to longer wavelengths. The magnitude of the shift in the latter case is not as large as that in the former. Table 4.4 shows the effect of solvent polarity on the λ_{max} and ε_{max} of different ketones. The long wavelength $n \rightarrow \pi^*$ absorption of 4-methyl-3-penten-2-one is blue shifted 24 nm in

TABLE 4.4. Comparison of absorption peak position λ_{max} and ε_{max} for different ketones

Name	Structure	λ_{max}(nm), $n \rightarrow \pi^*$	ε_{max}	λ_{max}(nm), $\pi \rightarrow \pi^*$	ε_{max}	Solvent
Acetone	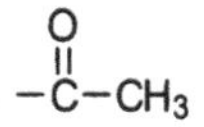 $-C(=O)-CH_3$	279	15	189	900	Hexane
4-methyl-3-penten-2-one	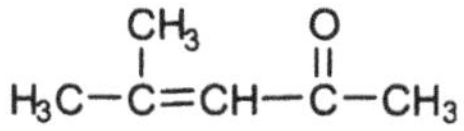 $H_3C-C(CH_3)=CH-C(=O)-CH_3$	329	41	230	12,600	Hexane
		305	60	243	10,000	Water

polarized water; thus, the excited state would appear to be less polar than the ground state. The red shift of 13 nm was observed in the $\pi \rightarrow \pi^*$ transition of this ketone with increasing solvent polarity, indicating that the excited state in this transition is more polar than the ground state. The λ_{max} and ε_{max} of acetone are shorter than that of 4-methyl-3-penten-2-one, because the π-electron of the latter is spread over at least four atomic centers and requires less energy to be excited. We called it a conjugated system. We will further discuss the absorption spectrum of conjugated systems later.

If the functional group of organic molecules exhibits characteristic UV-Vis absorption, we call it a chromophore. Table 4.5 lists some examples of chromophores. If a series of organic molecules has the same functional group and no complicating factors are present, all of the molecules will absorb at nearly the same wavelength and will have nearly the same molar extinction coefficient.

TABLE 4.5. UV absorption of single chromophores*

Chromophore	Example	λ_{max}, nm	ε_{max}	Solvent
>C=C<	Ethylene	171	15,530	Vapor
	1-Octene	177	12,600	Heptane
—C≡C—	2-Octyne	178	10,000	Heptane
		196	~2,100	Heptane
		223	160	Heptane
>C=O	Acetaldehyde	160	20,000	Vapor
		180	10,000	Vapor
		290	17	Hexane
>C=O	Acetone	166	16,000	Vapor
		189	900	Hexane
		279	15	Hexane
$-CO_2H$	Acetic acid	208	32	Ethanol
$-CONH_2$	Acetamide	178	9,500	Hexane
		220	63	Water
$-CO_2R$	Ethyl acetate	211	57	Ethanol
$-NO_2$	Nitromethane	201	5,000	Methanol
		274	17	Methanol
>C=N—	Acetoxime	190	5,000	Water
—C≡N	Acetonitrile	167	weak	Vapor
$-N_3$	Azidoacetic ester	285	20	Ethanol
—N=N—	Azomethane	338	4	Ethanol

*Data in the table are selected from reference [1].

If two or more chromophoric groups are present in a molecule and they are separated by two or more single bonds, the effect on the spectrum is usually additive; there is little electronic interaction between isolated chromophoric groups. On the other hand, if two chromophoric groups are separated by only one single bond (a conjugated system), a large effect on the spectrum results from the π-electron. When two chromophoric groups are conjugated (separate by only one single bond), the high-intensity ($\pi \rightarrow \pi^*$ transitions) absorption band is generally shifted about 15–45 nm to a longer wavelength with respect to the simple chromophore due to the ease of exciting conjugated π-electrons. The low intensity of the $n \rightarrow \pi^*$ transition band of certain chromophoric groups is also shifted to longer wavelengths; sometimes, this band is ob-

TABLE 4.6. UV absorption of conjugated chromophores*

Chromophore	Example	λ_{max}, nm	ε_{max}	Solvent
>C=C–C=C<	Butadiene	217	20,900	Hexane
>C=C–C≡C–	Vinylacetylene	219	7,600	Hexane
		228	7,800	Hexane
>C=C–C=O	Crotonaldehyde	218	18,000	Ethanol
		320	30	Ethanol
>C=C–C=O	3-Penten-2-one	224	9,750	Ethanol
		314	38	Ethanol
–C≡C–C=O	1-Hexyn-3-one	214	4,500	Ethanol
		308	20	Ethanol
>C=C–CO_2H	*cis*-Crotonic acid	206	13,500	Ethanol
		242	250	Ethanol
–C≡C–CO_2H	*n*-Butylpropiolic acid	~ 210	6,000	Ethanol
>C=C–C=N–	*N*-*n*-Butylcroton aldimine	219	25,000	Hexane
>C=C–C≡N	Methacrylonitrile	215	680	Ethanol
>C=C–NO_2	1-Nitro-1-propene	229	9,400	Ethanol
		235	9,800	Ethanol
O=C–C=O	Glyoxal	195	35	Hexane
		280	3	Hexane
		463	4	Hexane
HO_2C –CO_2H	Oxalic acid	~ 185	4,000	Water
		250	63	Water

*Data in the table are selected from reference [1].

scured by the peak with large extinction. Table 4.6 shows examples of UV absorption of conjugated chromophores. We would observe longer wavelength shifts with extended conjugation. The lengthened π-electron system results in greater delocalization of the π-electrons; the energy required for the $\pi \rightarrow \pi^*$ transitions is less, and the probability of these transitions is higher. Table 4.7 shows examples of this effect in a series of polyene aldehydes in alcohol solution.

Polythiophene is another good example of an extended conjugation system. The λ_{max} of $\pi \rightarrow \pi^*$ transition depends on the chain length of the polymer, as shown in Table 4.8. The chain length of the polymer can be expressed by average molecular weight (M_w, K dalton) and number of repeating units (n), as shown in the the second and third columns of Table 4.8. The polydispersity (PDI) describes the molecular weight distribution of the polymer; a good-quality polymer has a PDI less than 1.5. The PDI of high molecular weight polymers is usually higher due to a statistical distribution of kinetically controlled polymer synthesis. The λ_{max} is increased when the molecular weight of P3HT is increased due to the increase in conjugating length. For instance, the λ_{max} of a CF1 sample prepared from 8.2K dalton P3HT is red shifted by 40 nm as compared with that of a CF5 sample prepared from 50.4 K dalton P3HT. The amount of shift is also affected by the processing conditions. By increasing molecular weight from 8.2 to 50.4 K dalton, a red shift is larger for the film prepared from chloroform (CF sample) as compared with that from chlorobenzene (CB sample) (40 nm vs. 26 nm). When the same molecular weight of P3HT is used, the λ_{max} of the CB sample is longer than the λ_{max} of the CF sample. The high boiling point and slow evaporation rate of CB provide a longer time for P3HT to be self-organized and have a longer effective conjugation length, which results in a longer λ_{max}. Thermal annealing is more effective to increase the λ_{max} of the CF sample than that of the CB sample, because the CB sample is already self-organized during the solvent evaporation process. The λ_{max} of absorption being red shifted using different molecular weight P3HT and different solvents are shown in Figure 4.5(a). The effects of additional thermal annealing at 130°C for 12 hr on the UV-Vis absorption of different samples are plotted in Fig. 4.5(b). Both using CB solvent and thermal annealing are effective in promoting red shift of main-chain chromophores of P3HT and the $\pi \rightarrow \pi$ stacking between P3HT molecules, which is shown by the increase in the shoulder peak intensity of 625 nm.

TABLE 4.7. UV absorption of polyene aldehydes*

n	λ_{max}, nm	ε_{max}
1	217	15,650
2	270	27,000
3	312	40,000
4	343	40,000
5	370	57,000
6	393	65,000
7	415	63,000

*Data in the table are selected from reference [1].

TABLE 4.8. Comparison of absorption λ_{max} of different molecular weights of poly(3-hexyl thiophene) in different solvents and different fabrication conditions

				UV-Vis			
				λ_{max} CF sample		λ_{max} CB sample	
Polymer	M_w (K dalton)	n	PDI	Spin*	Thermal†	Spin	Thermal
1	8.2	32	1.54	513	510	529	519
2	10.1	47	1.27	516	511	534	525
3	17.3	90	1.15	557	559	561	563
4	27.9	140	1.20	551	554	560	560
5	50.4	160	1.90	553	554	555	553

*Thin-film sample was prepared by spin coating.
†Thin-film sample was prepared by spin coating plus thermal annealing at 130°C for 12 hr.

4.3.2 Fluorescence

An organic molecule is excited from ground state S_0 to an excited state S_n ($n > 1$) upon absorbing light. The excited molecule will return to the ground state following two successive steps as shown in Figure 4.6, which is called Jablonski electronic transition diagram. The molecule at S_n returns to the lowest excited state S_i by dissipating a part of its energy in the surrounding environment. This is called internal conversion. From the excited state S_i, the molecule will return to the ground state S_0 through different competitive processes:

1. Emission of a photon (fluorescence) with a radiative rate constant k_r
2. Part of the absorbed energy is dissipated in the medium as heat. This is called nonradiative energy and occurs at a rate constant of k_i
3. Some of the absorbed energy is released to nearby molecules by collisional quenching at a rate constant of k_q or by energy transfer at a rate constant of k_t
4. A transient passage occurs from S_i to a lower energy of excited triplet state T_i at a rate constant of k_{isc} (intersystem crossing). The triplet state can undergo similar processes as S_i. Photon can be emitted at a rate of k_p (phosphorescence). Some energy can be dissipated as nonradiative energy at rate constant of k_i' and can be transferred to another molecule at a rate constant of k_t', or quenched via collision at a rate of k_q'.

A chromophore that emits a photon is called a fluorophore. Many chromophores do not fluoresce because absorbed energy can be dissipated as heat or be transferred to another molecule. Figure 4.7 shows the spin configurations of the singlet and triplet states. The speed of electron rotation is very weak as compared with the light velocity, so the magnetic contribution to absorption is negligible as compared to the electric contribution. Therefore, during absorption, a displaced electron preserves the same

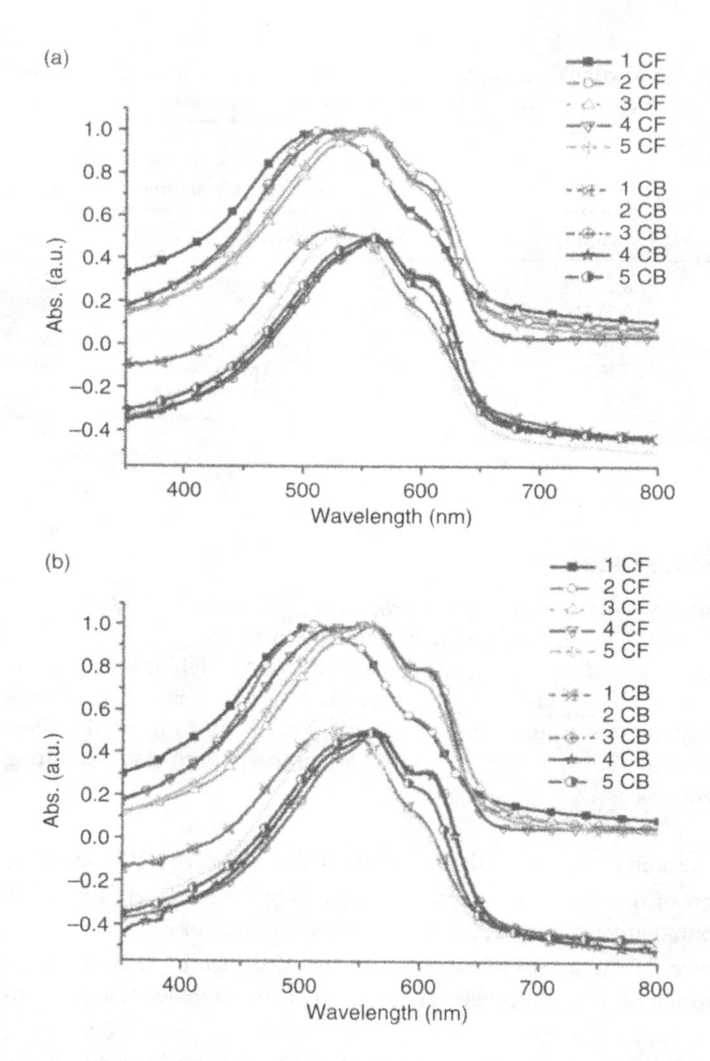

Figure 4.5. The absorption spectra of different molecular weights of poly(3-hexyl thiophene) in different solvents (CF: chloroform, CB: chlorobenzene) and different processing conditions, (a) as spun and (b) thermally annealed.

spin orientation. Thus, the transition of $S_0 \rightarrow S_n$ is allowed and the transition of $S_0 \rightarrow T_n$ is forbidden. In the ground state, all molecules except oxygen are in singlet form. Oxygen is in the triplet state. Upon light exposure, it reaches the singlet state, which is destructive to organic molecules.

From the electronic transition diagram, we know the absorption energy is higher than the emission energy and the fluorescence energy is higher than that of phosphorescence. Emission occurs from the excited state S_i independent of the excited wave-

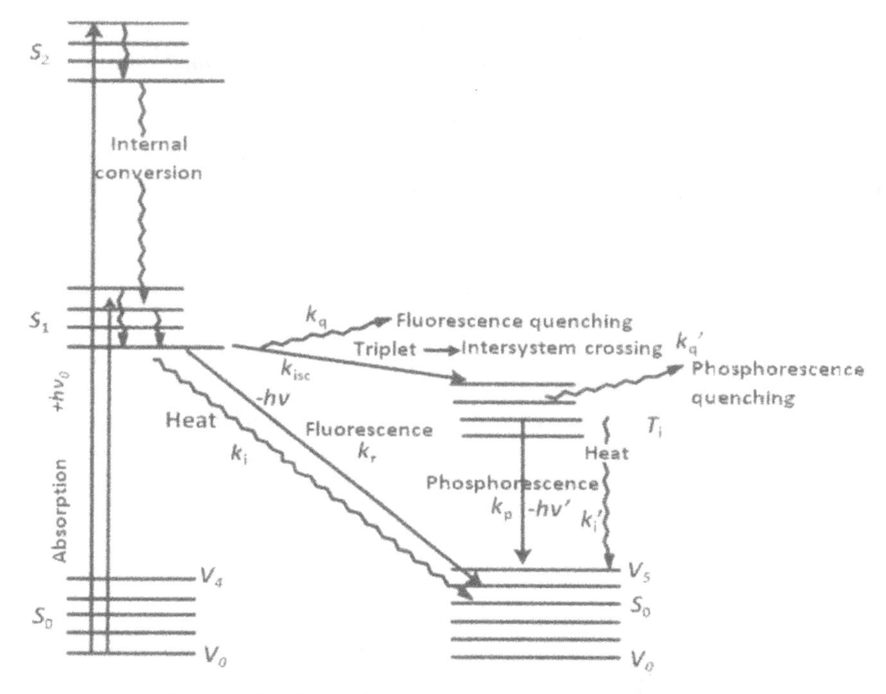

Figure 4.6. Jablonski electronic transition diagram [3].

length. Absorption and fluorescence do not require any spin reorientation. However, intersystem crossing and phosphorescence require a spin reorientation. Therefore, absorption and fluorescence are much faster than phosphorescence (10^{-15} s vs. 10^{-9} to 10^{-12} s).

We will focus only on the fluorescence of organic materials here. Generally, there is a maximum peak shift to longer wavelength from the absorption spectrum to the

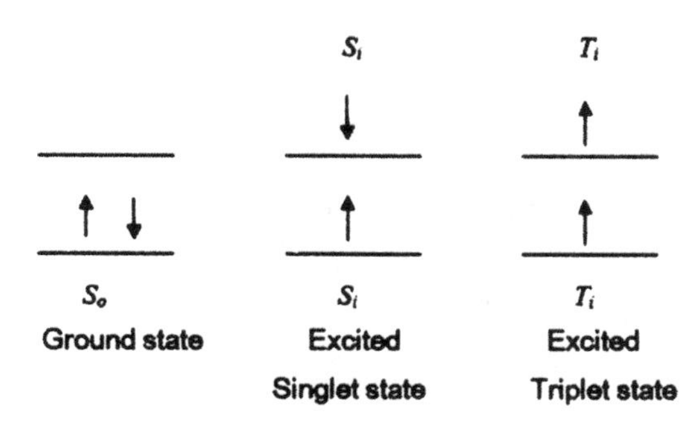

Figure 4.7. Spin configurations of the singlet and triplet states.

emission spectrum. It is called a stokes shift. The relationship between the absorption intensity and emission intensity can be described by the following equation:

$$I_F(\lambda) = 2.3 I_0 \varepsilon_{(\lambda)} C l \phi_F \tag{4-6}$$

where the I_0 is the intensity of incident light, ϕ_F is the fluorophore quantum yield, I_F is the intensity of fluorescence, and $\varepsilon_{(\lambda)}Cl$ is the optical density of the molecule. The quantum yield can be described by the following equation:

$$\phi_F = \frac{\text{emitted photons}}{\text{absorbed photons}} = \frac{k_r}{k_r + k_i + k_{isc}} \tag{4-7}$$

The ease of fluorescence of organic compounds depends also on their chemical structure. If the conjugating length is longer in the main chain of the molecule, the λ_{max} of fluorescence will be longer. That means the molecule can be easily excited and fluoresce. The organic compounds shown in Figure 4.8 are used as the standards for the

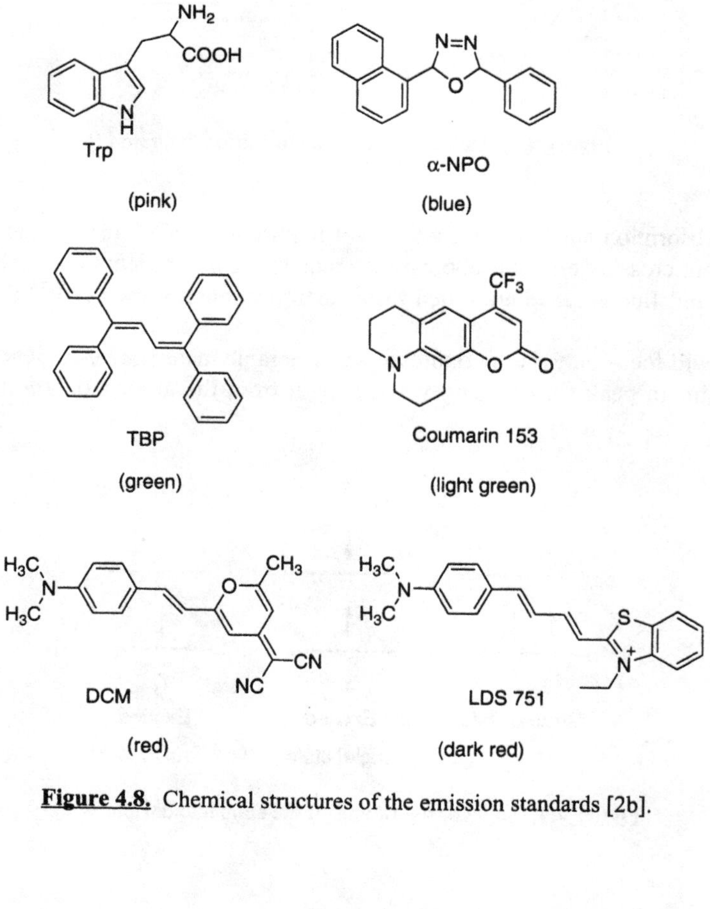

Trp (pink)

α-NPO (blue)

TBP (green)

Coumarin 153 (light green)

DCM (red)

LDS 751 (dark red)

Figure 4.8. Chemical structures of the emission standards [2b].

emission spectroscopy. They cover the whole range of the visible spectrum (Figure 4.9).

Effects of chemical structure on the absorption and emission of organic molecules are similar. The ease of electron delocalization in the molecule, the λ_{max} of molecule will be longer, as shown in Figure 4.10. Si has been used in organic molecules instead of carbon to improve their optical and electrical properties. The Si atom is larger than the C atom and the outer-shell electrons are easier to remove than those of the C atom. In general, organic molecules containing Si absorb at relatively longer wavelengths and exhibit higher conductivity [4]. The extent of conductivity of organic material is also directly related to the ease of electron delocalization in the molecule. This subject will be discussed in more detail in Section 4.5.

The fluorescence of organic materials is very close to their single molecules in the solution state. However, a red shift of fluorescence spectrum is usually observed in the solid state of organic materials due to the ease of energy transfer between molecules. The spectrum will also depend on the morphology of the polymer. Figure 4.11 shows the fluorescence spectra of different molecular weight poly(3-hexyl thiophene) films as spun and with thermal annealing using different solvents. The PL spectrum of polythiophene does not affect by its molecular weight. The PL spectrum of polythiophene is not influenced either by type of solvent if the solvent can dissolve the polymer well and is very dilute. However, the thermally annealed samples show a red shift due to strong intermolecular interactions. Table 4.9 summaries these results. The photoluminescence of a conducting polymer is dependent on its chemical structure, as shown in Figure 4.12. The alkyl chain of CH_3 or C_8H_{17} is an electron donating group that gives polythiopehene a higher band gap than aromatic substituted polythiophenes, in which the aromatic group is an electron withdrawing group. Aromatic substituted polythio-

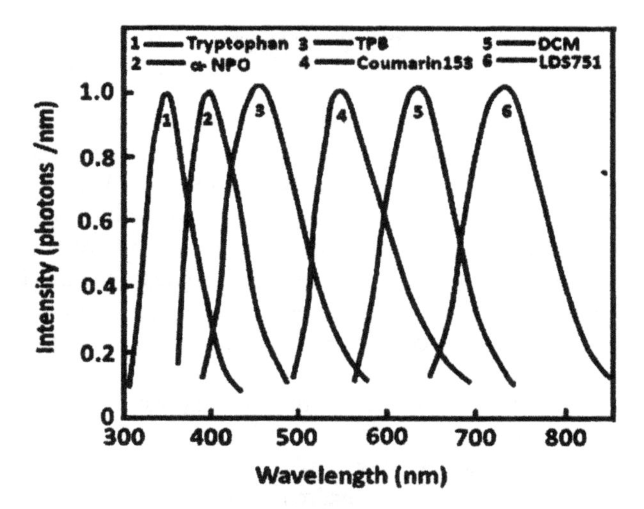

Figure 4.9. Fluorescence spectra of the emission standards [2b].

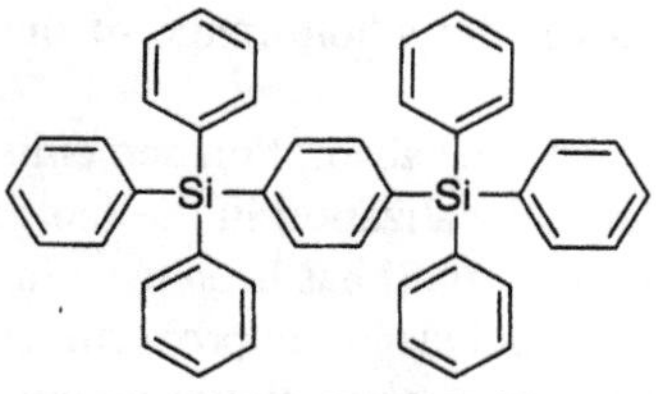

$\lambda_{max(abs)}$ = 265 nm(in CH_2Cl_2)
$\lambda_{max(exc)}$ = 296 nm(in CH_2Cl_2)

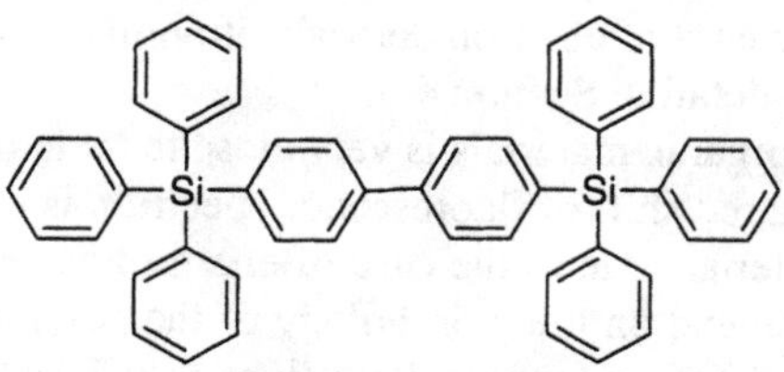

$\lambda_{max(abs)}$ = 271 nm(in CH_2Cl_2)
$\lambda_{max(exc)}$ = 324 nm(in CH_2Cl_2)

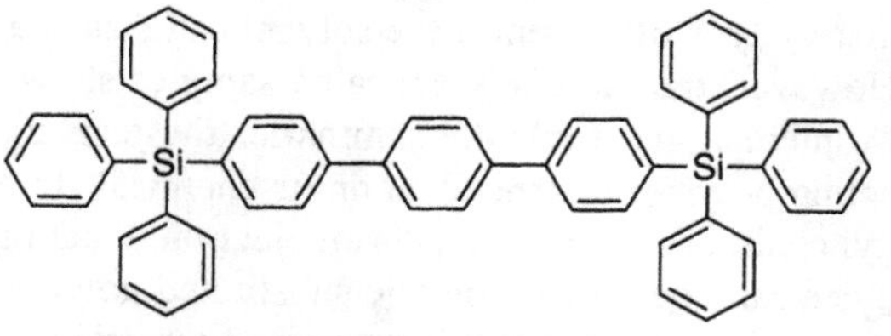

$\lambda_{max(abs)}$ = 296 nm(in CH_2Cl_2)
$\lambda_{max(exc)}$ = 358 nm(in CH_2Cl_2)

Figure 4.10. Effects of chemical structure on the optical properties of organic compounds.

phenes are donor—acceptor systems. These systems exhibit the characteristic of easily delocalized π electrons (i.e., extended conjugation length), which results in a lower band gap. The subject of band gap will be discussed in Section 4.4.

By increasing the size of side group, the solubility of polythiophene can be increased. However, there is a larger deviation from coplanarity of the adjacent thiophene units. This causes a diminished extended conjugation and, hence, a hypsochromic shift in the luminescence spectra, whereas the application of a phenyl ring as the soluble group enlarges the conjugated systems and causes a bathochromic shift in the luminescence spectra. This example shows that the control of conjugation length is a powerful tool for the band gap engineering of conjugated polymers (see further discussion in Section 4.4).

One kind of hybrid solar cell uses a mixture of polymers and semiconducting nanoparticles as its active layer. For example, the hybrid solar cell made from P3HT:TiO_2 nanorod has been extensively investigated (Section 7.3.4). Figure 4.13

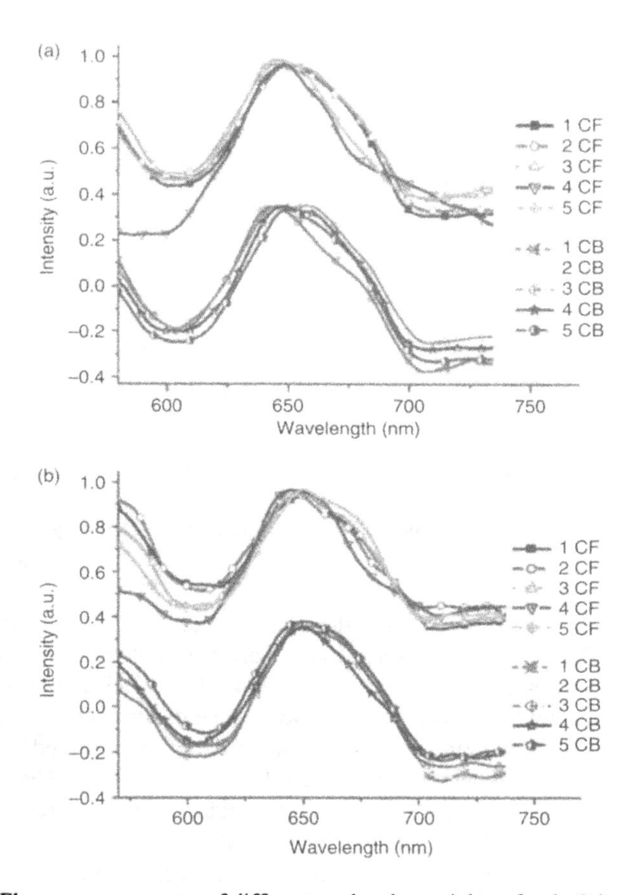

Figure 4.11. Fluorescence spectra of different molecular weights of poly(3-hexyl thiophene) in different solvents (CF = chloroform and CB = chlorobenzene). (a) film as spun and (b) film after thermal annealing at 130°C for 12 hours.

TABLE 4.9. Comparison of fluorescence λ_{max} of different molecular weights of poly(3-hexyl thiophene) in different solvents and under different process conditions.

			PL			
			λ_{max} in $CHCl_3$		λ_{max} in CB	
Polymer	M_w (K dalton)	PDI	Spin	Thermal	Spin	Thermal
1	8.2	1.54	646	651	644	648
2	10.1	1.27	648	648	651	658
3	17.3	1.15	644	644	648	647
4	27.9	1.20	649	652	646	651
5	50.4	1.90	644	652	648	650
	Averaged λ_{max}		646	649	647	650

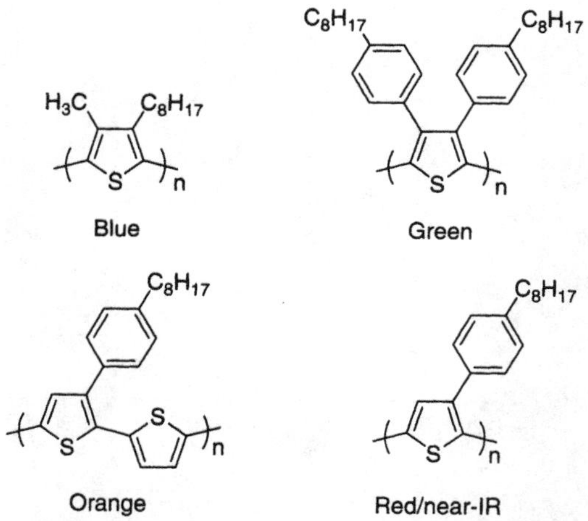

Figure 4.12. Structure and emission color of various substituted polythiophenes.

shows the effect of TiO_2 nanorod on the fluorescent intensity of P3HT. The P3HT is a p-type material. When the P3HT absorbs light, there are excitons formed at the higher energy level. The introduction of n-type TiO_2 nanorod into P3HT induces the charge separation of the excitons into electrons and holes. The electrons transport through n-type TiO_2, so a quench phenomenon of the fluorescence is observed. However, the amount of quenching is affected by the surface linker on the TiO_2. The TiO_2 is a hydrophilic inorganic material and P3HT is a hydrophobic organic material, so they are not compatible. To have a homogeneous hybrid P3HT:TiO_2 nanorod, the surface of TiO_2 needs to be modified with an organic linker. Oleic acid (OA) linker is used when the TiO_2 nanorods are synthesized. The long chain structure of oleic acid functions as a template to facilitate the growth of TiO_2 nanoparticles horizontally to obtain nanorod structure. However, the OA is an insulator and will not be a good linker to transport electrons. On the other hand, the linkers of pyridine (pyr), 2-bithienyl carboxylic acid (BCA), and 2-trithienyl phosphoric acid (TPA) contain π-electrons in their molecule, so they can more easily transport electrons compared to OA. Therefore, a large quench effect is observed for pyr, BCA, and TPA. The results indicate that the extent of active layer quenching can be used to gauge the extent of charge separation in solar cells.

4.4 BAND GAP OF ORGANIC MATERIALS

The principle of band gap formation in inorganic materials can be used to understand the band gap of organic materials. For inorganic materials, the band is formed by the overlapping of atomic orbitals in solid state. The band of organic materials is generated from the overlapping of molecular orbitals in solid state. The highest occupied molecu-

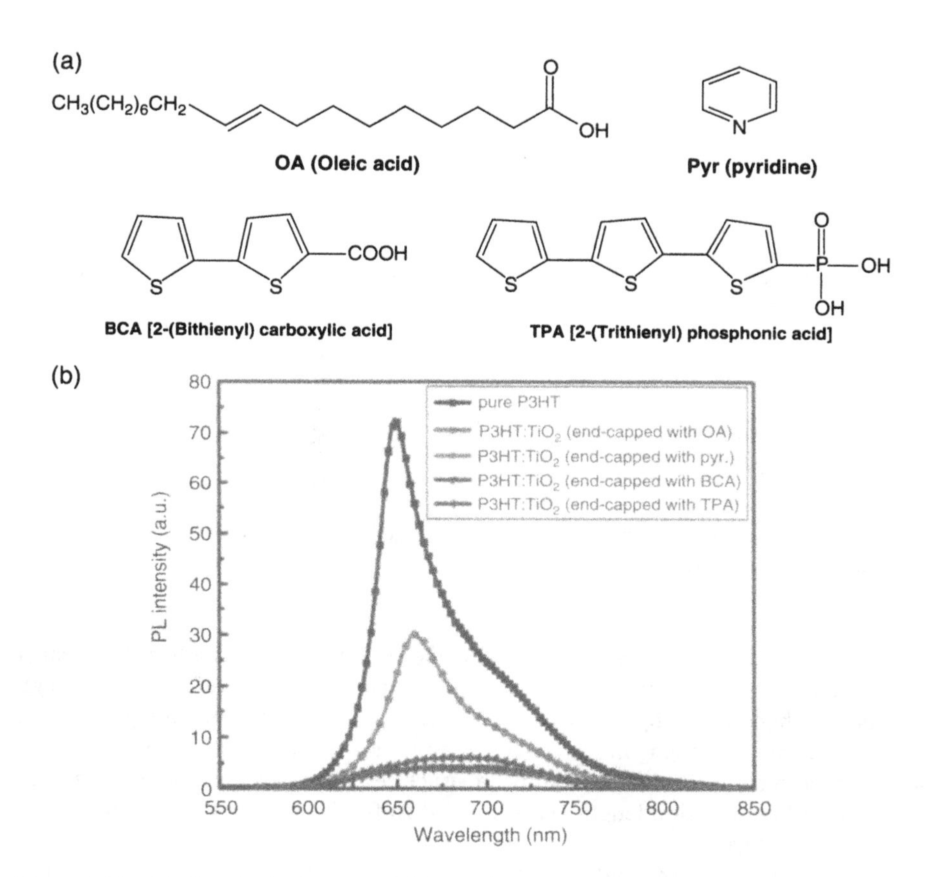

Figure 4.13. (a) Chemical structures of linkers. (b) Normalized fluorescence spectra of the thin films consisting of P3HT:TiO_2 nanorods (47:53 weight ratio) with different surface linkers end capped on TiO_2 nanorods.

lar orbital (HOMO) is equivalent to the valence band of inorganic materials, whereas the lowest unoccupied molecular orbital (LUMO) is similar to the conducting band. The difference in energy between these levels is called the band gap E_g. A majority of organic materials exhibit E_g greater than 4.0, which makes them insulators. Only the organic materials [5-8] containing conjugated structures with delocalized π-electrons show semiconducting properties. By manipulating the chemical structure of conjugated organic materials, the band gap of the materials can be engineered to have desired electrical and optical properties. Through such band gap engineering, it is possible to reduce the band gap to approximately zero, resulting in intrinsically conducting organic materials.

Figure 4.14 shows how the band structure of polythiophene originates from the interaction of the π-electron orbitals of repeating units throughout the chain (intramolec-

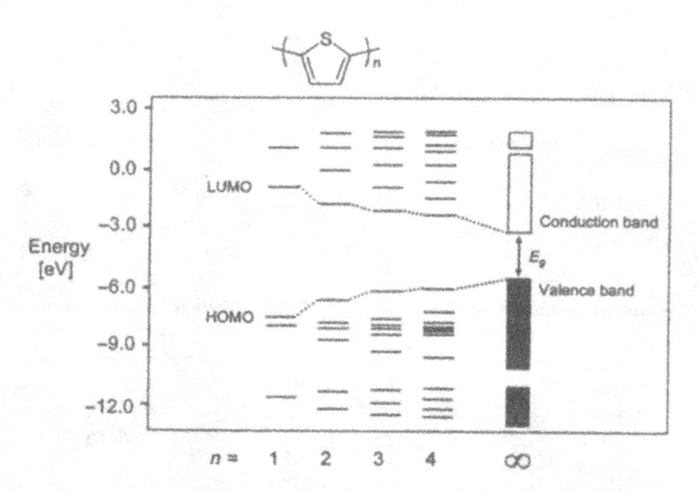

Figure 4.14. Calculated (frontier) energy levels of oligothiophenes with n = 1–4 of polythiophene, where E_g = band gap [6].

ular π conjugation) [6]. Each additional thiophene unit causes hybridization of each level, yielding more and more levels with reduced band gaps. Therefore, one strategy to reduce the band gap is by increasing the repeating unit of the conjugating structure, that is, the conjugating length. The π-electron orbitals of molecules can interact with each other to form three-dimensional band structures (intermolecular π conjugation). The increased conjugation length leads to a red shift of optical absorption due to the decrease of band gap.

Table 4.10 shows the band gap of different conducting polymers. Poly(*p*-phenylene vinylene) exhibits a lower band gap than poly(*p*-phenylene) because the vinylene structure increases the linearity of the molecule and enhances the probability of π-electron delocalization. Polyacetylene has the lowest band gap among the polymers shown because it is easy to be stretched out to obtain a linear conjugated path for fast charge transport. Polythiophene has a relatively low band gap because the π-electrons are eas-

TABLE 4.10. The band gap of different conducting polymers*

Polymer†	Band gap (eV)
Polyacetylene	1.5–1.7
Polythiophene	2.0–2.1
Poly(*p*-phenylene vinylene)	2.4
Poly(*p*-phenylene)	2.7
Poly(pyrrole)	3.2

*Data obtained from Reference [5,7].

†Chemical structures of polymers are shown in Table 4.11.

ier to move through the carbon atom. Polypyrrole has a relative high band gap. This can be attributed to the large π-donor strength of nitrogen [8].

4.5 ELECTRICAL CONDUCTING PROPERTIES OF ORGANIC MATERIALS

Electrical conductivity refers to the transport of charge carriers through a medium under the influence of an electric field and it is dependent on the number of charge carriers and their mobility. The charge carriers may be generated intrinsically or from impurities. They can be electrons, holes, or ions. Alternatively, electrons or holes may be injected from electrodes. We shall focus on the electronic conduction in organic materials here.

In general, organic materials such as polymers are insulating materials with conductivities ($\sigma < 10^{-8}$ S/cm) at least ten orders lower than metal ($\sigma > 10^2$ S/cm) and used in insulating and structural applications throughout the electronics industry. However, the discoveries of intrinsically conducting poly(sulfur nitride) $-(S{=}N)_n-$ in 1973 and I_2 doped poly(acetylene) $-(CH{=}CH)_n-$ in 1978 have stimulated major research efforts to elucidate the mechanism of conduction and to search for other conducting polymers for applications in batteries, transistors, light emitting diodes, and solar cells. Table 4.11 shows the comparison of the conductivities between metals and doped conducting polymers. A comparable conductivity to metal has been reached by doped conducting polymers. Therefore, the conducting polymers are also called synthetic metals.

Most conjugated polymers are *p*-type semiconductors. After doping with oxidants such as I_2, the electrons are released. Although a complete understanding of the conduction mechanism remains elusive, molecules must contain an extended conjugate backbone structure to release the delocalized π-electrons upon doping that result in conductivity. Kivelson proposed [11] the following conducting mechanism for charge transport involving mobile neutral solitons (Figure 4.15).

The dopant can be either electron acceptors (AsF_5 or acid) or electron donors (alkali metals, I_2). The configurational structure and conformational structure of polymers (i.e., polymer morphology) also have great influence on their conductivity. For instance, the conductivity of polyacetylene can reach as high as 1.5×10^5 S/cm when the film is properly oriented. Similar to semiconductor, the conductivity of polymers varies with the concentration and the type of dopant.

Another mechanism involving the formation of mobile charges [7], the polaron, was proposed as shown in Figure 4.16. The removal of one electron from the polythiophene chain (la) produces a mobile charge in the form of a radical cation (1b). The positive charge tends to induce local atomic displacements, leading to the polaronic behavior. Further oxidation can either convert the polaron into a spinless bipolaron (1c) or introduce another polaron (1d). In either case, introduction of each positive charge also means introduction of a negatively charged counter ion (Ox^-).

Organic molecules used in solar cell applications are usually undoped to retain their semiconducting properties. The relationship between the conductivity and mobil-

TABLE 4.11. Comparison of conductivities of metals and doped conjugated polymers

Material	Structure	Dopants	Conductivity (S/cm)
Copper[a]	Cu	none	5.8×10^5
Gold[a]	Au	none	4.1×10^5
Poly(acetylene)[a]	$-(C=C)_n-$	I_2	1.0×10^3–1.5×10^5
Poly(sulfur nitride)[a]	$-(S=N)_n-$	none	10^3–10^4
Poly(*p*-phenylene)[a]	$-(C_6H_4)_n-$	AsF_5	10^3
Poly(*p*-phenylene vinylene)[b]	$-(C_6H_4-CH=CH)_n-$	H_2SO_4	10^3
Polyaniline[b]	$-(C_6H_4-NH)_n-$	HCl(1M)	7.7
Polypyrrole[b]	[pyrrole ring, N–H]$_n$	PF_6	10^2–10^3
Poly(3-hexylthiophene)[b] P3HT	[thiophene ring, S; C_6H_{13}]$_n$	I_2	10^4
Poly(3,4-ethylenedioxythiophene)[c] PEDOT	[thiophene ring, S; O–CH₂CH₂–O]$_n$	$-(C-C)_n-$ with $C_6H_4SO_3H$	10–10^3

[a]Adapted from Reference [9].
[b]Adapted from Reference [5].
[c]Adapted from Reference [10].

ity of organic materials can be expressed by Eq. (4-8) which is adapted from the principle of semiconductor materials:

$$\sigma = ne\mu \tag{4-8}$$

Where σ(siemen/m) is the conductivity of the organic materials, n is the number of charge carriers, e is the elementary charge of 1.602×10^{-19} coulombs, and μ ($m^2/v \cdot s$) is the mobility of charge carrier. The number of charge carriers is important to the power conversion efficiency (PCE) of solar cell. The charge carrier mobility is even more im-

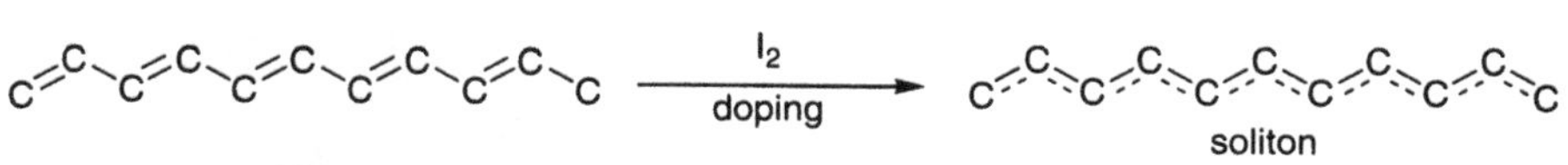

Figure 4.15. Kivelson mechanism for charge transport involving mobile neutral solitons.

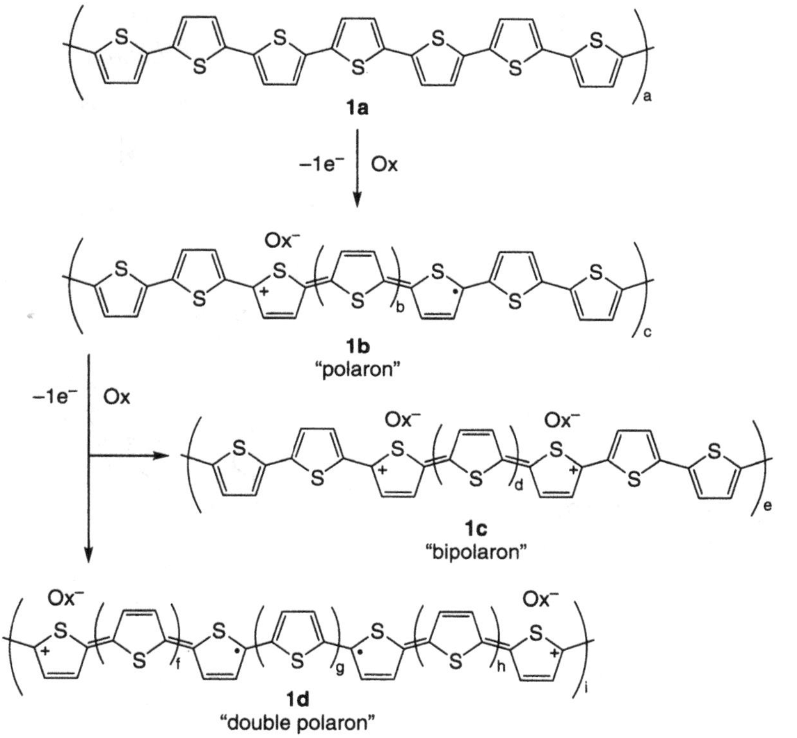

Figure 4.16. Structural changes in polythiophene upon doping with a suitable oxidant [7].

TABLE 4.12. State-of-the-art dyes used in dye-sensitized solar cells [12].

Dye	PCE %
N3	8.42
N749	11.1
C101	11.0
CYC-B1	12.5
YE05	10.1
C217	9.80

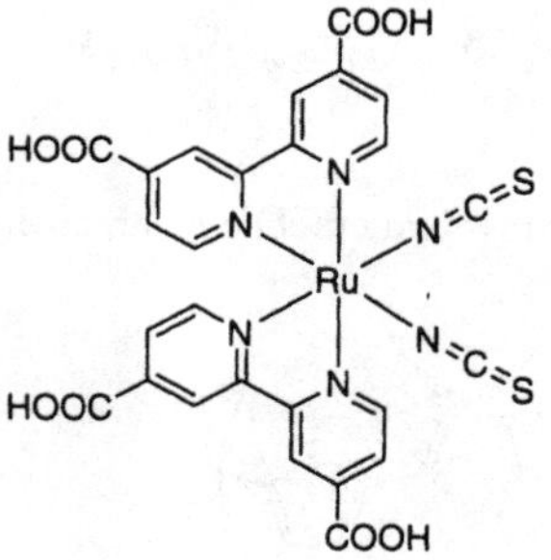

N3

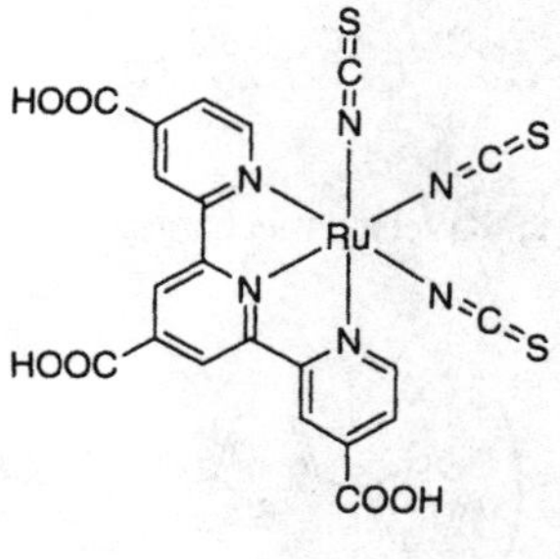

N749

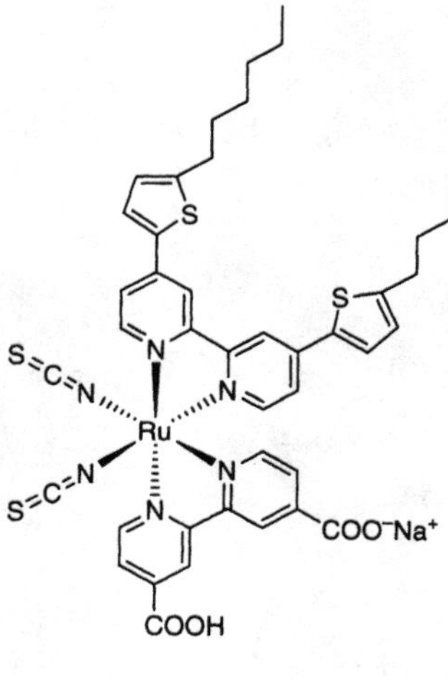

C101

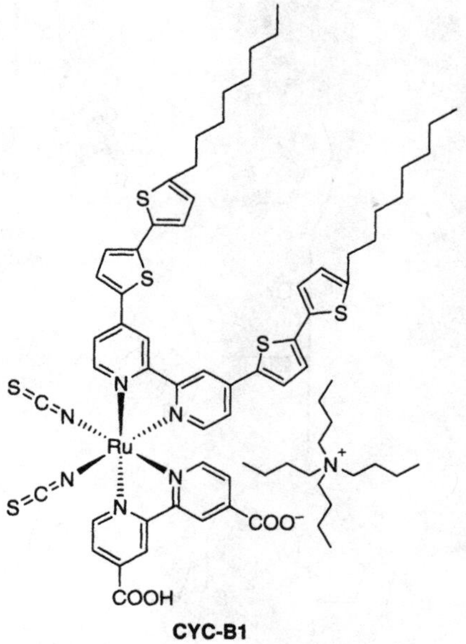

CYC-B1

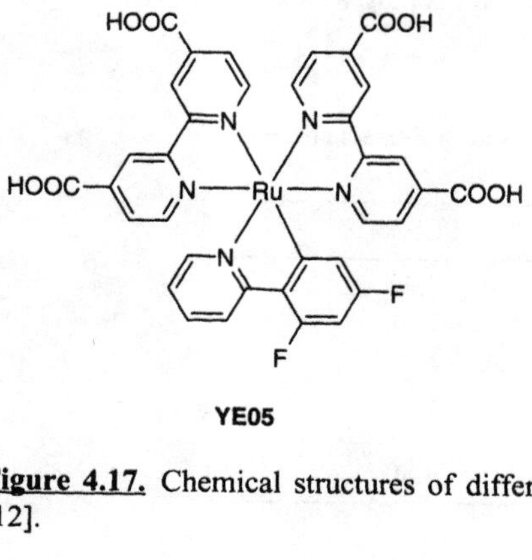

YE05

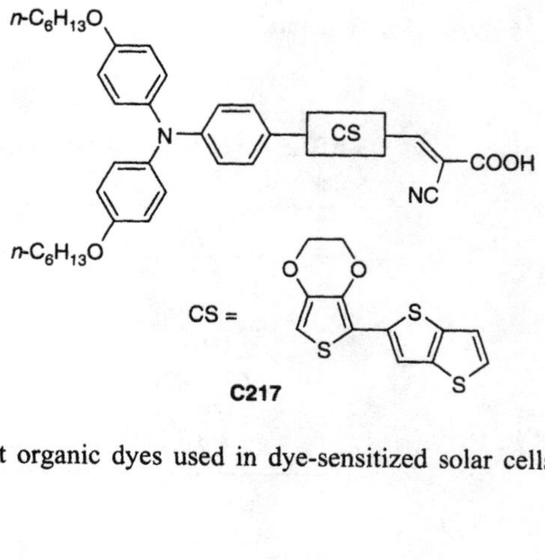

C217

Figure 4.17. Chemical structures of different organic dyes used in dye-sensitized solar cells [12].

portant to determine the PCE, because the charge carriers have to move to the opposite electrode in order to generate electricity. The mobility of poly(3-hexyl thiophene) can be increased by two to three orders when the polymer is annealed. The crystallinity of P3HT is increased after annealing, which facilitates the charge transport. Thus, in the application of poly(3-hexyl thiophene) in solar cells, an annealing process is usually applied after the fabrication of the device. The relationship between PCE and the mobility of solar cells will be discussed more detail in Chapter 7.

4.6 SUITABLE ORGANIC MATERIALS FOR SOLAR CELL APPLICATIONS

As we discussed before, organic molecules can absorb strongly in the visible range with high excitation coefficient through delocalized π-electrons and π-π stacking, making them useful for solar cell applications. Relatively small sized dye molecules as compared with polymers (molecular weight < 1000 versus molecular weight > 5000) have been used to fabricate dye-sensitized solar cells. Table 4.12 shows the state-of-the-art dyes used in dye-sensitized solar cells. Their chemical structures are shown in Figure 4.17. N749 has a higher PCE than that of N3 due to a higher incident-photon-to-current conversion efficiency (IPCE) of N749 [12] (Figure 4.18). Table 4.13 illustrates the state of the art of conducting polymers used in the polymer–fullerene solar cell. Their chemical structures are shown in Figure 4.19. A solar cell efficiency of 8.13% has been achieved by Solarmer Energy Company and certified by the U.S. National Renewable Energy Laboratory in 2010.

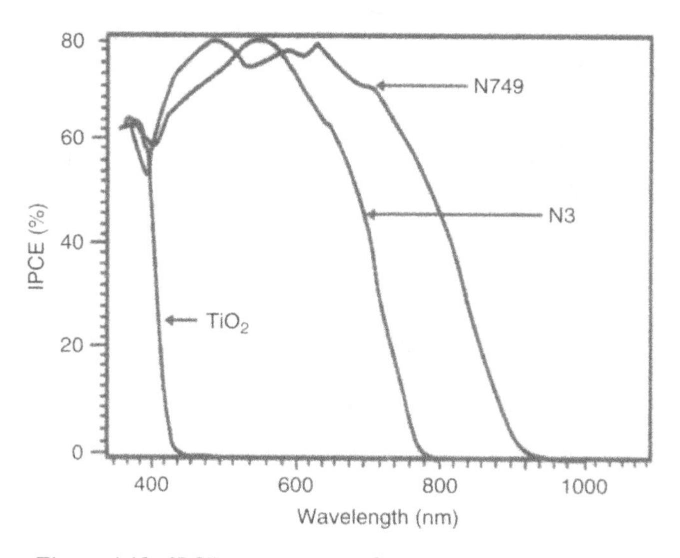

Figure 4.18. IPCE measurements for N3 and N749 dyes [12].

TABLE 4.13. State-of-the-art of conducting polymers used in the polymer–fullerene solar cell [13]

Polymer	Band gap (E_g)	PCE %
P3HT	2.00	5.1
PCDTBT	1.87	3.6–6.1
PCPDTBT	1.45	5.5
PSBTBT	1.45	5.1
PTPTBT	1.70	5.8
PTB7	1.70	7.4

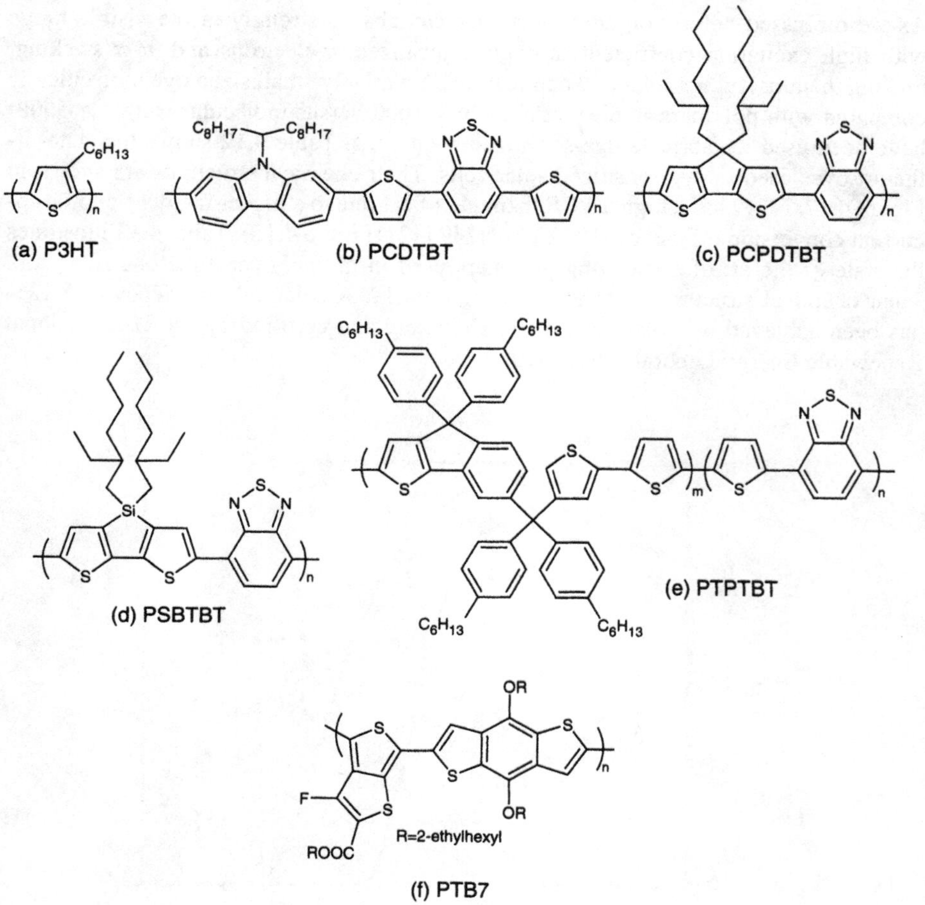

Figure 4.19. Chemical structures of conducting polymers used in the polymer–fullerence solar cell.

REFERENCES

1. J. R. Dyer (1965), *Applications of Absorption Spectroscopy of Organic Compounds,* Prentice-Hall.
2. (a) J. R. Albani (2007), *Principles and Applications of Fluorescence Spectroscopy,* Blackwell. (b) J. R. Lakowicz (2006), *Principles of Fluorescence Spectroscopy,* 3rd Ed., Springer.
3. A. Jablonsk (1935), *Zeitschrift fur Physik, 94,* 38–64.
4. J. A. Gardeck and M. Maroncelli (1998), *Appl. Spectroscopy, 52,* 1179–1189.
5. T. A. Skotheim, R. L. Elsenbaumber, and J. R. Reynolds (1998), *Handbook of Conducting Polymers,* Marcel Dekker.
6. U. Salzner, J. B. Lagowski, P. G. Pickup, and P. A. Porrier (1998), *Synth. Met., 96,* 177.
7. H. A. M. van Mullekom, J. A .J. M. Vekemans, E. E. Havinga, and E. W. Meijer (2001), *Materials Science and Engineering, 32,* 1–40.
8. G. Appel, O. Böhme, R. Mikalo, and D. Schmeiber (1999), *Chem. Phys. Lett, 313,* 411–415.
9. M. P. Stevens (1999), *Polymer Chemistry,* Oxford.
10. CLEVIOS Data Sheet, http://clevios.com.
11. S. Kivelson (1982), *Phys. Rev., B25,* 3798.
12. M. Gratzel (2009), *Acct. Chem. Res., 42,* 1788.
13. J. Chen and Y. Cao (2009), *Acct. of Chem. Res., 42*(11), 1709.

EXERCISES

1. Why do organic materials exhibit light weight, flexibility, low melting point, and ease of processing as compared to inorganic materials?
2. Define the functional group of organic molecules.
3. Explain why the amount of energy involved in the electronic transitions of organic molecules upon light absorption is in the decreasing order of $\sigma \rightarrow \sigma^* > n \rightarrow \sigma^* > n \rightarrow \pi^* > \pi \rightarrow \pi^*$. (Hint: draw the energy-level diagram of each bond formation.)
4. (a) Explain the increase of λ_{max} and ε_{max} of the following organic molecules when their bond length increases.

$H_2C{=}CH{-}CH{=}CH_2$

$\lambda_{max} = 220$ nm
($\varepsilon_{max} = 20{,}900$)

$H_2C{=}CH{-}CH{=}CH{-}CH{=}CH_2$

$\lambda_{max} = 257$ nm
($\varepsilon_{max} = 35{,}000$)

$H_2C{=}CH{-}CH{=}CH{-}CH{=}CH{-}CH{=}CH_2$

$\lambda_{max} = 287$ nm
($\varepsilon_{max} = 52{,}000$)

(b) Explain the increase of λ_{max} and ε_{max} of the following organic molecules when their ring number increases.

$\lambda_{max} = 255$ nm ($\varepsilon_{max} = 215$)

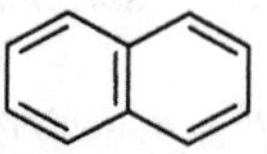

$\lambda_{max} = 314$ nm ($\varepsilon_{max} = 289$)

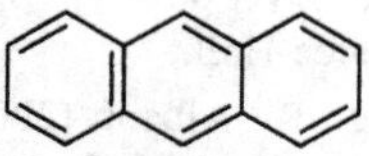

$\lambda_{max} = 380$ nm ($\varepsilon_{max} = 9{,}000$)

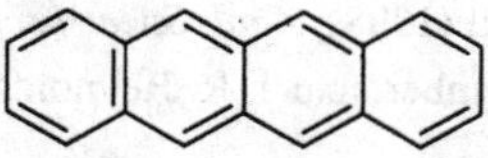

$\lambda_{max} = 480$ nm: a yellow compound ($\varepsilon_{max} = 12{,}500$)

5. (a) Which of the following aromatic compounds do you expect that will absorb at the longer wavelength?

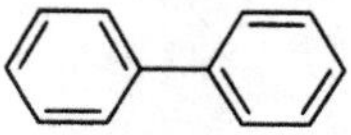

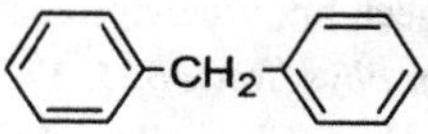

CH_2

(b) Naphthalene is colorless, but its isomer, azulene, is blue. Which compound has the lower-energy π-electronic transition?

naphthalene azulene

6. Compare the λ_{max} of absorption and fluorescence of the following organic molecules and arrange them in decreasing order (assume they all dissolve well in CH_2Cl_2). Explain your answer.

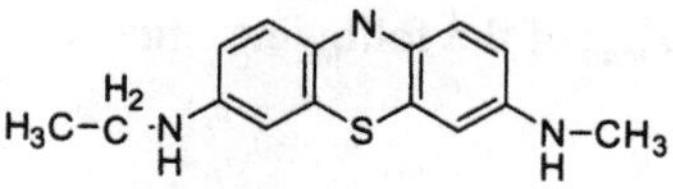

Azure B

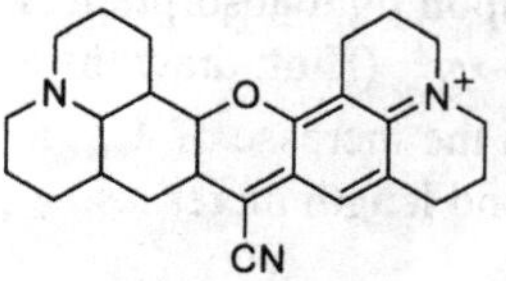

Rhodadmine800

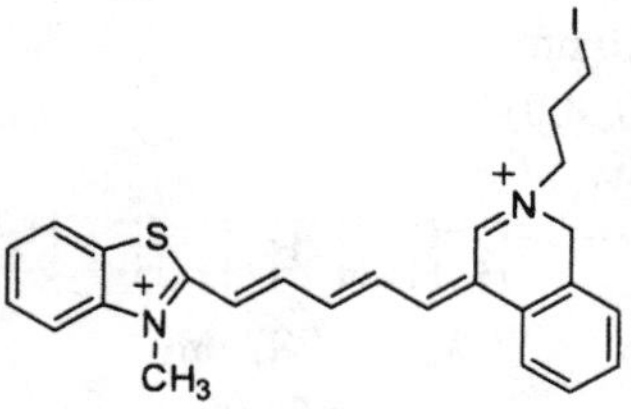

Thiazole Orange

7. Compare the band gap of the following organic molecules and arrange them in decreasing order.

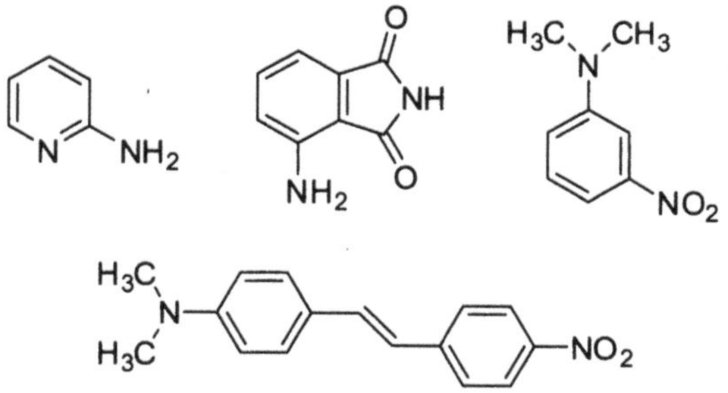

8. Select the organic molecules that exhibit semiconducting behavior from the following molecules and explain your answer.

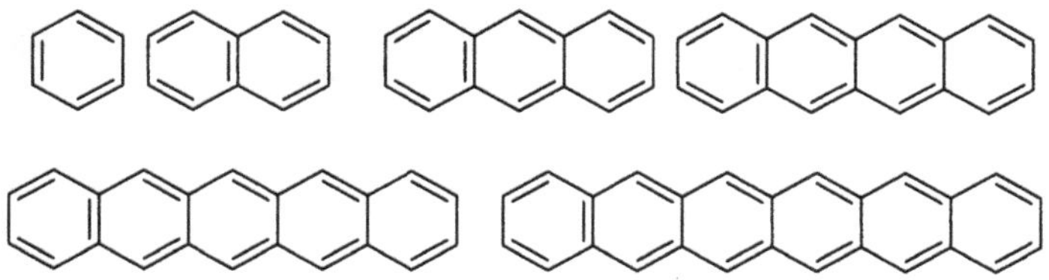

9. Arrange the following polymers in increasing order of expected power conversion efficiency for solar cell application and explain your answer. Assume that the molecular weight for each polymer is about the same—approximately 10,000.

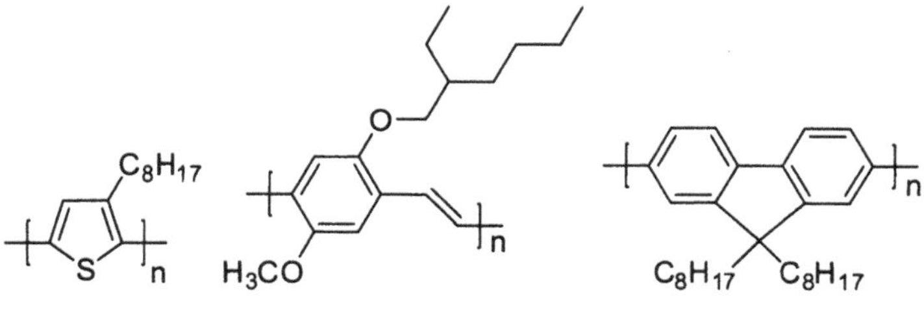

5

INTERFACE BETWEEN ORGANIC AND INORGANIC MATERIALS

Wei-Fang Su

Every real solid is bounded by its surface. However, most of physical properties of solid material can be well understood without considering the presence of its surface. The reason is, first, that one usually deals with properties such as transport and optical, mechanical, or thermal properties, to which all the atoms (for inorganic materials) or molecules (for organic materials) of the solid contribute more or less to the same extent. Second, there are many more atoms or molecules in the bulk of a solid sample than at its surface, provided the solid is of macroscopic size. In the case of an aluminum cube of 1 cm^3, for example, the ratio of surface atoms to the bulk atoms is $4 \times 10^{16}/6 \times 10^{22} = 6.7 \times 10^{-7}$. On the contrary, for the aluminum thin film of 2 mm × 2 mm × 100 nm, the ratio of surface atoms to the bulk atoms is $5.8 \times 10^{14}/2.4 \times 10^{16} = 2.4 \times 10^{-2}$. The amount of surface atoms in the thin film is almost five orders of magnitude more than in the cube. Thus, the property of thin film will be different from the bulk sample because there are more atoms interacting with their environment.

In the ambient condition, the real surface of a solid is far from ideal. A freshly prepared surface of a material is normally very reactive toward atoms and molecules in the environment. All kinds of strong chemical absorptions to weak physical absorptions give rise to an added layer on top of the solid. One example is the immediate formation of an extremely thin oxide layer on a freshly cleaved silicon crystal. Usually, the chemical composition and the geometrical structure of such a layer are not well defined. Thus, a well-defined thin passivation layer is usually deposited immediately after the silicon surface is prepared to avoid any contamination. The aluminum thin-film electrode used in devices has the same problem of oxidation, so the device is usually encapsulated right after the fabrication of the electrode.

Organic, Inorganic, and Hybrid Solar Cells. By C.-F. Lin, W.-F. Su, C.-I Wu, and I-C. Cheng

To study the interface properties of two materials, we need to understand the chemical and physical characteristics of each material first. Then we can predict their interface properties. If the two materials are compatible in nature, then we can assume that both of the materials are adapted to each other and in prefect contact. The interface of two materials is generated by bonding them through three kinds of bonding forces depending on the types of materials: (1) van der Waals attractive forces, (2) electrostatic forces, and (3) chemical bonding (including hydrogen bonding) forces. The van der Waals interfacial force can be very strong if the lattice constant and atom (molecule) radius constant are matched between two materials in prefect contact. This is an ideal case. In the real world, the van der Waals force of the interface is rather weak because each surface is far from perfect and will not achieve ideal contact. The necessary strong force can be achieved usually by electrostatic force or chemical bonding.

As we discussed in Chapter 4, organic materials consist of carbon and hydrogen atoms mainly and exhibit hydrophobic behavior. On the other hand, inorganic materials are made from other elements and have hydrophilic properties. In general, organic and inorganic materials are not compatible and will not form a good interface. In order to make them compatible, we need to modify each component chemically or physically. The optical and electronic properties of each component will be altered depending on the surface modification and the interface. In this chapter, we will discuss interfaces based on two kinds of junctions encountered in the organic–inorganic hybrid solar cells, as shown in Figure 5.1 [1]. The difference between two kinds of junctions is the structure of the active layer of the solar cell. The active layer consists of acceptors and donors which absorbs sun light to form excitons. The excitons dissociate at the interface to generate electrons and holes and then transport through acceptor and donor domains, respectively, to the opposite electrode and generate current. The planar heterojunction [Figure 5.1(a)] is made from the acceptor (e.g., C60 derivative, PCBM) deposited on top of the donor (e.g., P3HT), with a limited interface between two thin layers. On the other hand, the bulk heterojunction [Figure 5.1(b)] is made by blending an acceptor with a donor to form large interfaces dispersed throughout the bulk layer. The device with the structure of Figure 5.1(b) is a so-called forward-structure solar

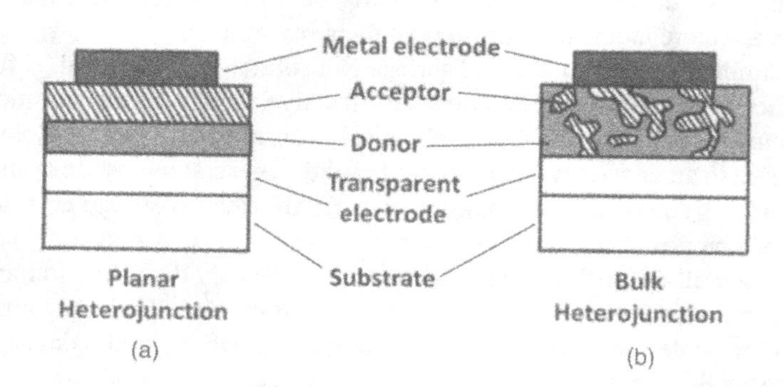

Figure 5.1. Schematic structures of forward-structure hybrid solar cell on a planar heterojunction (a) and on a bulk heterojunction (b) [1].

cell; the electron is transported out through the metal electrode and the hole is transported out through the transparent indium–tin oxide (ITO) electrode. When the transport of charge carriers is in the reversed direction, we call it an inverted solar cell.

Here, we will focus on the interfaces between of each layer the solar cell, including (a) transparent electrode (e.g., indium–tin oxide, ITO) and substrate, (b) transparent electrode and active layer, (c) donor and acceptor of active layer, (d) active layer and metal electrode, and (e) impedance characteristics at the interface. We will address the problems encountered at the interfaces between the layers of solar cell and will provide solutions to the problems. The detailed working principle of different solar cells will be discussed in the next three chapters. In brief, each layer has to have a compatible interface in terms of chemical similarity and matched band gap to facilitate the charge transport. Figure 5.2 shows the band diagram of an organic base solar cell, which is in cascade fashion for ease of charge transport. Figure 5.2(a) is an example of a forward-structure solar cell of ITO/P3HT:PCBM/Al. The energy level of ITO is lower than that of Al, so the hole is transported through the ITO electrode and the electron is transported out of the Al electrode to form current. Figure 5.2(b) is an example of inverted solar cell of ITO/P3HT:PCBM/Au. The energy level of ITO is higher than that of Au, so the direction of charge carrier transport is in the reverse order of the forward-structure solar cell.

5.1 INTERFACE BETWEEN TRANSPARENT ELECTRODE AND SUBSTRATE

The transparent substrate can be either glass or plastic. Polyethylene terephthalate (T_g = 80°C), polyethylene naphthalate (T_g = 155°C), and polycarbonate (T_g = 149°C) are commonly used organic substrates due to their high transparency (> 80% in visible

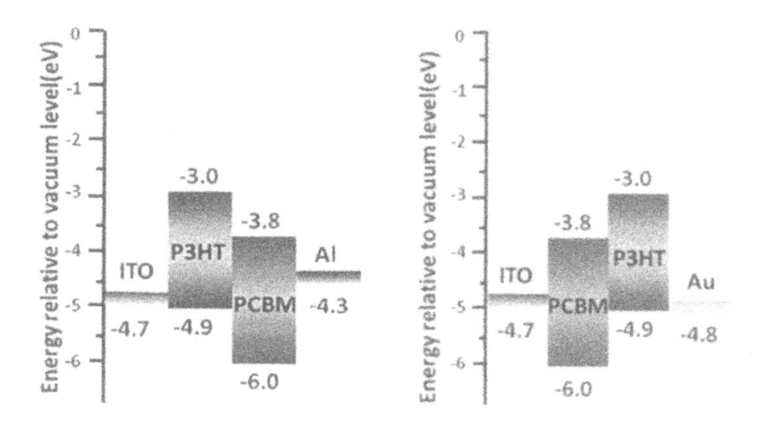

Figure 5.2. Band gap diagram of an organic-based solar cell. (a) Forward structure. (b) Inverted structure.

wavelength) and high glass-transition temperature (T_g). The glass transition temperature is the temperature at which the material becomes soft and losses its mechanical strength. Plastic is flexible and light weight. On the other hand, glass is a rigid substrate but more dimensionally stable and provides a hermetic seal against oxygen and moisture. Glass is routinely used in the fabrication of solar cells. The plastic substrates are still in the development stage because the transparent anode materials such as indium–tin oxide (ITO) coated plastics exhibit relatively low conductivity.

The transparent electrode material is required in solar cell fabrication so that sunlight can pass through the electrode and substrate. The light can be absorbed by the active layer. Transparent inorganic oxides such as fluoro-doped tin oxide (FTO) or ITO are used extensively in organic or hybrid solar cells. They are deposited on either glass or plastic by a sputtering process. Because the plastic substrate exhibits a low softening temperature of ~200°C as compared with ~600°C for glass, the sputtering power used is relatively low for plastic. Thus, the small crystalline ITO is obtained on plastic substrates with an order lower of conductivity as compared to ITO on glass.

The compatibility between ITO and glass is generally good because both are inorganic. The thickness of ITO is in the range of 150 nm, which is much smaller than the thickness of glass (~1.5 mm), so the ITO can be layered on the glass rather well with van der Waals forces. Table 5.1 summarizes the properties of transparent electrodes used in the fabrication of hybrid solar cells.

The compatibility between ITO and plastic is expected to be not good due to their different chemical structures and characteristics. Details on the thickness of the ITO coating on plastic is not available from the manufacturers but we can assume it is in the range of 150 nm. As compared with the thickness of plastic (0.15 mm), it is still far thinner than the plastic. We know that when the material becomes thinner, it becomes flexible. Thus, the ITO can be layered on the plastic rather well with van der Waals forces. The sputtering process can also create a rough surface on the plastic that can create mechanical bonding between the plastic and the ITO.

TABLE 5.1. Properties and suppliers of ITO electrodes on different substrates

Sample	Substrate thickness (mm)	CTO* thickness (nm)	Transparency at 550 nm (%)	Resistance (ohm/sq)	Supplier
ITO/glass (CF TOP ITO)	0.4-1.1	150 ± 20	≥90	≤15	Aim Core Technology (Taiwan)
FTO/glass (Pilkington TEC7)	2.2–4.0	500 ± 25	80–82	6-8	Hartford Glass (USA)
ITO/PET (ITOPET50)	0.0125–0.25	—	85	40–60	Visiontek System Ltd. (USA)
ITO/PEN	0.2	—	≥80	≤15	Visiontek System Ltd. (USA)

*CTO = conducting transparent oxide

Graphene has emerged as an alternative transparent electrode material (Figure 5.3). It is a one atom thick and two-dimensional crystalline arrangement of carbon atoms, so it exhibits extremely high carrier mobility (15,000 $cm^2V^{-1}s^{-1}$), high optical transparency, low sheet resistance, and great flexibility. Intrinsic graphene is a semi-metal or zero-gap semiconductor. It can be applied either by vacuum or solution processes. The quality of the graphene film is usually better with the vacuum process compared to the solution process. The graphene solution is prepared by adding exfoliated graphite sheet to strong acid to produce graphene oxide, which then is reduced to graphene using a strong reduction agent such as hydrazine. Uniform size single-sheet graphene is rather difficult to prepare using current solution processes, so the quality is poor. There is ongoing intensive research to develop solution-processable graphene for the transparent electrodes of flexible electronics [2].

5.2 INTERFACE BETWEEN TRANSPARENT ELECTRODE AND ACTIVE LAYER

The inorganic ITO layer is not compatible with the organic-based active layer. The surface of ITO is rather rough at about ~40–50 nm. The contact between ITO and active layers such as P3HT:PCBM is poor, resulting in inefficient charge transport. The chemistry and physics of P3HT:PCBM will be discussed in Chapter 7. For the forward-structure solar cell, a transparent conducting polymer, PEDOT:PSS, about 40–50 nm is usually coated on the ITO and planarize the ITO to improve its contact with P3HT:PCBM. The PEDOT:PSS also serves as an electron-blocking layer, which reduces the recombination of electrons and holes and facilitates hole transport [3]. The band gap of the transparent electrode has to be matched with the band gap of the

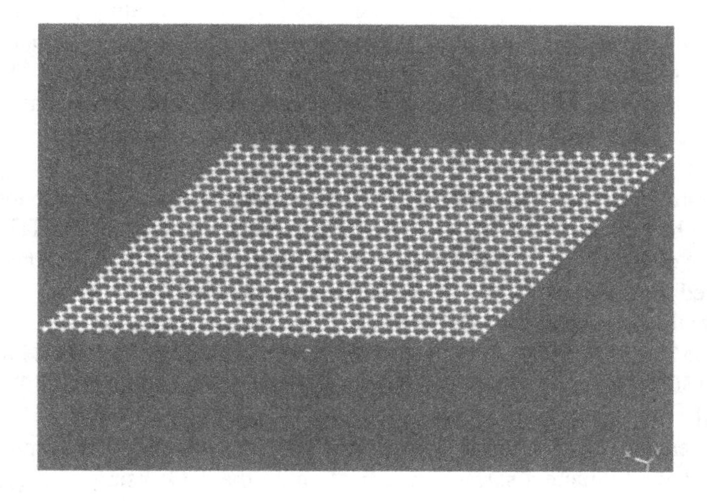

Figure 5.3. Graphene has a two-dimensional-carbon network structure.

active layer to facilitate the charge transport, as shown in Figure 5.2. The introduction of an electron blocking layer of PEDOT:PSS can reduce the energy gap between the transparent electrode layer and the active layer so the transport of holes is improved.

On the other hand, for the inverted solar cell, a sol–gel ZnO layer is used to replace PEDOT:PSS for matched band alignment [4]. The ZnO layer is compatible chemically with the ITO. It also has the advantage over PEDOT:PSS of being nonacidic, so it will not corrode the ITO. The chemical compatibility between the ZnO layer and the organic active layer is not good. However, the organic active layer can be layered on top of the ZnO layer smoothly using a relatively low surface energy processing solvent such as chlorobenzene for making the organic active layer.

The interface between the graphene transparent electrode and active layer of P3HT:PCBM is compatible due to their similar chemical characteristics. Chemical-vapor-deposition-prepared graphene film exhibits a surface roughness of ~0.9 nm and sheet resistance of 230 ohms/sq (at 72% transparency). When solar cells with chemical-vapor-deposition (CVD) graphene and ITO electrodes were fabricated side-by-side on flexible PET, they have exhibited power conversion efficiencies of 1.18 and 1.27% respectively [5].

5.3 INTERFACE BETWEEN DONOR AND ACCEPTOR OF AN ACTIVE LAYER

For polymer-based solar cells, the most widely used donor is P3HT and the acceptor is PCBM. The two materials can be intermixed very well because they both are organic materials and are compatible. However, the PCBM tends to aggregate upon heating. For bulk heterojunction solar cells of P3HT:PCBM, thermal annealing is usually a required step to yield high-performance solar cells [6]. The annealing process promotes the crystallization of P3HT and induces the formation of a bicontinuous phase to facilitate charge transport, as shown in Figure 5.1(b). However, the control of annealing temperature and time is very critical to form optimized bicontinuous phases. Overannealed active layers will form large aggregated PCBM domains and disrupted bicontinuous phases, which will reduce the charge transport efficiency [7].

To improve the thermal stability of organic-based solar cells, the donor material, PCBM, can be replaced by TiO_2 nanorods. The thermal stability of P3HT:TiO_2 is better than that of P3HT:PCBM [8] because of the larger size and higher density of TiO_2 as compared with that of PCBM (5×20 nm versus 0.7 nm; 4.23 versus 1.72 g/cm^3). The thermal decomposition temperature has been increased by 36°C from P3HT:PCBM to P3HT:TiO_2 [Figure 5.4(a)]. The morphology of P3HT:TiO_2 did not change at 150°C [Figure 5.4(b)] but a large aggregate is formed for P3HT:PCBM [Figure 5.4 (c)].

However, the TiO_2 nanorod is inorganic material, which is not compatible with P3HT. We have to make a surface modification on the TiO_2 using an organic linker. The linker can be a small molecule [9], dye [10], or oligomer [11]. Three criteria have

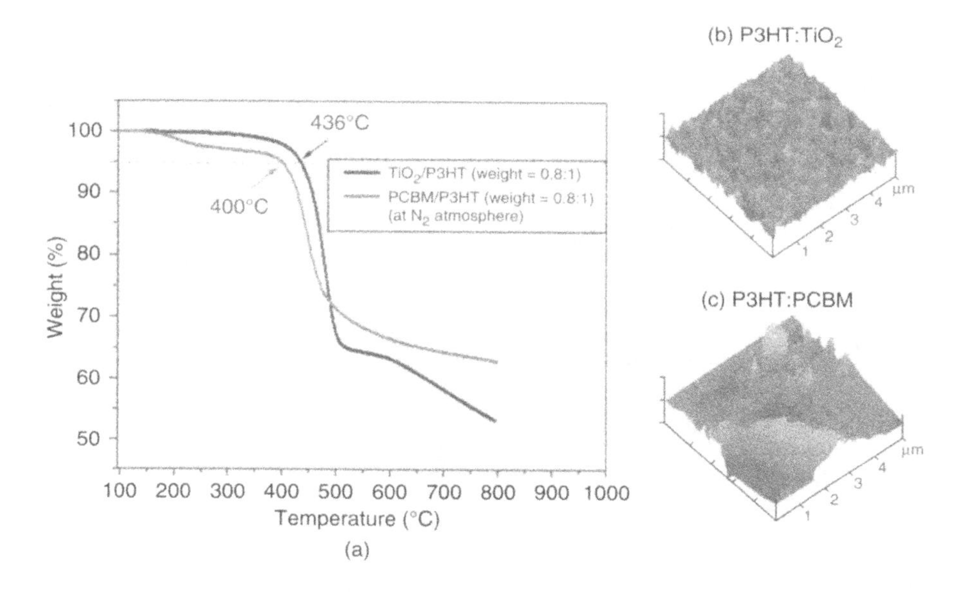

Figure 5.4. (a) Thermal stability study of a polymer–nanoparticle hybrid by thermal gravimetric analysis; Thermal AFM study at 150°C for (b) P3HT:TiO_2 and (c) P3HT:PCBM.

to be met to be an ideal linker: (1) amphiphilic functionality, (2) aligned band gap between polymer and TiO_2, and (3) adequate size. The amphiphilic functionality can link both TiO_2 and polymer simultaneously to achieve compatible properties. The linker usually has –COOH functionality which produces very strong bonding to the TiO_2 from the chelating bond, as shown in Figure 5.5. The aligned band gap is necessary to afford good carrier transport and reduced charge recombination, as explained earlier. The adequate size can cover the surface of TiO_2 completely to reduce surface defects and to decrease carrier trapping.

Figure 5.6 shows four different kinds of surface linkers that have been employed to modify the surface of TiO_2 nanorods for P3HT:TiO_2 solar cells. Figure 5.6(a) shows the chemical structure of the linkers. They all can bond well on that TiO_2 surface

Figure 5.5. Formation of chelating bond between TiO_2 and RCOOH.

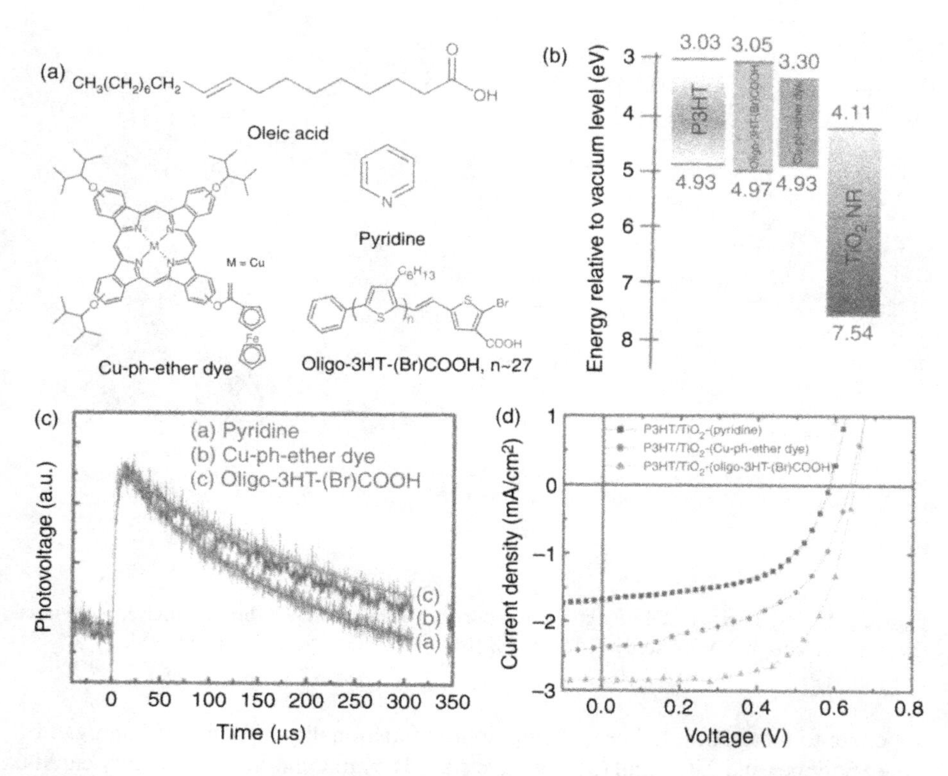

Figure 5.6. Amphiphilic surface linkers for P3HT:TiO_2 hybrid solar cells. (a) Chemical structures of surface linkers for TiO_2 nanorods. (b) Band diagram of different linkers. (c) Exciton lifetime of different linkers by transient photo-voltage measurement. (d) Solar cell performance of different linkers [11].

through –COOH chelating or an unshared electron (oxygen or nitrogen). Figure 5.6(b) indicates that the band gap of each linker is aligned between P3HT and TiO_2 for ease of charge transport. Figure 5.6(c) shows transient photo-voltage data for hybrid cells fabricated from surface modified TiO_2 and P3HT. The data was fit into a single-exponential line decay, which falls within the expected range except for a very small voltage perturbation that can be ignored. For P3HT/TiO_2 thin films, the decay of the voltage signal is dependent on the rate of carrier recombination at the P3HT/TiO_2 interface [11]. The charge carrier lifetime can be obtained via the exponential fitting line. Thus, we can figure out the recombination rate constant (k_{rec}) from the inverse of the charge carrier lifetime [12]. Table 5.2 summarizes the data.

These results demonstrate that both modifiers enhance the charge carrier lifetimes and reduce the charge recombination rate constants. Additionally, oligo-3HT-(Br)COOH exhibits a longer lifetime than Cu-ph-ether dye. The oligo-3HT-(Br)COOH is not as flat as Cu-ph-ether dye and its size is larger (2652.6 Å^3 versus 1125.3 Å^3) so it

TABLE 5.2. Charge carrier lifetime and recombination constants of a P3HT/TiO_2-(surface modifier) system [11]

Surface modifier	Charge carrier lifetime, t(μs)	Recombination rate constant, k_{rec}(S^{-1})
TiO_2-(pyridine)	190.2	5.26×10^{-3}
TiO_2-(Cu-ph-ether dye)	259.5	3.85×10^{-3}
TiO_2-(oligo-3HT-(Br)COOH)	300.9	3.32×10^{-3}

acts as a more effective energy barrier and spatial obstacle for charge recombination. Moreover, oligo-3HT-(Br)COOH is a conducting material; the polar groups of carboxylic acid and Br facilitate efficient charge transport to reduce charge recombination. The result is consistent with the dipole moment of the modifiers that were calculated by molecular modeling to be 1.5 debye, 2.4 debye, 2.6 debye, and 13.7 debye for oleic acid, pyridine, Cu-ph-ether dye, and oligo-3HT-(Br)COOH, respectively. Figure 5.6(d) shows that the expected solar cell performance is in the decreasing order of oligomer > dye > pyridine > oleic acid. The oligomer provides the best performance for solar cells because it is conducting, large enough to act as a shelter for the reduction of charge carrier recombination, and adheres well to TiO_2 through –COOH functionality. The dye is better than pyridine due to the shelter effect of its large size dye molecule. The pyridine is better than oleic acid because of the aromatic characteristics of pyridine.

5.4 INTERFACE BETWEEN ACTIVE LAYER AND METAL ELECTRODE

The thin layer of metal can be layered on top of the active layer. However, the organic characteristics of the active layer are not compatible with metal. Better contact has been achieved by thermal annealing after the deposition of metal on the active layer. In order to improve the contact and reduce the resistance between the active layer and metal electrode, a thin (several nm) electron transport layer has been used in between the two.

By placing a thin layer of LiF (2.90 eV, 0.6 nm) or Ca (2.89 eV, 20 nm), which are less conductive than that of Al (4.28 eV, 100 nm) between metal electrode and active layer of P3HT:PCBM, the efficiency of solar cell can be increased from 1.53% to 3.08% for LiF and 3.69% for Ca due to the formation of ohmic contact [13]. Calcium is a chemically active metal that reacts with the sulfur atom of P3HT during the deposition of Ca film and forms up to 3 nm of CaS clusters [14].

A thin TiO_2 nanorod layer of about 20 nm is placed between the aluminum electrode and MEHPPV:TiO_2 active layer to form a large network between the active layer and electrode to improve the electron transport efficiency, as shown in Figure 5.7. A 2.5 times power conversion efficiency is observed [15]. This electron transport layer is

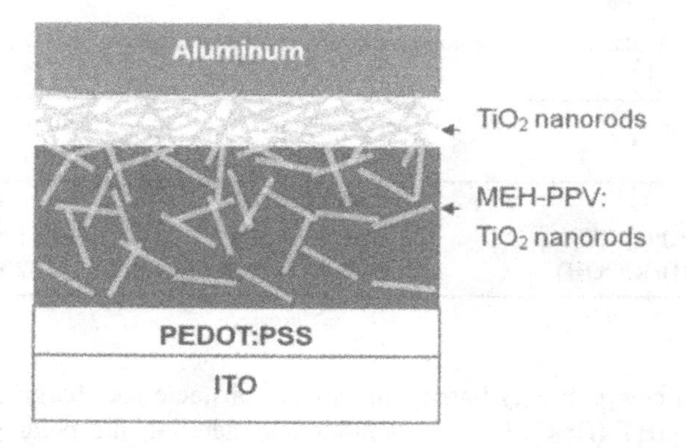

Figure 5.7. Large interconnecting network formed in hybrid MEH-PPV:TiO_2 nanorod solar cells by placing a TiO_2 nanorod electron transport layer between the active layer and the Al.

also used in P3HT:TiO_2 hybrid solar cells with similar results [16]. This technique has been applied to inverted solar cell as well, which has improved their power conversion efficiency from 4.4% to 5.6% [17].

5.5 IMPEDANCE CHARACTERISTICS AT THE INTERFACE

If a potential is applied across a device, a current is caused to flow through the device, with a value determined by the interface characteristics of each component. The relationship between the applied potential and the current flow is known as the impedance. If the applied potential is a sinusoid ($\Delta E \sin \omega t$), then the subsequent current will also be sinusoidal with a value $\Delta i \sin(\omega t + \phi)$. Harmonics of this current (2ω, 3ω, ..., etc.) will also flow. The impedance (Z) has a magnitude $\Delta E/\Delta i$ and phase (ϕ), and is thus a vector quantity.

In a real solar cell, power is dissipated through the resistance of the contacts and leakages at the edges of devices. To account for these effects, a simple resistor–capacitor (RC) equivalent circuit in dark conditions is often used, as shown in Figure 5.8(a) [18]. The series resistance R_s is the sum of interfacial resistance (contact resistance) and bulk resistance of the photoactive layer. A lower series resistance means that higher current will flow through the device. R_{sh} shunt resistance occurs in real devices across the surface, at pin holes in the p–n junction, or at grain boundaries. High shunt resistance corresponds to fewer shorts or leaks in the device. An ideal cell would have an R_s approaching zero and R_{sh} approaching infinity. The behavior of this simple circuit can be determined by an impedance analyzer. Under certain bias voltage (V_{bias}), one inputs a small perturbation alternating voltage signal, at time t; the voltage of the circuit $V(t)$ can be expressed by Eq. (5-1). The corresponding current can be measured by an impedance analyzer, as shown in Eq. (5-2). Then, the impedance can be deduced from Eq. (5-3).

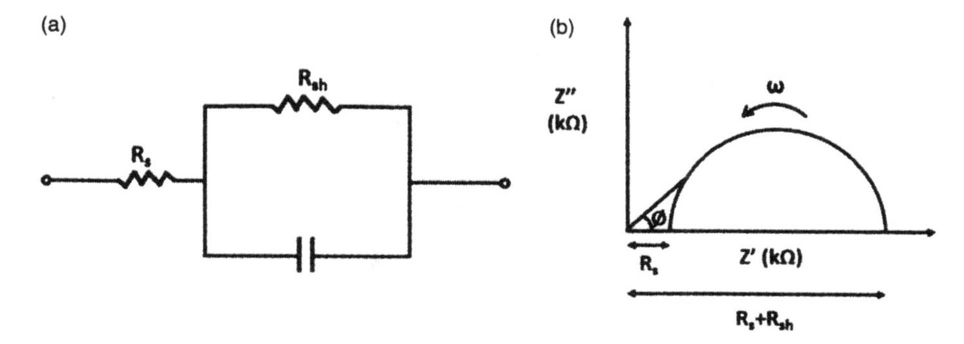

Figure 5.8. (a) Simple equivalent circuit and (b) complex plane impedance spectrum.

$$V(t) = V_{bias} + \hat{V}e^{i\omega t} \tag{5-1}$$

$$I(t) = I_0 + \hat{I}e^{i(\omega t+\phi)} \tag{5-2}$$

$$Z(\omega) = -\frac{\hat{V}}{\hat{I}} = Z'^{(\omega)} + iZ''(\omega) \tag{5-3}$$

where $e^{i\omega t} = \cos(t) + i\sin(t), \omega = 2\pi f$ is angular frequency, and f is frequency. The impedance $Z(\omega)$ contains real part Z' and imaginary part Z''. Figure 5.8(b) shows an example of the impedance spectrum of an ideal RC circuit, either as the real and imaginary parts against frequency, or the magnitude and phase against frequency, allows the characteristics of the system to be established. A predominant RC (a resistor–capacitor) semicircle arc is observed in this complex plot. It is also called Nyquist plot. At high frequencies, a much smaller RC arc is found and the current flows easily (effectively shorting out the R_{sh}, leaving only the effect of the R_s and the transport). As the frequency decreases, the current becomes less and less and the response due to R_{sh} increases. As the frequency approaches zero, there is no current flow and the cell impedance is a function only of R_{sh} and R_s.

This technique has been employed to study the effectiveness of the interface between donors and acceptors in hybrid solar cells of ZnO/P3HT [19]. Figure 5.9 shows, through modification at ZnO nanocolumns, that the shunt resistance for a ZnO/P3HT solar cell can be increased in the impedance spectrum of the devices under illumination. The experimental data can be fitted relatively well into the simple equivalent circuit model (inset). The R_{sh} extracted from the intercepts at the low-frequency end of the plot are 4.6×10^4, 7.6×10^5, and 1.6×10^6 ohm for the reference device, APS-CdSe, and APhS-CdSe modified devices, respectively. In the modified devices, the increased shunt resistance is due to the reduction of electron–hole recombination at the interfaces. APS (3-aminopropyl trimethoxysilane) is less conducting compared with APhS (p-aminophenyl trimethoxysilane), so the APhS exhibits the highest R_{sh}.

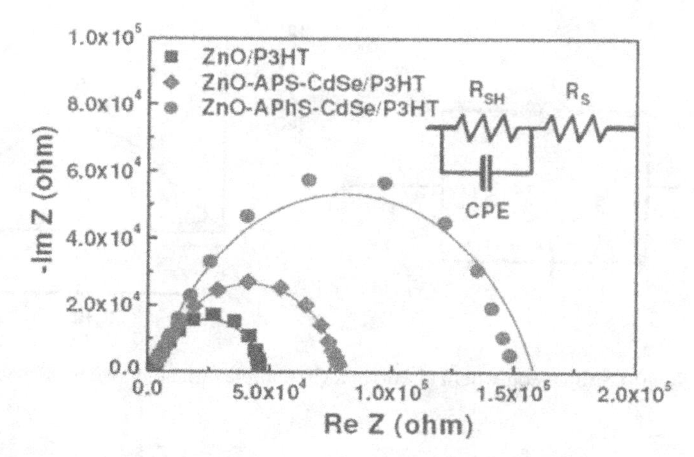

Figure 5.9. Electrochemical impedance spectroscopy for devices of ITO/ZnO nanocolumn/P3HT/Ag without and with a CdSe quantum-dot layer attached with different linkers under illumination (the equivalent circuit is shown in the inset) [19].

Impedance spectroscopy has been used to determine the charge carrier mobility and lifetime of ITO/P3HT:PCBM/Al solar cells [20]. Reverse-bias capacitance exhibits Mott–Schottky-like behavior, indicating the formation of a Schottky junction (band bending) at the P3HT:PCBM/Al contact. Impedance measurements at different biases were performed and showed that minority carriers (electrons) diffuse out of the depletion zone and their accumulation contributes to the capacitive response at forward bias. A simple equivalent circuit accounting for the diffusion-recombination mechanism was proposed, which can fit the experimental data well, as shown in Figure 5.10. This model comprises a distributed resistor r_t (standing for the electron transport), distributed chemical capacitance $c_n = C_{\mu}$, and r_{rec} (accounting for the elec-

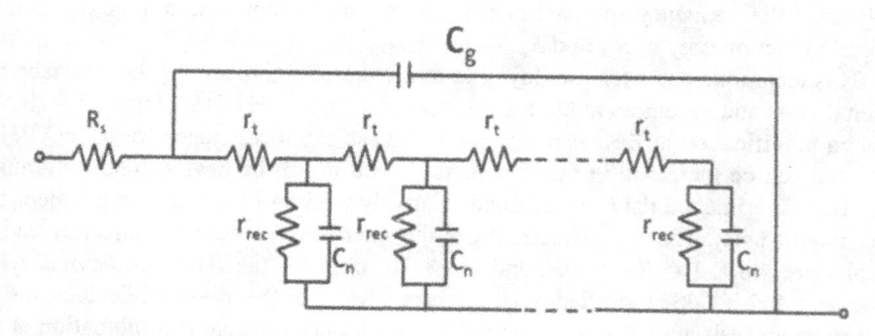

Figure 5.10. Equivalent circuit accounting for the diffusion–recombination mechanism proposed for the ITO/P3HT:PCBM/Al solar cell [20].

tron recombination resistance). The series resistance R_s accounts for the resistance from contact and wire. Finally, a capacitor, $C_g \sim \varepsilon_0\varepsilon A/L$, represents the dielectric contribution of the solar cell diode, ε_0 is the permittivity of the vacuum, A is the device area and L is the thickness of active layer. This impedance contains two characteristic times related to the electron diffusion $\tau_d = r_t c_n$ (transit time), and the effective lifetime $\tau_n = r_{\text{rec}} c_n$.

Figure 5.11 shows that the experimental data fit the model well. The position of characteristic times τ_d and τ_n are marked. The R_s is about 100 Ω and C_g is about 1.0 nF when ε is assumed to be 3 for P3HT:PCBM, $A = 9$ mm^2, and $L = 200$ nm. Because the diffusion–recombination of charge-carrier mechanism is proposed for this solar cell, the diffusion coefficient can be calculated by Eq. (5-4):

$$D_n = \frac{L^2}{\tau_d} \tag{5-4}$$

Assuming that the electron statistics slightly depart from the dilute concentration conditions, the electron mobility can be calculated by using the Nernst–Einstein relationship:

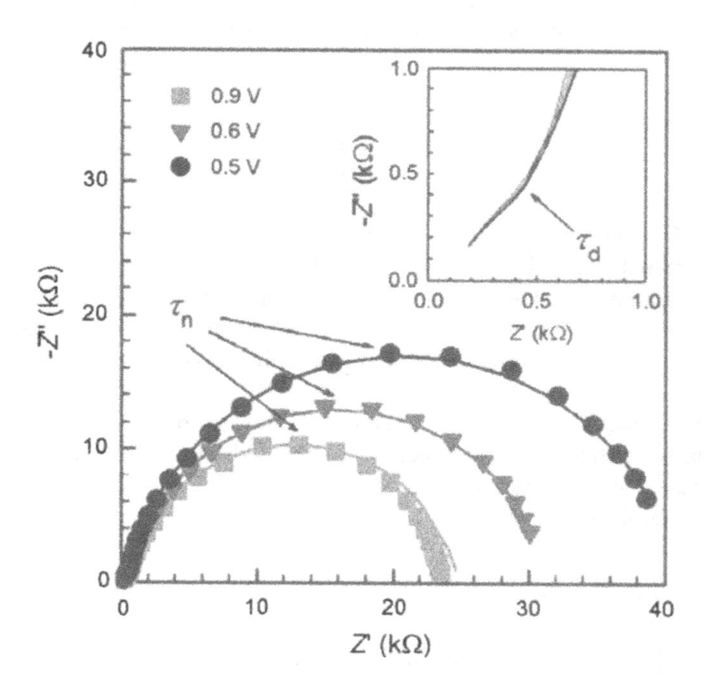

Figure 5.11. Impedance spectra measured in the dark at different bias as indicated. Experimental points and fitting results were obtained using the equivalent circuit model of Figure 5.10. Characteristic times are marked. Inset: detail of the high-frequency interval that signals the transition between diffusion and recombination responses [20]. (Reprinted with permission.)

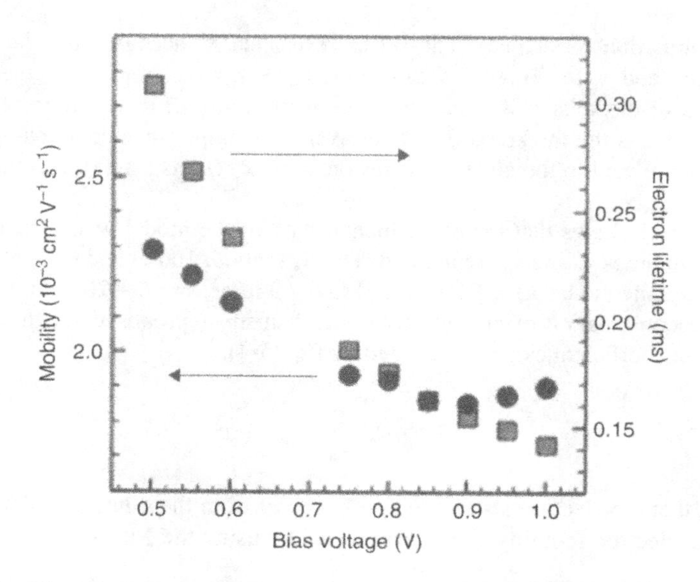

Figure 5.12. Electron mobility and electron recombination time determined from fitting parameters of an diffusion–recombination equivalent circuit model [20]. (Reprinted with permission.)

$$\mu_n = \frac{eD_n}{k_B T} \tag{5-5}$$

where the k_BT is the thermal energy. Figure 5.12 summarizes the results. The electron mobility extracted from this exhibits a nearly constant value close to 2×10^{-3} $cm^2V^{-1}s^{-1}$. Electron effective lifetime results within the time interval 0.3–0.1 ms decrease as the bias increases. Such times refer to losses produced by electron–hole bimolecular recombination along the active layer, although surface recombination cannot be excluded.

Impedance analysis has been well established and routinely used to investigate the interface characteristics inside solar cells. The above discussions have shown that the impedance characteristics are important to understand the interface behaviors of multi-junctions in hybrid solar cells.

REFERENCES

1. W. J. Potscavage, Jr., A. Sharma, and B. Kippelen (2009), *Acct. Chem. Res., 42*(11), 1758–1767.
2. G. Eda, Y. Y. Lin, S. Miller, C. W. Chen, W. F. Su, and M. Chhowalla (2008), *Applied Physics Letters, 92,* 233305.

3. C. J. Brabec and J. R. Durrant (2008), *MRS Bulletin, 33,* 670–675.
4. Y. H. Lin, P. C. Yang, J. S. Huang, G. D. Huang, I. J. Wang, W. H. Wu, M. Y. Lin, W. F. Su, and C. F. Lin (2011), *Solar Energy Materials and Solar Cells, 95,* 2511–2515.
5. C. W. Zhou, L. G. De Arco, Y. Zhang, C. W. Schlenker, K. Ryu, and M. E. Thompson (2010), *ACS Nano, 4,* 2865.
6. H. C. Liao, C. S. Tsao, T. H. Lin, C. M. Chuang, C. Y. Chen, U. S. Jeng, C. H. Su, Y. F. Chen, and W. F. Su (2011), *J. Am. Chem. Soc., 133,* 13064–13073.
7. Y. C. Huang, S. Y. Chuang, M. C. Wu, H. L. Chen, C. W. Chen, and W. F. Su (2009), *Journal of Applied Physics, 106,* 034506.
8. Y. C. Huang, W. C. Yen, Y. C. Liao, Y. C. Yu, C. C. Hsu, M. L. Ho, P. T. Chou, and W. F. Su (2010), *Applied Physics Letters, 96,* 123501.
9. Y. Y. Lin, T. H. Chu, C. W. Chen and W. F. Su (2008), *Applied Physics Letter, 92,* 053312.
10. Y. Y. Lin, T. H. Chu, S. S. Li, C. H. Chuang, C. H. Chang, W. F. Su, C. P. Chang, M. W. Chu, and C. W. Chen (2009), *Journal of The American Chemical Society, 131,* 3644.
11. Y. C. Huang, J. H. Hsu, Y. C. Liao, W. C. Yen, S. S. Li, S. T. Lin, C. W. Chen, and W. F. Su (2011), *Journal of Materials Chemistry, 21*(12), 4450–4456.
12. B. C. O'Regan, and F. J. Lenzmann (2004), *J. Phys. Chem B, 108,* 4342–4350.
13. M. O. Reese, M. S. White, G. Rumbles, D. S. Ginley, and S. E. Shaheen (2008), *Appl. Physics Letters, 92,* 053307.
14. J. Zhu, F. Bebensee, W. Hieringer, W. Zhao, J. H. Baricuatro, J. A. Farmer, Y. Bai, H-P. Steinrück, J. M. Gottfried, and C. T. Campbell (2009), *J. Am. Chem. Soc., 131,* 13498.
15. T. W. Zeng, Y. Y. Lin, H. H. Lo, C. W. Chen, C. H. Chen, S. C. Liou, H. Y. Huang, and W. F. Su (2006), *Nanotechnology, 17,* 5387–5392.
16. T. W. Zeng, H. H. Lo, C. H. Chang, Y. Y. Lin, C. W. Chen, and W. F. Su (2009), *Solar Energy Materials and Solar Cells, 93,* 952–957.
17. Y. H. Lin, P. C. Yang, J. S. Huang, G. D. Huang, I. J. Wang, W. H. Wu, M. Y. Lin, W. F. Su, and C. F. Lin (2011), *Solar Energy Materials & Solar Cells, 95,* 2511–2515.
18. A. L. Fahrenbrach and R. H. Bube (1983), *Fundamentals of Solar Cells,* Academic Press, New York.
19. T. W. Zeng, I. S. Liu, F. C. Hsu, K. T. Huang, H. C. Liao, and W. F. Su (2010), *Optics Express, 18,* S3, A357–365.
20. G. Garcia-Belmonte, A. Munar, E. M. Barea, J. Bisquert, I. Ugarte, and R. Pacios (2008), *Organic Electronics, 9,* 847.

EXERCISES

1. Calculate the number of atoms in bulk and on the surface for 1 cm^3 of silicon and 1 cm^3 of calcium, respectively.
2. Draw the energy band diagrams of ITO/PEDOT:PSS/P3HT:PCBM/Al and ITO/P3HT:PCBM/PEDOT:PSS/Au. Which solar cell has a faster charge carrier transport rate and why?

3. Arrange the following linkers for TiO_2 nanorods in the decreasing order of P3HT:TiO_2 solar cell performance and explain your answer. (Hint: see reference 10.)

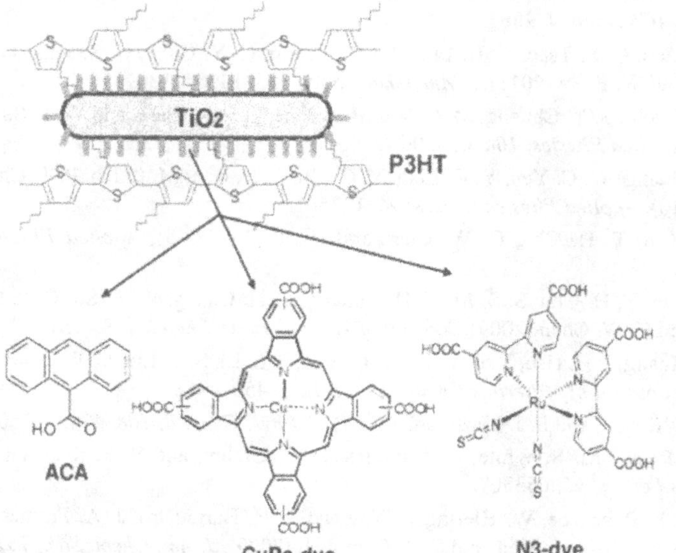

4. Compare and explain the differences in electrical and optical properties between chemical-vapor-deposition-prepared graphene and reduced graphene oxide from a solution process. (Hint: see reference 5 and Valentini, L., Cardinali, M., Bon, S. B., Bagnis, D., Verdejo, R., Lopez-Manchado, M. A., and Kenny, J. M. *J. Mater. Chem.* 2010, *20*, 10943.)
5. Design two forward-structure P3HT:PCBM solar cells with different electron transport layers and hole transport layers, then compare the expected differences in power conversion efficiency between these two solar cells.
6. Calculate the electron mobility and lifetime of aITO/P3HT:PCBM/Al solar cell using the literature data. (Hint: see reference 20.)

6

INORGANIC SOLAR CELLS

I-Chun Cheng

6.1 INTRODUCTION

The first diffused silicon monocrystalline p–n junction was discovered by R. Ohl at Bell Laboratories in 1940 and this light-sensitive electrical device was patented in 1941 [1]. In 1954, D. Chapin, C. S. Fuller and G. Pearson reported an efficiency of 4.5% and then 6% for a silicon monocrystalline solar cell with a diffused p–n junction [2], which was an important milestone in the development of the modern photovoltaic cell. In 1963, Sharp Corporation in Japan developed the first silicon solar cell module and used it for terrestrial applications. The energy crisis in 1973 stimulated work on thin-film solar cells as a path to reducing the cost of photovoltaic electricity. In 1976, D. Carlson and C. Wronski at RCA Research Laboratory demonstrated the first amorphous solar cell with a conversion efficiency about 2.4% [3]. By 1980, amorphous silicon and CdS/CdTe thin-film solar cells had been used in consumer application. In 1985, a crystalline-Si solar cell with conversion efficiency higher than 20% was reported by M. Green at the University of New South Wales. Since the 1990s, the development of photovoltaic technology has focused on large-scale production. In the last decade, global solar cell production has been increasing by an average of more than 20% each year and reached 10.7 GW [4] and 20.5 GW [5] in 2009 and 2010, respectively. In this chapter, we will go over several solar cell technologies based on inorganic semiconductors. First, the basics of p–n homojunction solar cell operation are introduced. Next, crystalline silicon solar cells are described. Finally, thin-film solar cell technologies, including amorphous silicon-based, CdTe, and $CuInSe_2$-based cells, are discussed.

Organic, Inorganic, and Hybrid Solar Cells. By C.-F. Lin, W.-F. Su, C.-I Wu, and I-C. Cheng

6.2 BASIC PRINCIPLES

6.2.1 p–n Junction in Equilibrium

The basis of a typical solar cell is a p–n junction. Let us consider joining a uniformly doped p-type semiconductor region and a uniformly doped n-type semiconductor region together, which is called a step junction. Before contact, as illustrated in Figure 6.1(a), the p-type semiconductor contains a large concentration of holes, whereas the n-type semiconductor contains a large concentration of electrons. Upon bringing these two regions into intimate contact, as illustrated in Figure 6.1(b), diffusion of carriers takes place because of the large concentration gradients existing between these two regions. When holes diffuse from the p-type region into the n-type region, and electrons diffuse from *n* to p, a space-charge region of width *W* is formed due to the uncompensated negative acceptor ions N_a^- in the p side and the uncompensated positive donor ions N_d^+ in the *n* side near the junction. Hence, an electric field *E* is built up and directed from the positive charge toward the negative charge, which prevents further net flow of carriers.

Since in equilibrium the net flux of both electrons and holes is equal to zero, the drift current due to *E* must cancel the diffusion current for both types of carriers. For instance, the net hole current at location *x* in Figure 6.2 can be expressed as:

$$J_p(x) = -qD_p \frac{dp(x)}{dx} + q\mu_p p(x)E(x) = 0 \tag{6-1}$$

where q is the magnitude of electron charge, D_p is the diffusion coefficient for holes, μ_p is the hole mobility, and $p(x)$ is the hole concentration at location x. After rearranging Equation 6-1, writing the electric field in terms of the potential gradient, $E(x) = -dV(x)/dx$, and incorporating Einstein's relationship, $\mu_p = qD_p/kT$, we can obtain

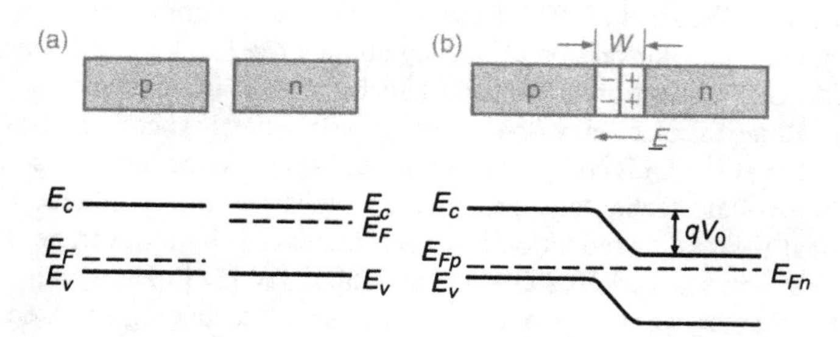

Figure 6.1. Schematic illustration and corresponding energy band diagram of a p–n junction in equilibrium. (a) Isolated p-type and n-type semiconductor materials, where E_C, E_F, and E_V denote the conduction band edge, Fermi level, and valence band edge of the semiconductor, respectively. (b) Junction between the p-type and n-type semiconductor materials, where E_{Fp} and E_{Fn} are the quasi-Fermi levels of p-type and n-type semiconductors, and V_0 is the built-in potential.

$$-\frac{q}{kT}\frac{dV(x)}{dx}=\frac{1}{p(x)}\frac{dp(x)}{dx} \tag{6-2}$$

Integrating Equation 6-2 over the step junction gives

$$-\frac{q}{kT}V_0=\ln\frac{p_n}{p_p} \tag{6-3}$$

where V_0 is the potential difference between the n side and the p side, $V_n - V_p$, called the built-in potential, and p_n and p_p are the hole concentrations in the n-side and p-side neutral regions, respectively. Equation 6-3 can be further written as

$$\frac{p_p}{p_n}=e^{qV_0/kT} \tag{6-4}$$

Similarly, we can get the expression for electron concentration on either side of the junction as follows:

$$\frac{n_n}{n_p}=e^{qV_0/kT} \tag{6-5}$$

where n_n and n_p are the electron concentrations in the n-side and p-side neutral regions, respectively.

By applying the equilibrium condition $p_p n_p = p_n n_n = n_i^2$, $p_p \approx N_a$, and $n_n \approx N_d$, the built-in potential V_0 can be related to the doping concentrations on each side of the junction as

$$V_0=\frac{kT}{q}\ln\frac{N_a N_d}{n_i^2} \tag{6-6}$$

where n_i is the intrinsic carrier concentration in the semiconductor material.

To obtain the characteristics of the space charge region, we assume that the carrier concentration within it is negligible and only ionized acceptors and donors are considered. The assumption of carrier depletion over the space charge region is called the depletion approximation. The space charge density in the p side and n side are $-qN_a$ and $+qN_d$, respectively. The penetration of the space charge region into the p side and the n side are denoted as x_p and x_n with total width $(x_p + x_n)$ equal to W, as illustrated in Figure 6.2. Since the junction must have equal number of charges on either side, we obtain

$$qN_a x_p = qN_d x_n \tag{6-7}$$

To calculate the distribution of electrical field $E(x)$, we apply Poisson's equation as follows:

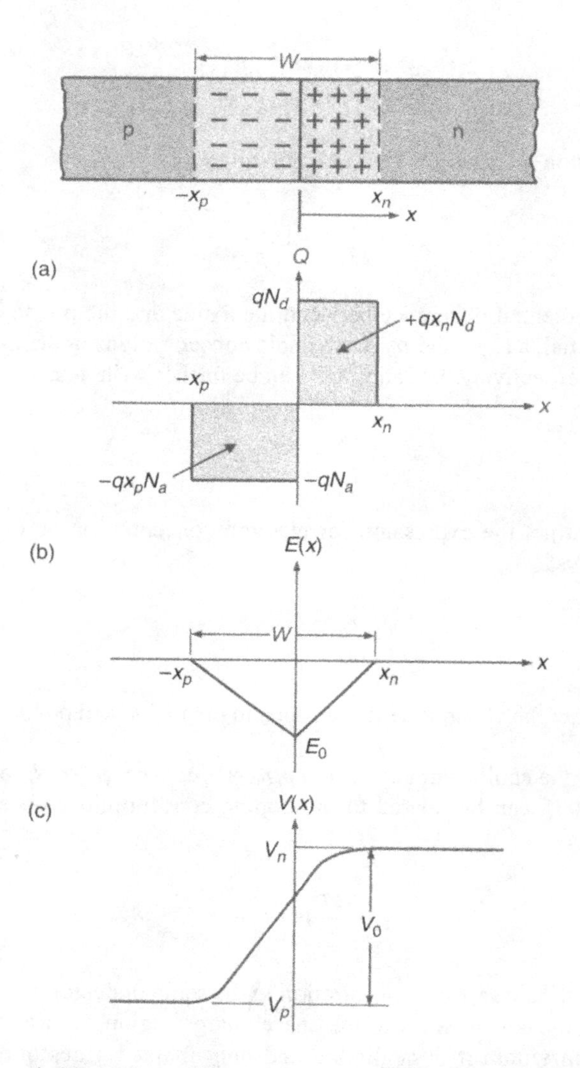

Figure 6.2. Schematic illustrations of the (a) charge density distribution, (b) electrical field distribution, and (c) potential distribution within the space charge region of a step p–n junction in equilibrium.

$$\frac{dE(x)}{dx} = \frac{q}{\varepsilon}\left(p - n + N_d^+ - N_a^-\right) \tag{6-8}$$

where ε is the dielectric constant of the semiconductor material. Outside the space charge region (neutral regions) the net charge density is zero. Within the space charge region the contribution of the carriers to the space charge is neglected and complete ionization of the dopants is assumed, and Poisson's equation can be approximated as

$$\frac{dE(x)}{dx} = -\frac{q}{\varepsilon}N_a \qquad -x_p < x < 0 \tag{6-9}$$

$$\frac{dE(x)}{dx} = +\frac{q}{\varepsilon}N_d \qquad 0 < x < x_n \tag{6-10}$$

Equations 6-9 and 6-10 tell us that the electrical field $E(x)$ is negative everywhere within the space charge region. Its magnitude increases as x increases in the p side and then decreases as x increases in the n side, as shown in Figure 6.2(b). At $x = 0$, maximum value of electrical field E_0 is reached, which can be obtained by integrating Equations 6-9 and 6-10 with integration limits of $E(-x_p) = 0$ and $E(0) = E_0$ for Eq. 6-9 and $E(0) = E_0$ and $E(x_n) = 0$ for Eq. 6-10. Then E_0 can be expressed as

$$E_0 = -\frac{qN_a x_p}{\varepsilon} = -\frac{qN_d x_n}{\varepsilon} \tag{6-11}$$

and the built-in potential V_0 is related to the width of depletion region as

$$\begin{aligned} V_0 &= -\int_{-x_p}^{x_n} E(x)\,dx = -\frac{1}{2}E_0 W \\ &= \frac{1}{2}\frac{q}{\varepsilon}N_a x_p W = \frac{1}{2}\frac{q}{\varepsilon}N_d x_n W \end{aligned} \tag{6-12}$$

Since $N_a x_p = N_d x_n$ and $x_p + x_n = W$, we can substitute x_n and x_p in Equation 6-12 with $x_n = [N_a/(N_a + N_d)]W$ and $x_p = [N_d/(N_a + N_d)]W$ to obtain V_0 in terms of N_a, N_d, W, q, and ε as

$$V_0 = \frac{1}{2}\frac{q}{\varepsilon}\frac{N_a N_d}{N_a + N_d}W^2 \tag{6-13}$$

Then the width of the space charge region can be expressed in terms of built-in potential, doping concentration, and constants q and ε as

$$W = \left[\frac{2\varepsilon V_0}{q}\left(\frac{1}{N_a} + \frac{1}{N_d}\right)\right]^{1/2} \tag{6-14}$$

The equation above indicates that the width of the space charge region varies as the square root of the potential across the p–n junction.

6.2.2 Current–Voltage Characteristics

In this section, current–voltage (I–V) characteristics of a p–n junction will be discussed. If we apply a voltage bias V to the p side with respect to n side and neglect the voltage drop in the neutral region, the electrostatic potential barrier across the junction is lowered by a forward bias $V = V_F$ from the built-in potential V_0 to $(V_0 - V_F)$ and increased by a reverse bias $V = -V_R$ from the built-in potential V_0 to $(V_0 + V_R)$, as illustrated in Figure 6.3. Therefore, the electric field within the space charge region decreases with forward bias and increases with reverse bias.

Since change in the electric field can result in change in the width of the space charge region, we can expect W to decrease under forward bias and increase under reverse bias. The value of W can be obtained by replacing V_0 in Equation 6-14 with new potential barrier of $(V_0 - V_F)$ under forward bias and $(V_0 + V_R)$ under reverse bias:

$$W(V = V_F) = \left[\frac{2\varepsilon(V_0 - V_F)}{q}\left(\frac{1}{N_a} + \frac{1}{N_d}\right)\right]^{1/2} \tag{6-15}$$

$$W(V = -V_R) = \left[\frac{2\varepsilon(V_0 + V_R)}{q}\left(\frac{1}{N_a} + \frac{1}{N_d}\right)\right]^{1/2} \tag{6-16}$$

The electron energy barrier reduces from qV_0 to $q(V_0 - V_F)$ under forward bias, whereas it increases from qV_0 to $q(V_0 + V_R)$ under reverse bias.

When voltage bias V is applied, the ratio of hole concentrations on the two sides of the space charge region becomes:

$$\frac{p(-x_p)}{p(x_n)} = e^{q(V_0 - V)/k_BT} \tag{6-17}$$

Considering the case with low-level injection, where the changes in the majority carrier concentrations can be neglected, $p(-x) \approx p_p$, the ratio of minority carrier (hole) con-

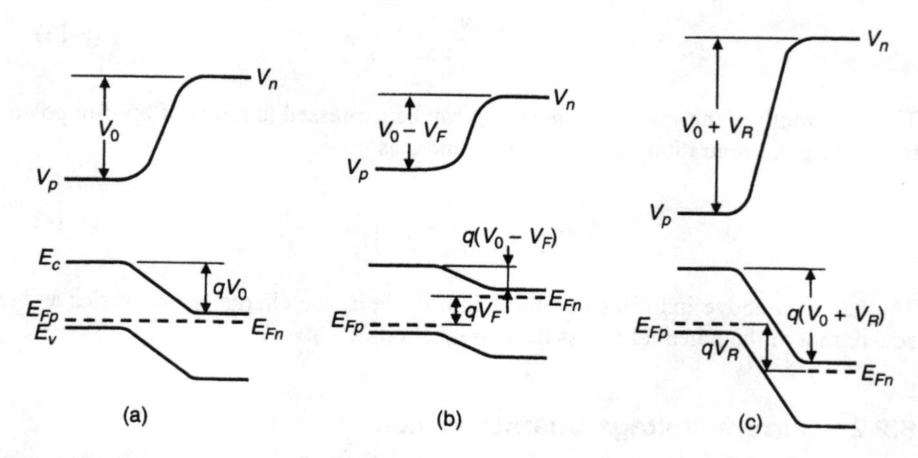

Figure 6.3 Schematic illustrations of space charge region width, electrostatic potential, and energy band diagram of a p–n junction (a) in equilibrium, (b) under forward bias, and (c) under reverse bias.

centrations at the edge of the space charge region on the n-side under bias and at equilibrium can be obtained by taking the ratio of Equation 6-4 and Equation 6-17 as follows:

$$\frac{p(x_n)}{p_n} = e^{qV/k_BT} \tag{6-18}$$

Similarly, the ratio of minority carrier (electron) concentrations at the edge of the space charge region on the p-side under bias and at equilibrium can be expressed as follows:

$$\frac{n(-x_p)}{n_p} = e^{qV/k_BT} \tag{6-19}$$

Therefore, the minority carrier concentrations at the edges of the space charge region are raised under forward bias and reduced under reverse bias. The excess hole and electron concentrations at the edge of the space charge region under forward bias can be described by Equations 6-20 and 6-21, respectively:

$$\Delta p_n(x_n) = p(x_n) - p_n = p_n\left(e^{qV/k_BT} - 1\right) \tag{6-20}$$

$$\Delta n_p(-x_p) = p(-x_p) - n_p = n_p\left(e^{qV/k_BT} - 1\right) \tag{6-21}$$

Assuming that the lengths of the p-side and n-side regions are longer than the minority carrier diffusion lengths L_e (electron diffusion length) and L_h (hole diffusion length), the excess minority carrier concentration distribution can be expressed as

$$\Delta p_n(x) = \Delta p_n(x_n)e^{-(x-x_n)/L_h} \qquad \text{for } x \geq x_n \tag{6-22}$$

$$\Delta n_p(x) = \Delta n_p(-x_p)e^{-(-x_p-x)/L_e} \qquad \text{for } x \leq -x_p \tag{6-23}$$

The hole diffusion current $I_{p,diff}(x)$ and electron diffusion current $I_{n,diff}(x)$ can be calculated as

$$I_{p,diff}(x) = -qAD_h\frac{d\Delta p_n(x)}{dx} = qA\frac{D_h}{L_h}\Delta p_n(x_n)e^{-(x-x_n)/L_h} = qA\frac{D_h}{L_h}\Delta p_n(x) \tag{6-24}$$

$$I_{n,diff}(x) = -qAD_e\frac{d\Delta n_p(x)}{dx} = -qA\frac{D_e}{L_e}\Delta n_p(-x_p)e^{-(-x_p-x)/L_e} = -qA\frac{D_e}{L_e}\Delta n_p(x) \tag{6-25}$$

where A is the area of the p–n junction and D_h and D_e denote the diffusion coefficient of holes and electrons, respectively.

Then we can obtain the hole and electron diffusion currents at the edges of the space charge region as follows:

$$I_{p,diff}(x_n) = qA\frac{D_h}{L_h}\Delta p_n(x_n) = qA\frac{D_h}{L_h}p_n\left(e^{qV/k_BT}-1\right) \quad (6\text{-}26)$$

$$I_{n,diff}(-x_p) = -qA\frac{D_e}{L_e}\Delta n_p(-x_p) = -qA\frac{D_e}{L_e}n_p\left(e^{qV/k_BT}-1\right) \quad (6\text{-}27)$$

If the recombination in the space charge region is neglected, the electron current at $-x_p$ must equal the electron current at x_n. The total current I is given by

$$I = I_{p,diff}(x_n) - I_{n,diff}(-x_p) = qA\left(\frac{D_h}{L_h}p_n + \frac{D_e}{L_e}n_p\right)\left(e^{qV/k_BT}-1\right) = I_0\left(e^{qV/k_BT}-1\right) \quad (6\text{-}28)$$

where I_0 is called reverse saturation current, because when a reverse bias $V = -V_R$ much greater than the thermal voltage of k_BT/q is applied across the junction, the total current I equals $-I_0$.

In reality, some of the minority carriers can recombine within the space charge region, W. The recombination current is proportional to n_i and increases with forward bias as $e^{qV/2kBT}$ approximately. In general, the I–V characteristic of a diode can be written as

$$I = I_0\left(e^{qV/nk_BT}-1\right) \quad (6\text{-}29)$$

where n is often called the ideality factor, which varies between 1 and 2. The diode characteristic is diffusion-controlled when $n = 1$ and is controlled by recombination in the space charge region when $n = 2$.

6.2.3 Photovoltaic Current-Voltage Characteristics

Figure 6.4 shows the photovoltaic operation principle of a p–n junction under open- and short-circuit conditions. When photons with energy greater than the energy band gap of the semiconductor material are absorbed by the cell, they create electron–hole pairs. Electron–hole pairs generated in the space charge region can drift apart immediately due to the built-in electrical field, E_0. Electron–hole pairs generated within the minority carrier diffusion lengths, L_h and L_e, of both sides can diffuse to the edges of the space charge region and then are separated by the built-in electrical field, too. Once the electrons drift and reach the n-type neutral region and the holes drift and reach the p-type neutral region, an open circuit voltage V_{oc} builds up when the front and back terminals are opened, as illustrated in Figure 6.4 (a). The quasi-Fermi levels of n-region and p-region are separated by V_{oc}. When the two terminals are shorted, the excess electrons can travel through the external circuit to recombine with the excess holes in the p-side and short-circuit current I_{sc} is measured, as illustrated in Figure 6.4 (b).

In real operation, the two terminals are connected though an external resistive load R_L, as depicted in Figure 6.5. When the cell is under illumination, the current I though the cell can be divided into two parts: (1) the photocurrent current, $-I_{ph}$, due to the flow

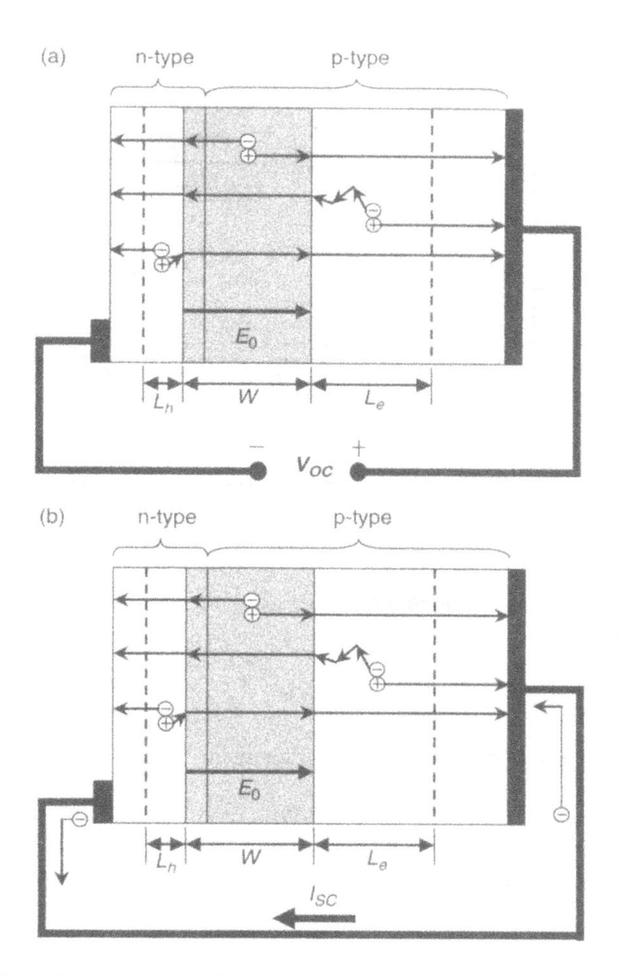

Figure 6.4. Schematic illustrations of photovoltaic operation principle of a p–n junction under (a) open-circuit condition and (b) short-circuit condition. Where W, L_e, and L_h denote the width of the space charge region, electron diffusion length within the p-type region, and hole diffusion length within the n-type region, respectively.

of the photo-generated carriers, which depends on the incident light intensity ϑ, and (2) the forward diode current, I_d, due to the voltage developed across the external load R_L. The photocurrent can be described by

$$I_{ph} = qAG(L_h + W + L_e) \tag{6-30}$$

where A (unit: cm^{-2}) is the area of the diode and G (unit: $cm^{-3}sec^{-1}$) is the electron–hole pair generation rate per unit volume. Since G is proportional to light in-

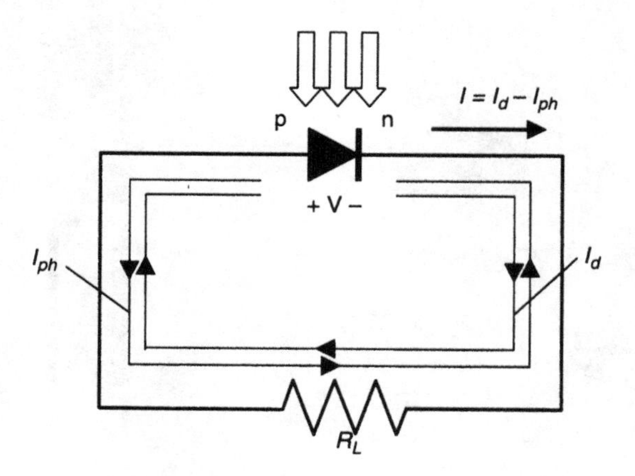

Figure 6.5. Photocurrent I_{ph} and diode current I_d in a p–n junction solar cell driving an external resistive load R_L.

tensity, the photocurrent can be related to the light intensity with a proportionality constant K as follows:

$$I_{ph} = K\vartheta \tag{6-31}$$

The diode current is given by the diode characteristics as

$$I_d = I_0\left(e^{qV/nk_BT} - 1\right) \tag{6-32}$$

where n is the ideal factor of a diode. Therefore, the total current through the cell is

$$I = -I_{ph} + I_d = -K\vartheta + I_0\left(e^{qV/nk_BT} - 1\right) \tag{6-33}$$

The I–V characteristics of a typical p–n junction solar cell under illumination is shown in Figure 6.6. It can be viewed as diode characteristics measured in the dark being shifted down by I_{ph}. By setting $I = 0$, we can obtain the open circuit voltage V_{oc} of the cell as

$$V_{OC} = \frac{nk_BT}{q}\ln\left(\frac{I_{ph}}{I_0} + 1\right) = \frac{nk_BT}{q}\ln\left(\frac{K\vartheta}{I_0} + 1\right) \tag{6-34}$$

It can be seen that it depends on the intensity of illumination. If $V_{oc} \gg nk_BT/q$, the open-circuit voltage can be approximated as

$$V_{OC} \approx \frac{nk_BT}{q}\ln\left(\frac{K\vartheta}{I_0}\right) \tag{6-35}$$

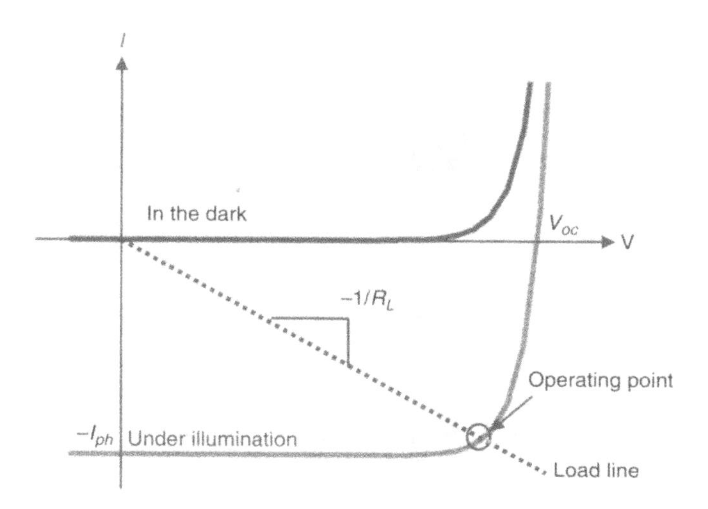

Figure 6.6. I–V characteristics of a typical p–n junction solar cell. The short-circuit current is equal to $-I_{ph}$, which depends on the intensity of illumination. The I–V characteristics measured under illumination can be viewed as the I–V characteristics measured in the dark being shifted down by I_{ph}.

Next, we consider the current and voltage under operation. The I–V characteristics of the external resistive load can be described by

$$I = -\frac{V}{R_L} \tag{6-36}$$

Since the actual current and voltage in the circuit must satisfy the I–V characteristics of the cell, described by Equation 6-33, and that of the load resistance, described by Equation 6-36, simultaneously, the operating point can be determined by the intersect of these two characteristic curves, as shown in Figure 6.6.

6.2.4 Series and Shunt Resistances

The I–V characteristic of a practical solar cell can deviate from that of an ideal p–n junction solar cell due to the presence of internal series and/or shunt resistances. The series resistance, R_{series}, can arise from the resistance of the device material and the electrodes, whereas the shunt resistance, R_{shunt}, can be caused by the leakage current flowing around the edges of the devices or through the cell, such as across grain boundaries in a polycrystalline device. Figure 6.7 illustrates the equivalent circuits of an ideal p–n junction solar cell and a p–n junction solar cell with series and shunt resistances. The series resistance can cause a voltage drop in the output of the cell and result in a smaller voltage across the load resistance, whereas the shunt resistance can divert

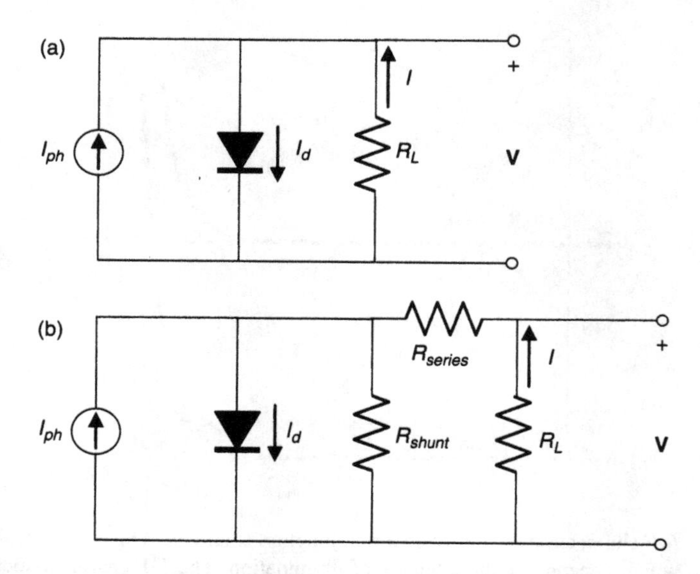

Figure 6.7. Equivalent circuits of (a) an ideal p–n junction solar cell and (b) a p–n junction solar cell with series and shunt resistances.

the photocurrent away from the load resistance. When both resistances are present, the I–V characteristic of a p–n solar cell becomes

$$I = -I_{ph} + I_0\left(e^{q(V-IR_{series})/nk_BT} - 1\right) + \frac{V - IR_{series}}{R_{shunt}} \quad (6\text{-}37)$$

To realize an efficient cell, R_{series} should be as small as possible and R_{shunt} should be as large as possible.

6.3 CRYSTALLINE SILICON SOLAR CELLS

Crystalline silicon solar cells and modulus have been dominant photovoltaic technology for a long period of time. Table 6.1 shows the photovoltaic cell production based on various technologies from 2007 to 2009. Although technologies other than crystalline silicon cells have gradually played important roles in the photovoltaic market in the past couple of years, 81% of cell production was still based on crystalline silicon technology in 2009. In early days, p-type Czochralski silicon wafers were the major substrates used in solar cell production. Today, multicrystalline (or polycrystalline) silicon cells have become an alternative due to their lower cost.

A conventional crystalline silicon solar cell structure is illustrated in Figure 6.8. The shallow p–n junction diode is usually achieved by diffusion or implantation of n-type dopants, usually phosphorus, onto p-type monocrystalline or multicrystalline sil-

TABLE 6.1. Photovoltaic Cell Production

Technology	Solar cell production (unit: MW)		
	2007	2008	2009
Standard crystalline silicon	3,054	5,637	8,020
Super monocrystalline silicon	260	447	653
CdTe	207	504	1,019
CIGS	28	74	166
Amorphous silicon	198	388	796
Total	3,746	7,049	10,655

Source: GTM Research.

icon wafers of ~300 μm thickness. The n-type region, usually called the emitter, of 0.2 to 1 μm thickness is heavily doped; therefore, the space charge region extends primarily into the p-side. Metallic finger and bus electrodes are deposited onto the n-side, as depicted in Figure 6.8. The electrode pattern is used to allow light to enter the cell while keeping the series resistance small at the same time. The patterns are usually realized by screen printing. A thin antireflection layer, usually silicon nitride, is coated on top of the n-layer to increase the amount of light transmittance into the cell. When photons with energy greater than the silicon band gap of 1.12 eV enter the cell, they create electron–hole pairs and contribute to the energy conversion process.

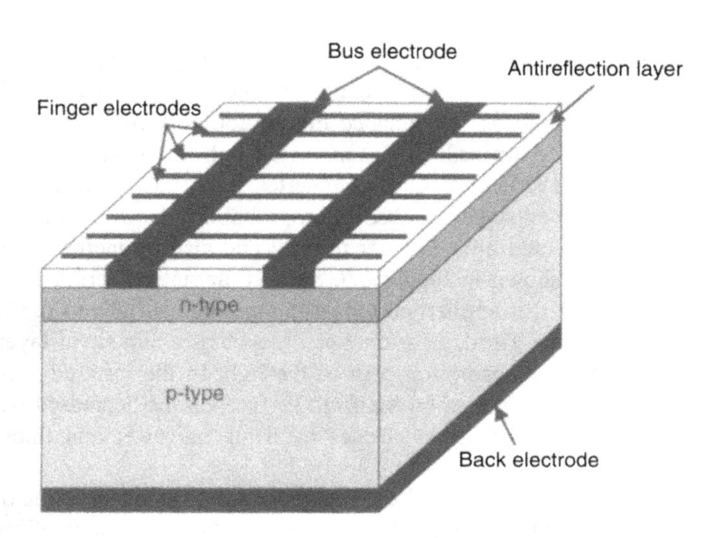

Figure 6.8. A schematic illustration of a conventional crystalline silicon solar cell structure. The n-region is narrow and heavily doped. The finger electrodes on the front surface of the cell are used to reduce the series resistance. The carriers collected by the finger electrodes are then transferred to the bus electrodes.

Typical crystalline silicon-based solar cell efficiency ranges from 15–18% for multicrystalline (or polycrystalline) cells to 22–24% for high-efficiency single-crystal cells with special structures to increase the amount of photons absorbed. The world record efficiency for the best silicon solar cell of 24.7% was reported by University of New South Wales' ARC Photovoltaic Centre of Excellence. Losses commonly seen in silicon solar cells include:

1. Photons with energies less than the silicon band gap of 1.12 eV cannot be absorbed.
2. Carriers generated by absorption of photons with energies larger than 1.12 eV may loose their energies due to the thermalization process.
3. The open-circuit voltage is determined by the separation of quasi-Fermi levels of p- and n-sides instead of the energy band gap of silicon.
4. Optical loss (or photon collection loss) is caused by surface reflection and electrode shading.
5. Charge recombination occurs in the bulk and on the front and rear surfaces of the cells.
6. Series resistance due to front or rear current collection patterns and contacts causes a voltage drop in the output of the cell.
7. Shunt resistance due to leakage current flow around the edges or through the grain boundaries of the cell can divert the photocurrent away from the external load.

Losses 1–3 are due to the fundamental limit of silicon material, whereas 4–7 can be minimized by special structure design or process improvements. Because silicon has a high refractive index—3.5 at wavelength of 700 nm—the surface reflection is high. To reduce the optical loss or enhance the photon collection efficiency, light-trapping techniques can be utilized. Commonly used light-trapping techniques are as follows. First, a textured structure, such as inverted pyramids, is introduced on the top surface of the cell to enhance the absorption probability by allowing a second or third chance for the reflected light to enter the cell and also by increasing the effective absorption path, as shown in Figure 6.9. Second, an antireflection coating, usually an SiN_x or TiO_x layer, is applied on top of the textured surface to further reduce the reflected light intensity. Third, a back reflector, such as a thin silver layer, is coated to increase the effective absorption path of the light in the infrared spectrum. In addition, the electrode shading area on the front surface can be decreased without deteriorating the conductance of the electrodes by using narrower and thicker finger electrodes.

Next is the charge recombination loss. Recombination can occur at the defects inside the bulk. Surfaces and interfaces also contain significant amounts of recombination centers. For instance, photons with high energies are absorbed strongly near the top surface and the electron–hole pairs generated can be lost by the recombination on the front surface. The recombination loss can be reduced by impurity gettering, surface

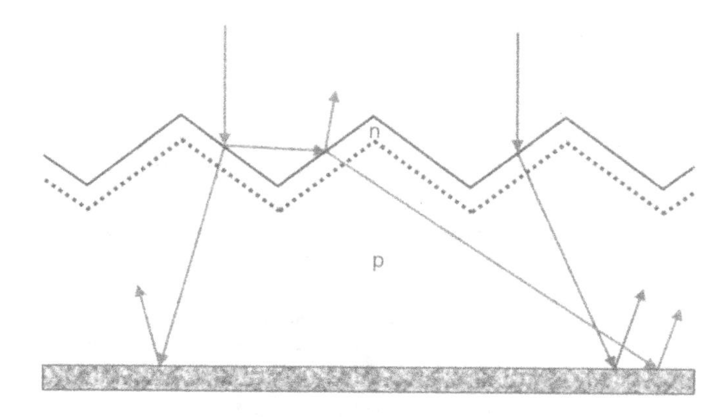

Figure 6.9. Surface texturing increases the optical path length relative to the light normally incident on a flat surface and reduces the reflection by allowing a second or third chance for the reflected light to enter the cell.

passivation, and back-surface field introduction. The impurity gettering is used to reduce the bulk defects. The front and rear surfaces can be passivated with dielectric material, typically a thin SiO_2 layer, to minimize the surface recombination. Introducing heavily doped region between the bulk semiconductor and the majority carrier contact at the back, usually called the back surface field (BSF), is also a way to reduce the recombination on the rear surface.

Series and shunt resistance are the major causes of resistive loss. To reduce the contact resistance between the n-type emitter and front contact without introducing high surface recombination loss, selective emitter structures can be employed, in which the doping concentration and thickness of the emitter under the electrodes is kept high and thick while the doping concentration and thickness of the emitter in the illumination region is kept low and thin. Buried contact technology, developed at the University of New South Wales [6,7], is another way to reduce the contact resistance. The buried contact, with high aspect ratios, can be achieved by forming grooves on the surface and then depositing metal contacts by electroless plating. To prevent the shunting problem, low-quality feedstock silicon wafers should be avoided.

The typical process flow for an industrial crystalline silicon solar cell fabrication is shown in Figure 6.10. Phosphorus is usually used as an n-type dopant for the emitter and aluminum is commonly used as a p-type dopant for the back surface field.

To enhance the performance of the crystalline silicon solar cells, some novel cells have been demonstrated. Here we discuss two examples: the passivated emitter and rear locally diffused (PERL) cell from the University of New South Wales and the heterojunction with intrinsic thin layer (HIT) cell from SANYO Electric Co. Ltd.

PERL cells [8] are the world's highest performing silicon solar cells. The PERL cells, as shown in Figure 6.11, are produced from silicon wafer with high purity. The front and rear surfaces of the cells are passivated with SiO_2 layers to minimize the re-

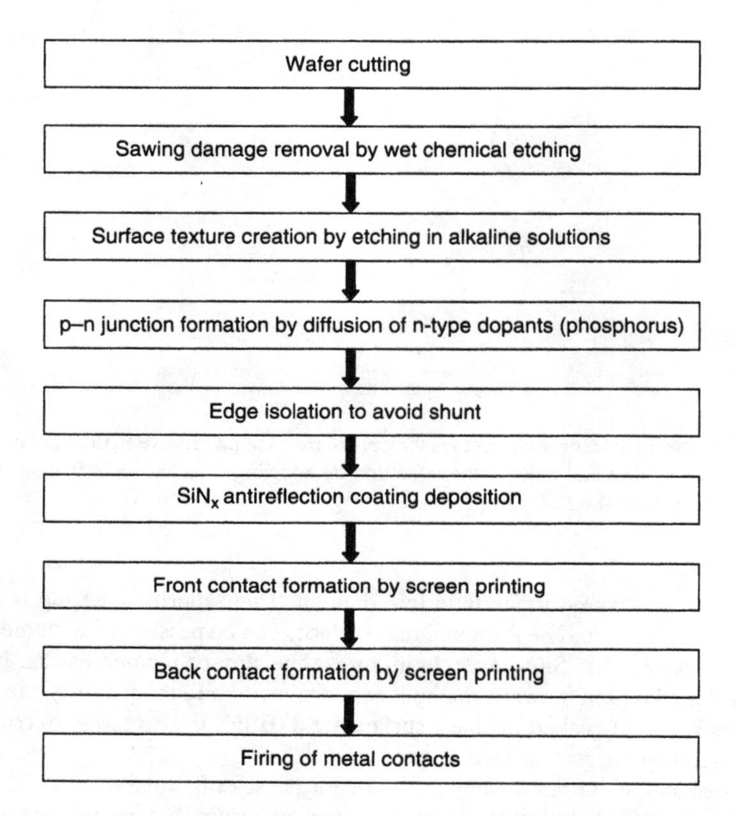

Figure 6.10. Typical process flow for industrial crystalline silicon solar cell fabrication.

combination loss. Inverted-pyramid texturing and antireflection coatings are applied on the top surface to enhance the light-trapping efficiency. A locally boron-diffused region underneath the aluminum contacts on the back side is introduced to decrease the contact resistance.

HIT cells were introduced by Sanyo Electric Co. Ltd in the 1990s. Efficiencies of 22% and 23% were achieved in 2007 and 2009, respectively. The HIT cell is composed of a textured, thin (<100 μm), low-doped, n-type monocrystalline silicon wafer sandwiched between a p-type amorphous silicon layer as the emitter and an n-type amorphous silicon layer as the back surface field. Ultrathin, high-quality intrinsic amorphous silicon layers are inserted between the doped amorphous silicon layers and the crystalline substrate. Because the band gaps of amorphous silicon (a-Si:H) and crystalline silicon (c-Si) are ~1.8 eV and 1.12 eV, respectively, two heterojunctions are presented in this structure. The excellent a-Si:H/c-Si heterointerface enables a high open circuit voltage and results in a better temperature coefficient compared to conventional p–n homojunction crystalline silicon solar cells [9]. A schematic cross section of the HIT cell is illustrated in Figure 6.12. The textures on the top and bot-

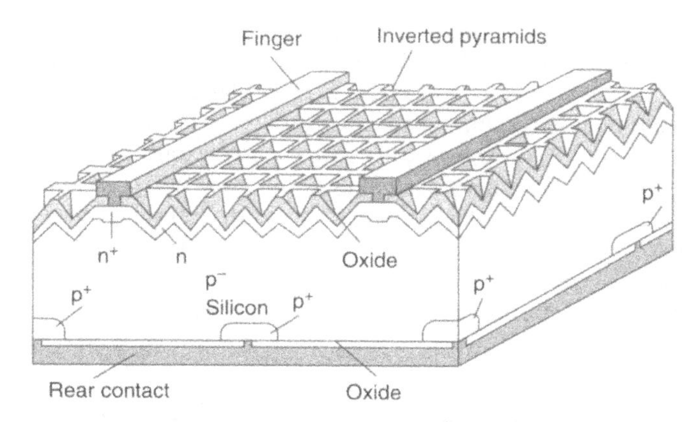

Figure 6.11. Passivated emitter rear locally diffused (PERL) cell [8].

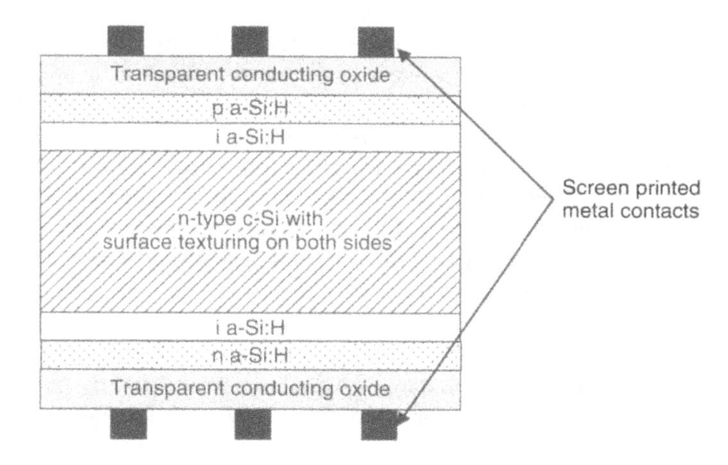

Figure 6.12. Schematic cross section of a heterojunction with intrinsic thin layer (HIT) cell. The good performance is achieved by the improvements in a-Si:H/c-Si heterointerfaces and the optical confinement. The surface texturing is not shown here.

tom surfaces can assist in light trapping. Grid electrodes are used on both sides, making it a bifacial cell.

6.4 THIN-FILM SOLAR CELLS

The development of thin-film solar cell originated in the 1960s with CdS/Cu_2S. The energy crisis in 1973 stimulated work on thin-film solar cells as a path to reducing the cost of photovoltaic electricity. a-Si:H, CdTe, and $CuInSe_2$ became attractive materials for solar cell applications. In 1976, D. Carlson and C. Wronski at RCA Research Laboratory reported the first solar cell based on a-Si:H with a conversion efficiency about 2.4% [3]. In the early 1980s p^+–i–n^+ a-Si:H /ITO solar cells were made on organic polymer foil substrates by Okaniwa and coworkers [10]. To enhance the performance of a-Si:H-based solar cells, stacked cells and surface texture substrates were introduced. At the same time, the development of CdTe- and $CuInSe_2$-based solar cell advanced gradually. In the 1990s, the performance of thin-film solar cells reached a conversion efficiency of >10%, >16%, and >19% for a-Si:H-based, CdTe-based, and $CuInSe_2$-based cells, respectively. At the same time, much effort was devoted to commercialization of these thin-film solar cells. In this section, we will focus on these three commonly used thin-film solar cells.

6.4.1 Amorphous Silicon-Based Thin-Film Solar Cells

Since the first a-Si:H solar cell reported in 1976 with a conversion efficiency of 2.4%, a-Si:H solar cell technology has improved considerably and has become one of the industrially mature solar cell technologies today. Different from crystalline silicon, amorphous silicon behaves like a direct-band-gap material with an optical band gap of 1.7 to 1.8 eV. It can be deposited at relatively low temperature on various types of substrates. The basic device structure is p–i–n with p- and n-layers of ~20 nm and i-layer of ~ 200nm. The p-layer serves as the window layer. However, due to the lack of long-range ordering, it suffers from light-induced degradation, called the Staebler–Wronski effect [11]. To minimize this effect, the intrinsic layer is made into a thin [12,13] multijunction structure by using alloys with different band gaps and introduced to compensate the insufficient absorption.

The p–i–n structure is designed such that most of the photon absorption takes place in the i-layer. The reason why a-Si:H solar cell uses a p–i–n structure instead of simple p–n junction is because doping in a-Si:H can create defects and the lifetimes of minority carriers in doped a-Si:H are very short. Insertion of an intrinsic layer between p- and n-layers can not only increase the region for photon absorption but also ensure that the lifetimes of the photogenerated carriers are sufficiently large. Assuming that the i-layer is truly intrinsic (in amorphous silicon, the unintentionally doped i-layer is usually slightly n-type), the built-in electric field in the i-layer is uniform, as shown in Figure 6.13, which is different from that in a simple p–n junction. Even though the photon absorption mainly occurs in the i-layer, the i-layer may introduce series resistance due to its poor conductivity compared to the doped regions. In addition, if the i-layer is not truly intrinsic, the electrical field may fall to zero within in the i-layer due to the charged impurities. Therefore, the typical i-layer thickness is approximately 20 nm.

In a-Si:H p–i–n, illumination through the p-layer is favorable, as illustrated in Figure 6.14. For illumination though the n-layer, a pronounced drop in the power can occur because the holes have to travel for a longer distance before being collected and the electric field collapses due to the build-up of the slowly drifting holes. The recombina-

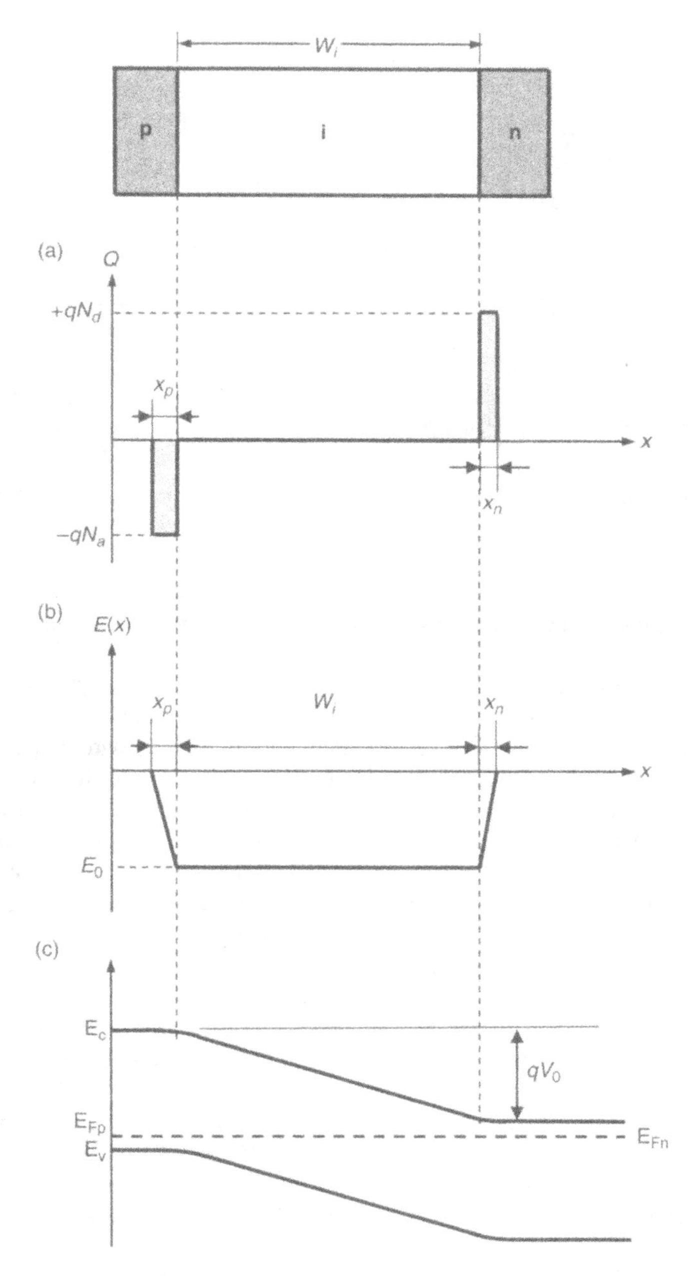

Figure 6.13. Schematic illustrations of the (a) charge density distribution, (b) electrical field distribution, and (c) energy band diagram of a p–i–n junction in equilibrium.

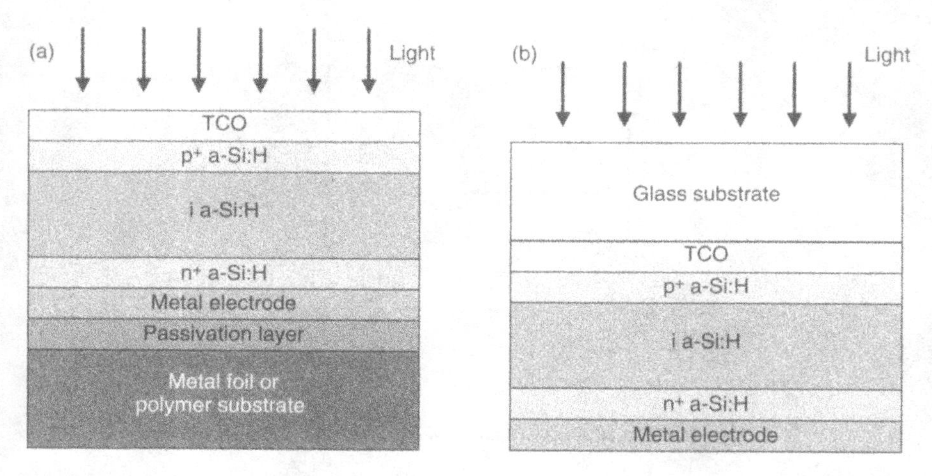

Figure 6.14. a-Si:H thin-film transistor with (a) substrate configuration and (b) superstrate configuration.

tion of electrons and holes happens predominately in the weak field regions near the n-layer [14].

Two device configurations—substrate and superstrate configurations—are commonly used for a-Si:H-based thin-film solar cells. In the substrate configuration, as shown in Figure 6.14(a), light is incident upon the cell directly without passing through the substrate. Therefore, flexible metal foils or organic polymer foils can be used. In the superstrate configuration, as illustrated in Figure 6.14(b), light needs to pass through the substrate before reaching the cell. Glass coated with a transparent conducting oxide (TCO) layer, such as SnO_2 or ZnO, is often used as the substrate. The surface texturing of the TCO can be obtained by applying proper growth conditions [15] or produced by postdeposition etch in alkaline solution [16].

The typical conversion efficiency of a single-junction a-Si:H solar cell is around 8%. To achieve the best performance, a multijunction structure is used, in which the a-Si:H cell often serves as the top cell and a-SiGe:H cells are used as the bottom cells. The band gap of the a-Si:H can be modulated to a lower value by alloying with Ge to form a-SiGe:H or to a higher value by applying C to form a-SiC:H. Cells with different band-gap material cover the absorption of different portion of the solar spectrum. The conversion efficiency of a triple-junction a-Si:H/a-SiGe:H/a-SiGe:H solar cell can reach 12%. However, there are three major problems encountered in the multijunction with a-Si:H and a-SiGe:H absorbers: (1) it is difficult to produce a-SiGe:H with a band gap less than 1.5 eV having good stability and sufficient optoelectronic properties, (2) a large band offset occurs at the interface between the doped layer and a-SiGe:H absorber, and (3) the cost of the germane source gas is relatively high. Therefore, microcrystalline silicon (μc-Si:H), also known as nanocrystalline silicon (nc-Si:H), has been investigated as an alternative absorber in conjunction with a-Si:H.

In early days, the application of μc-Si:H in photovoltaics was limited to p-type window layers in p–i–n a-Si:H cells. It was first studied as the absorption layer for solar cell application by the University of Neuchatel in early 1990s. μc-Si:H is a complex material that consists of silicon nanocrystals embedded within amorphous silicon "tissue." Because of the existence of a crystalline phase, it suffers from much milder light-induced degradation compared to a-Si:H and it has a much higher doping efficiency than a-Si:H. It can be produced in essentially the same manner as the a-Si:H thin film. The most common way to deposit μc-Si:H thin film is by introducing a sufficient amount of hydrogen into source gases during the plasma-enhanced chemical-vapor deposition process. The optical band gap of μc-Si:H is ~1.1 eV, which is lower than that of a-Si:H. Because of this property, it can be used to absorb and convert the light in the near infrared region of the solar spectrum. However, similar to crystalline silicon, it has an indirect band gap. Its absorption coefficient in the visible region is lower than that of a-Si:H; therefore, a much thicker absorption layer of 1.5 μm to 2 μm is required in comparison with ~200 nm in the a-Si:H case to insure sufficient absorption. For industrialization, great efforts are being made to μc-Si:H thin films with high deposition rate. The world record initial efficiency of 14.7% was reported by Yamamoto and coworkers with a "micromorph" structure [17], which was first introduced by University of Neuchatel in 1994. A schematic cross section of a micromorph solar cell, where a μc-Si:H bottom cell is used in conjunction with an a-Si:H top cell, is shown in Figure 6.15. An intermediate transparent layer was inserted between the a-Si:H and μc-Si:H layers to enhance partial reflection of light back into the top cell. The reflection was achieved by the difference in the refractive index between the interlayer and the a-Si:H. Currently, commercial photovoltaic modules based on this tandem cell approach have shown stabilized efficiencies about 10%.

a-Si:H-based thin-film solar cells are commonly produced by two approaches: the multichamber process and roll-to-roll manufacturing process. The roll-to-roll process was developed for cells with substrate-type configuration. An example process flow of a-Si:H-based triple-junction on steel foil solar cells made using roll-to-roll deposition is as follows [18]: (1) washing of the web, (2) deposition of a thin film of textured Ag/ZnO back reflector, (3) deposition of the a-Si:H/a-SiGe:H/a-SiGe:H triple-junction solar cell stack, (4) deposition of a TCO layer that serves as a transparent, conducting antireflection coating. The roll is then cut into individual cells and wires and bus bars are applied for current collection. A picture of the cell processor is shown in Figure 6.16.

6.4.2 CdTe Thin-Film Solar Cells

Cadmium telluride (CdTe) is a *II–VI* compound semiconductor with a direct ban dgap of ~1.45 eV. It has a high absorption coefficient of > 5×10^5 cm^{-1} at its band gap. CdTe has a zincblende crystal structure. Thin-film CdTe is polycrystalline in nature. CdTe heterojunction solar cells have been investigated since 1960. Superstrate polycrystalline CdTe/CdS thin-film solar cells were first demonstrated by Adirovich and coworkers in 1969 with an efficiency of ~2%. In early 2000, efficiency of 16.5% was demonstrated in a laboratory device.

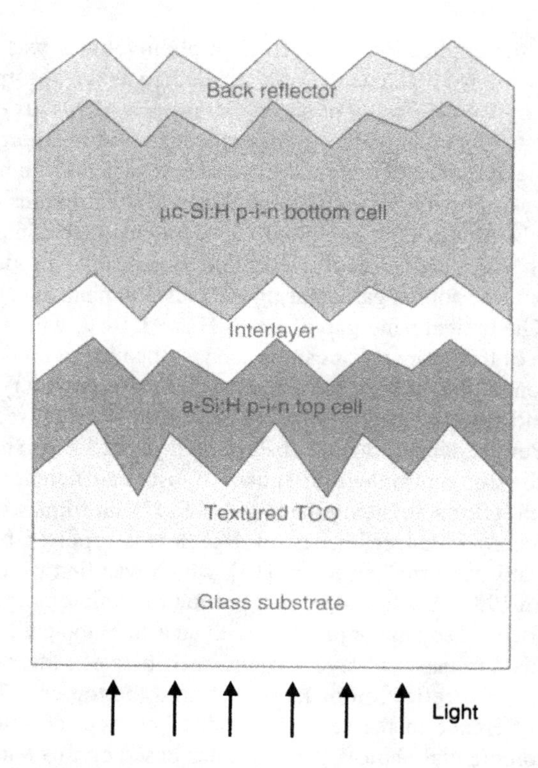

Figure 6.15. Schematic cross section of a micromorph solar cell, a μc-Si:H/a-Si:H stacked cell with a transparent interlayer as the intermediate reflector.

Figure 6.16. Roll-to-roll a-Si:H-based triple-junction cell processor [18].

Most of the high-efficiency CdTe solar cells reported are made with a superstrate configuration. To fabricate high-performance CdTe solar cells with substrate configuration, transferring the cell from a carrier superstrate onto a substrate has been demonstrated. The schematic cross section of a typical CdTe solar cell is shown in Figure 6.17. The solar cell is a heterojunction device with CdS as the n-type window layer and CdTe as the p-type absorber. The TCO on the superstrate serves as the front contact to the device. The basic requisition for the TCO is high optical transparency with low sheet resistance. Because the CdS window is typically thin and tends to have pinholes, a high-resistance oxide layer is grown prior to the deposition of CdS to avoid shunting. A variety of resistive materials have been used for the high-resistance oxide layer, such as SnO_2, In_2O_3, Ga_2O_3, and Zn_2SnO_4. Several methods are available for depositing a high-quality n-type CdS window layer, including physical vapor deposition, chemical bath deposition, and close-spaced sublimation. After the deposition of CdS, the film is often annealed in $CdCl_2$ at 400°C to minimize the interdiffusion between CdTe and CdS in the subsequent process steps, which may reduces the optical transmission of the window layer and result in lowering of the quantum efficiency in the short wavelengths. A large number of techniques with substrate temperatures (T_{sub}) ranging from room temperature to 600°C can be used for the deposition of device-quality CdTe thin films [19,20]. For example: close-space sublimation (T_{sub} = 600°C), chemical spraying (T_{sub} > 500°C), vapor-transport deposition (T_{sub} = 400°C–600°C), physical vapor deposition (T_{sub} = 400°C), metal organic chemical vapor deposition (T_{sub} = 200°C–400°C),

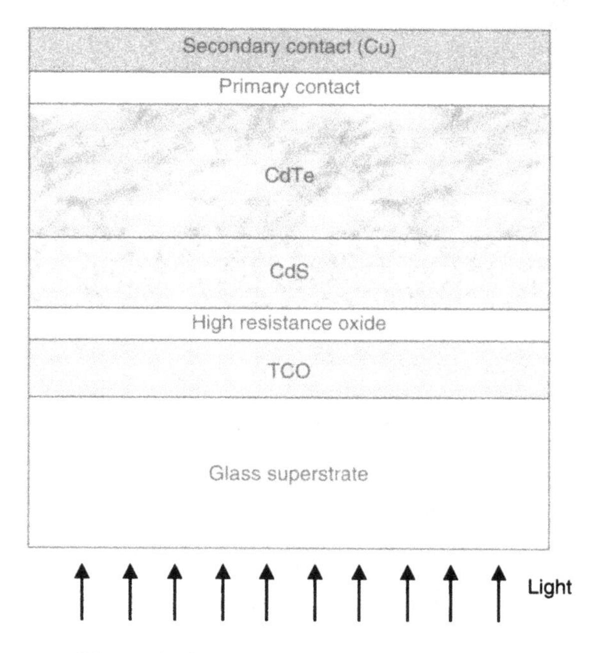

Figure 6.17. Basic CdTe solar cell structure.

sputtering (T_{sub} = 200°C), galvanic deposition (T_{sub} = 90°C), and screen printing (T_{sub} = 25°C), followed by a heat treatment at 600°C. Close space sublimation is favored by industry because of its high deposition rate of 1–10 μm/min. After the deposition, postdeposition $CdCl_2$ treatment is performed on the CdTe thin film at a temperature of ~400°C in air to promote the recrystallization and grain growth. Afterward, a surface treatment is used to form the tellurium-containing p^+ layer as the primary back contact followed by the deposition of the copper layer as the secondary contact.

CdS_xTe_{1-x} alloy has been observed at the CdS/CdTe interface due to interdiffusion after the $CdCl_2$ heat treatment. The formation of the interfacial layer provides both beneficial and detrimental effects. Figure 6.18 shows the optical band gap of CdS_xTe_{1-x} thin film. The diffusion of CdTe into CdS causes undesirable reduction in the optical transmission of the window layer. On the contrary, the diffusion of CdS into CdTe narrows the optical band gap of the absorber, which results in a rise of the quantum efficiency in the long wavelengths. The formation of the CdS_xTe_{1-x} alloy reduces the interfacial strain caused by the large lattice mismatch between CdTe and CdS, and possibly passivates the interface defects [21,22]. Interdiffusion consumes the CdS film, which can be beneficial to the window transmission. However, as the CdS thickness decreases, the probability of pinhole formation increases, which may cause a reduction in open-circuit voltage.

6.4.3 $CuInSe_2$-Based Thin-Film Solar Cells

$CuInSe_2$-based solar cells have been considered as the most promising low-cost photovoltaic technology for power generation. Very high efficiencies of 20.3% and 13.5% have been demonstrated in $Cu(In,Ga)Se_2$ solar cells and $Cu(In,Ga)(Se,S)_2$ solar cell

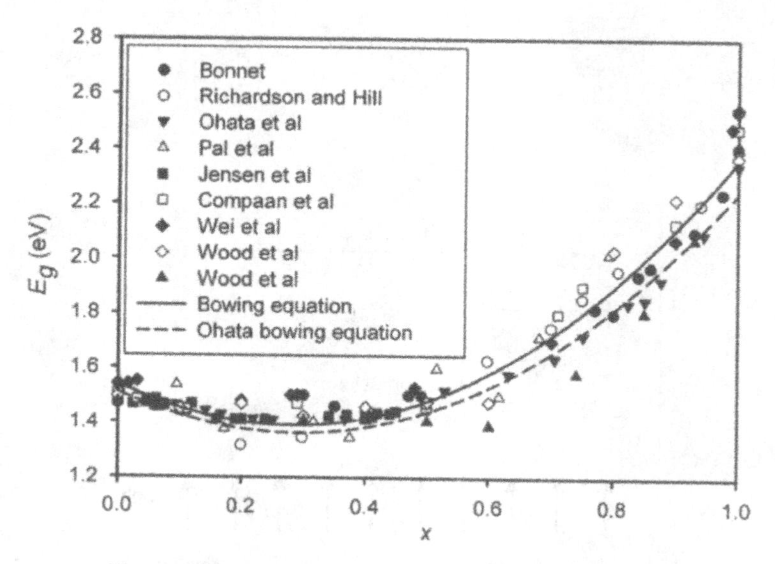

Figure 6.18. Optical band gap of thin-film CdS_xTe_{1-x} [23].

TABLE 6.2. Band gaps of $CuInSe_2$-based alloys

Material	Band gap (eV)
$CuInSe_2$	1.02
$CuGaSe_2$	1.68
$CuAlSe_2$	2.67
$CuInS_2$	1.53
$CuGaS_2$	2.53
$CuAlS_2$	3.49
$Cu(InGa)Se_2$	1.1–1.2
$Cu(InGa)(SeS)_2$	1.5–1.8

modules. The study of $CuInSe_2$-based solar cells dates back to the 1970s, when the first $CuInSe_2$ solar cell was demonstrated at Bell Laboratories by evaporating n-type CdS onto p-type $CuInSe_2$ crystals. The $CuInSe_2$-based alloy includes $CuInSe_2$, $CuGaSe_2$, $CuAlSe_2$, $CuInS_2$, $CuGaS_2$, $CuAlS_2$, and $Cu(In,Ga)(Se,S)_2$. These alloys are direct band gap materials and their room-temperature band gaps range from 1.02 to 3.49 eV, as listed in Table 6.2. For photovoltaic device application, alloys with band gaps less than 1.3eV are more suitable, as shown in Figure 6.19 [24].

$CuInSe_2$-based solar cells are essentially p–n heterojunction devices with an n-type CdS layer as the window layer and a p-type $CuInSe_2$-based alloy as the absorber. These alloys are polycrystalline in nature and have a chalcopyrite crystal structure. $CuInSe_2$-related solar cells can be implemented in both substrate and superstrate configurations. However, most of the high-efficiency $CuInSe_2$-based solar cells are made with substrate configurations, as shown in Figure 6.20.

First, Mo is sputtered as the back ohmic contact, followed by the deposition of $CuInSe_2$-based alloys. $CuInSe_2$ and $Cu(InGa)Se_2$ have a broad single-phase region, giving them such that they have wide tolerances to variation in composition. For instance, high-efficiency $CuInGaSe_2$ solar cells can be fabricated with Cu/(In + Ga) ratio ranging from 0.7 to ~1.0 and Ga/(Ga + In) < 0.3. $CuInSe_2$-based alloys can be deposited by a wide variety of methods. For commercial manufacturing, there are two approaches commonly used to deposit $CuInSe_2$-based thin films. The first approach is coevaporation, a single-step process with Cu, In, Ga, and Se being simultaneously evaporated onto a substrate heated to 400°C to 600°C. Efficiency of 19.9% for $Cu(InGa)Se_2$ cell has been reported by the National Renewable Energy Laboratory using three-stage coevaporation with modified surface termination in 2008 [25]. Scientists at the Zentrum für Sonnenenergie- und Wasserstoff-Forschung, Baden-Württemberg, Germany (Centre for Solar Energy and Hydrogen Research, ZSW) have achieved a champion efficiency of 20.3% for a $Cu(In,Ga)Se_2$ cell with an area of 0.5 cm^2 in July 2010 produced by modified coevaporation [26]. The second approach to produce device-quality $CuInSe_2$-based film is a two-step process in which Cu, Ga, and In are deposited using a low-cost, low-temperature technique followed by an annealing step in an Se-containing and/or S-containing atmosphere, usually H_2Se and/or H_2S, at 400°C to 600°C. This method is more amenable to batch processing but the annealing step

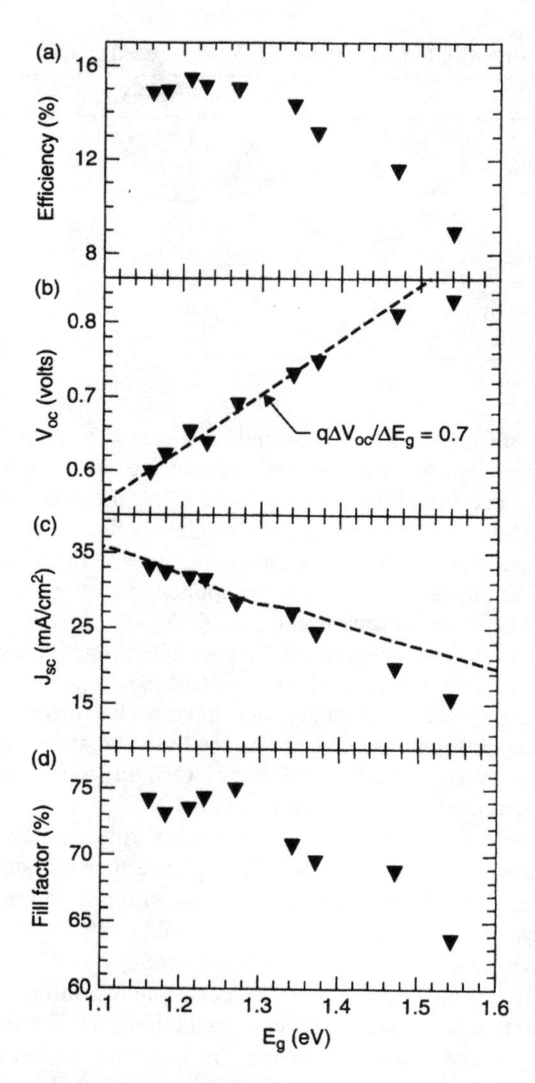

Figure 6.19. Band-gap dependence of J-V parameters of $Cu(In,Ga)Se_2$ solar cells: (a) efficiency, (b) open-circuit voltage, (c) short-circuit current, and (d) fill factor [24].

takes longer time and H_2Se and H_2S are toxic, highly flammable and environmental unfriendly. To improve the material usage and reduce the process cost for mass production, $CuInSe_2$ and $Cu(InGa)Se_2$ nanoparticle-based inks or colloids have been developed and vacuum-free processes, such as roll-to-roll printing or spraying, are adopted for the deposition of the absorber layer [27–31]. Similar to the CdTe solar cell, CdS is used as the n-type window layer. The most common method used to deposit CdS is

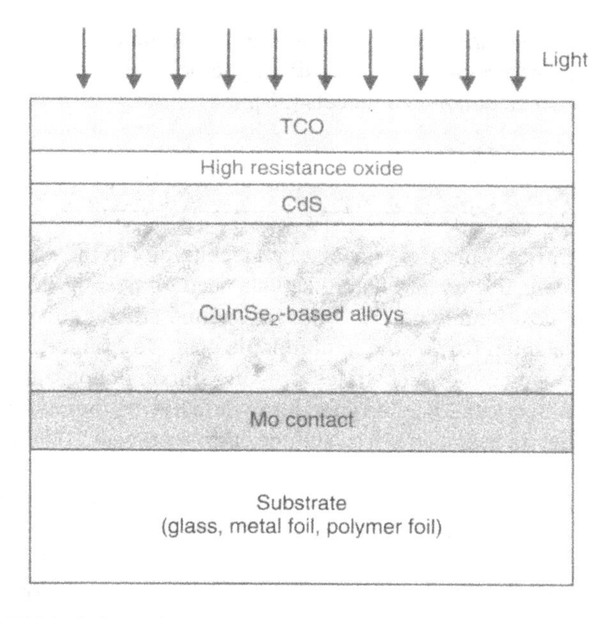

Figure 6.20. Schematic cross section of a typical $CuInSe_2$-based solar cell.

chemical bath deposition, which can provide a conformal coating. In recent years, significant research effort has been devoted to replacing CdS with a Cd-free window layer. Next, a high resistance oxide layer, usually intrinsic ZnO, is grown prior to the deposition of the top contact. This highly resistive oxide eliminates the subsequently deposited TCO in contact with the $CuInSe_2$-based absorber through the pinholes in CdS. Finally, a TCO, usually n-doped ZnO or ITO, is deposited as the top contact.

An important issue in the growth of $CuInSe_2$-based alloys is the incorporation of sodium. Soda lime glass is commonly used as the substrate for $CuInSe_2$-based solar cells. It has been reported that the alkali impurities, such as sodium (Na) or potassium (K), inside the glass can diffuse through the Mo contact and produce beneficial effects for the $CuInSe_2$-based absorbers. For instance, the presence of Na can enhance the grain sizes and the crystal preferred orientation of (112) perpendicular to the substrate in $CuInSe_2$ [32]. The catalytic effect of Na on oxidation can enhance the oxidation-related passivation of Se vacancies present at $Cu(In,Ga)Se_2$ surfaces and grain boundaries, consequently improving the cell efficiency [33,34]. In practice, when an alkali-free substrate is used or better control of Na supply is required, Na is introduced by diffusion from sodium-containing precursor layers, such as Na_2Se, NaF, or alkali-silicate glass, deposited prior to the growth of $Cu(In,Ga)Se_2$ or before the sputtering of the Mo back contact [35–37].

Another important feature of $CuInSe_2$-based solar cells is their insensitivity to defects at the junction, which allows the exposure of the $CuInSe_2$-based absorber to the air prior to the deposition of CdS layer. A post-deposition anneal in air or oxygen atmosphere at 200°C is commonly performed to enhance the cell performance. It is believed that oxygen incorporation can passivate the selenium vacancies on the grain boundaries, which act as donor-like defects [38].

6.5 OUTLOOK

There has been significant progress in solar cell technology in the past 40 years. To be competitive, solar cells for terrestrial applications need to meet the criteria of high performance and low cost simultaneously. For crystalline silicon solar cell technology, wafer cost is a substantial part of the total module cost. To bring the cost down, technologies involving using thinner substrates, growing silicon ribbons, and making silicon wafers that have monocrystalline characteristics while having the cost- and capital-efficiency benefits of the traditional multicrystalline casting process [39] are developed. For thin-film solar cells, cell production is responsible for a great portion of the total cost. At present, high efficiency cells are typically fabricated by batch vacuum processes, such as plasma-enhanced chemical vapor deposition for a-Si:H-based cells and evaporation for $CuInSe_2$-based cells. To reduce the production cost, vacuum-based roll-to-roll manufacturing processes on flexible foil substrates are being developed. However, expensive facilities and high operation cost are still inevitable when a vacuum process is used. The development of vacuum-free printing or spray-deposition processes and solution-processable organic polymers with sufficient photovoltaic performance offers the prospect of realizing cost-effective solar cells for terrestrial applications.

REFERENCES

1. M. Riordan and L. Hoddeson (1997), "The Origins of the pn Junction," *IEEE Spectrum, June 1997,* pp. 46–51.
2. D. Chapin, C. Fuller and G. Pearson (1954), "A New Silicon p–n Junction Photocell for Converting Solar Radiation into Electrical Power," *J. Appl. Phys., 25,* pp. 676–677.
3. D. E. Carlson and C. R. Wronski (1976), "Amorphous Silicon Solar Cell," *Appl. Phys. Lett., 28,* pp. 671–673.
4. S. Mehta, GTM Research (2010), "2009 Global PV Cell and Module Production Analysis."
5. http//:www.solarbuzz.com/facts-and-figures/market-facts/global-pv-market.
6. S. Wenham (1993), "Bruied Contact Silicon Solar Cells," *Prog. Photovolt., 1,* pp. 3–10.
7. M. A. Green and S. R. Wenham (1984), Australian Patent No. 5703309, Australia.
8. M. Green, J. Zhao, A. Wang, and S. R. Wenham (1999), "Very High Efficient Silicon Solar Cell - Science and Technology," *IEEE TED, 46,* pp. 1940–1947.
9. D. Fujishima, H. Inoue, Y. Tsunomura, T. Asaumi, S. Taira, T. Kinoshita, M. Taguchi, H. Sakata, and E. Maruyama (2010), "High-Performance HIT Solar Cells for Thinner Silicon Wafers," in *Proceedings of 35th IEEE Photovoltaic Specialists Conference,* pp. 3137–3140.
10. H. Okaniwa, K. Nakatani, M. Yano, M. Asano, and K. Suzuki (1982), "Preparation and Properties of a-Si:H Solar Cells on Organic Polymer Film Substrate," *Japanese Journal of Applied Physics, 21,* pp. 239–244.

11. D. Staebler and C. Wronski (1977), "Reversible Conductivity Changes in Discharge-Produced Amorphous Si," *Appl. Phys. Lett., 31,* pp. 292–294.
12. L. Yang and L. Chen (1991), "Thickness Dependence of Light Induced Degradation in a-Si:H Solar Cells," *Journal of Non-Crystalline Solids 137–138,* pp. 1189–1192.
13. A. Klaver, R.A.C.M.M. van Swaaij (2008), "Modeling of Light-Induced Degradation of Amorphous Silicon Solar Cells," *Solar Energy Materials and Solar Cells 92,* pp. 50–60.
14. X. Deng and E. A. Schiff (2003), "Amorphous Silicon-Based Solar Cells," in *Handbook of Photovoltaic Science and Engineering,* Edited by A. Luaue and S. Hegedus, Wiley, pp. 530–534.
15. J. Meier, U. Kroll, S. Dubail, S. Golay, S. Fay, J. Dubail, and A. Shah (2000), "Efficiency Enhancement of Amorphous Silicon p–i–n Solar Cells by LP-CVD ZnO," in *Proceedings of 28th IEEE Photovoltaic Specialist Conference,* pp. 746–749.
16. O. Kluth, B. Rech, L. Houben, S. Wieder, G. Schöpe, C. Beneking, H. Wagner, A. Löffl, and H. W. Schock (1999), "Texture Etched ZnO:Al Coated Substrates for Silicon Based Thin Film Solar Cells," *Thin Solid Films, 351,* pp. 247–253.
17. K. Yamamoto, A. Nakajima, M. Yoshimi, T. Sawada, S. Fukuda, T. Suezaki, M. Ichikawa, Y. Koi, M. Goto, H. Takata, T. Sasaki, and Y. Tawada (2003), "Novel Hybrid Thin Film Silicon Solar Cell and Module," in *Proceedings of 3rd World Conference on Photovoltaic Energy Conversion,* pp. 2789–2792.
18. A. Banerjee, G. DeMaggio, K. Lord, B. Yan, F. Liu, x. Xu, K. Beernink, G. Pietka, C. Worrel, B. Dotter, J. Yang, and S. Guha (2009), "Advances in Cell and Module Efficiency of A-Si:H Based Triple-Junction Solar Cells Made Using Roll-to-Roll Deposition," in *Proceedings of 34th IEEE Photovoltaic Specialists Conference,* pp. 116–119.
19. D. Bonnet and P. Meyers (1998), "Cadmium-Telluride—Material for Thin Film Solar Cells," *Journal of Materials Research, 13,* pp. 2740–2753.
20. H. R. Moutinho, F. S. Hasoon, F. Abulfotuh, and L. L. Kazmerski (1995), "Investigation of Polycrystalline CdTe Thin Films Deposited by Physical Vapor Deposition, Close-Spaced Sublimation, and Sputtering," *Journal of Vacuum Science and Technology A, 13,* pp. 2877–2883.
21. D. Wang, Z. Hou, and Z. Bai (2011), "Study of Interdiffusion Reaction at the CdS/CdTe Interface," *Journal of Material Research, 26,* pp. 697–705.
22. C. J. Bridge, P. Dawson, P. D. Buckle, and M. E. Ozsan (2000), "Photoluminescence Spectroscopy and Decay Time Measurements of Polycrystalline Thin Film CdTe/CdS Solar Cells," *Journal of Applied Physics, 88,* pp. 6451–6456.
23. D. W. Lane (2006), "A Review of the Optical Band Gap of Thin Film CdS_xTe_{1-x}," *Solar Energy Materials and Solar Cells, 90,* pp. 1169–1175.
24. W. N. Shafarman, R. Klenk, and B. E. McCandless (1996), "Characterization of $Cu(InGa)Se_2$ Solar Cells with High Ga Content, in *Proceedings of 25th IEEE Photovoltaic Specialist Conference,* pp. 763–768.
25. I. Repins, M.A. Contreras, B. Egaas, C. DeHart, J. Scharf, C.L. Perkins, B. To, and R. Noufi (2008), "19.9% Efficient $ZnO/CdS/CuInGaSe_2$ Solar Cell with 81.2% Fill Factor," *Progress in Photovoltaics: Research and Applications, 16,* pp. 235–239.
26. http://www.zsw-bw.de/1/topics/pv-materials-research/pv-materials/.
27. V. K. Kapur, A. Bansal, P. Le, O. I. Asensio (2003), "Non-Vacuum Processing of $CuIn_{1-x}Ga_xSe_2$ Solar Cells on Rigid and Flexible Substrates Using Nanoparticle Precursor Inks," *Thin Solid Films,* 431–432, pp. 53–57.

28. S. Yoon, T. Yoon, K.-S. Lee, S. Yoon, J. M. Ha, S. Choe (2009), "Nanoparticle-Based Approach for the Formation of CIS Solar Cells," *Solar Energy Materials and Solar Cells, 93,* pp. 783–788.
29. US Patent 7663057—CIGS nanoparticle ink for photovoltaics.
30. E.-J. Bae, J.-M. Cho, J.-D. Suh, C.-W. Ham, K.-B. Song (2010), "Preparation of CIGS Nanoparticles and Thin Films by One Step Non-Vacuum Deposition Methos," in *Proceedings of 35th IEEE Photovoltaic Specialists Conference,* pp. 3391–3393.
31. D. L. Schulz, C. J. Curtis, R. A. Flitton, H. Wiesner, J. Keane, R. J. Matson, K. M. Jones, P. A. Parilla, R. Noufi, and D. S. Ginley (1998), "Cu-In-Ga-Se Nanoparticle Colloids as Spray Deposition Precursors for Cu(In,Ga)Se_2 Solar Cell Materials," *Journal of Electronic Materials, 27,* pp. 433–437.
32. S.-H. Wei, S. G. Zhang, and A. Zunger (1999), "Effects of Na on the Electrical and Structural Properties of $CuInSe_2$," *Journal of Applied Physics, 85,* pp. 7214–7218.
33. L. Kronik, D. Cahen, and H. W. Schock (1998), "Effects of Sodium on Polycrystalline Cu(In,Ga)Se_2 and its Solar Cell Performance," *Advanced Materials, 10,* pp. 31–36.
34. D. Rudmannm A. F. da Cunha, M. Kaelin, F. Kurdesau, H. Zogg, A. N. Tiwari, and G. Bilger (2004), "Efficiency Enhancement of Cu(In, Ga)Se_2 Solar Cells Due to Post-Deposition Na Incorporation," *Applied Physics Letters, 84,* pp. 1129–1131.
35. T. Nakada, D. Iga, H. Ohbo and A. Kunioka (1997), "Effects of Sodium on Cu(In,Ga)Se_2-Based Thin Films and Solar Cells," *Japanese Journal of Applied Physics, 36,* pp. 732–737.
36. D. Rudmann, G. Bilger, M. Kaelin, F.-J. Haug, H. Zogg, and A. N. Tiwari (2003), "Effects of NaF Coevaporation on Structural Properties of Cu(In,Ga)Se_2 Thin Films," *Thin Solid Films, 431–432,* pp. 37–40.
37. S. Ishizuka, A. Yamada, P. Fons, and S. Niki (2009), "Flexible Cu(In,Ga)Se_2 Solar Cells Fabricated Using Alkali-Silicate Glass Thin Layers as an Alkali Source Material," *Journal of Renewable and Sustainable Energy, 1,* p. 013102.
38. D. Cahen and R. Noufi (1989), "Defect Chemical Explanation for the Effect of Air Anneal on CdS/$CuInSe_2$ Solar Cell Performance," *Applied Physics Letters, 54,* pp. 558–560.
39. "Getting on the Grid," (2007), *Frontiers, BP's Magazine of Technology and Innovation, 19,* pp. 14–18.

EXERCISES

1. What is the difference between a junction photodiode and a solar cell?
2. An Si p–n junction has the following properties: $N_a = 10^{16}$ cm^{-3}, $N_d = 10^{18}$ cm^{-3}. Calculate the maximum electric field and width of the depletion region at room temperature when (a) no bias is applied, (b) forward bias of 1V is applied, and (c) reverse bias of –5 V is applied.
3. The reverse saturation current and short-circuit current of a Si p–n junction solar cell at room temperature are 10 pA and 10 mA, respectively. Assume the cell has an area of 1 cm^2.

   (a) Find the open-circuit voltage at room temperature when the junction is diffusion controlled.

(b) If the illumination intensity is doubled, find the short-circuit current and open-circuit voltage at room temperature.

(c) Under the same illumination intensity, find the reverse-saturation current and open-circuit voltage at 330K and 350K, respectively.

4. A p–n junction solar cell has an open-circuit voltage of 0.5 V and short-circuit current of 10 mA. Another p–n junction solar cell has an open-circuit voltage of 0.6 V and short-circuit current of 8 mA. If the ideal factor of the two cells is one. Find the open-circuit voltage and short-circuit current when the two cells are connected (a) in series and (b) in parallel.

5. Consider a multicrystalline (polycrystalline) silicon solar cell that has an ideal factor $n = 1.5$, a reverse saturation current of 1 μA, and a photo current of 10 mA under illumination.

   (a) Plot the I–V characteristics when shut resistance is infinite, 1 kΩ, and 100 Ω, respectively.

   (b) Plot the I–V characteristics when series resistance is 0, 10 Ω and 100 Ω, respectively.

6. What are the factors that limit the conversion efficiency of a c-Si:H solar cell?

7. Thin-film solar cells can be made in substrate and superstrate configurations. Describe the difference between these two configurations.

8. The p–i–n structure is often used for a-Si:H thin-film solar cells. Can we use a p–n junction structure instead? Why or why not? Draw the absorption spectra of amorphous silicon and crystalline silicon. A much thinner absorber is often used in a-Si:H thin-film solar cells compared to that in c-Si solar cells. Why?

9. Compared to a-SiGe:H, what are the advantages of using μc-Si:H as an absorber in conjunction with a-Si:H?

10. n^+ CdS is used as the window layer in CdTe and $Cu(InGa)Se_2$ solar cells. What are the advantage(s) and disadvantage(s) when it is made thin? What do you do to overcome the disadvantage(s)?

11. What are the influences of sodium (Na) and oxygen (O) on the performance of $Cu(InGa)Se_2$ solar cells?

7

ORGANIC SOLAR CELLS

Wei-Fang Su

The driving force for the organic solar cell is the prospect of having very low cost solar cells (<0.5 USD/W_p) due to the use of less material as compared with silicon solar cells (submicron thickness versus 100 micron thickness) and the ease of using low-energy-consumption fabrication processes such as printing and dip coating.

At present, there are three kinds of organic solar cells under development: (1) dye-sensitized solar cells, (2) organic molecule solar cells, and (3) polymer solar cells. The principle and performance of each type of organic solar cell are not the same and will be discussed in detail in the following.

7.1 DYE-SENSITIZED SOLAR CELL

The dye-sensitized solar cell (DSSC) was invented by Michael Grätzel in 1985, so it is also called the Grätzel cell [1]. A power conversion efficiency of 12.5% has been reached using the newly developed dye CYC-B1 [2] (Figure 4.17), which is among the highest for an organic solar cell.

7.1.1 Structure of the DSSC

The structure and operation principle of the DSSC are illustrated in Figure 7.1 [3]. It is an electrochemical cell. The anode is made from mesoporous TiO_2 coated on a fluorine-doped SnO_2 transparent electrode (FTO). The dye is absorbed on the mesoporous TiO_2 layer. Upon exposure to sunlight, the dye is oxidized and generates excitons. The excitons are separated at the interface of the dye and TiO_2. The electrons are injected through the TiO_2 anode and the holes are transported by the redox electrolyte to the cathode made from a Pt-coated, fluorine-doped SnO_2 transparent electrode, thus generating current.

The redox electrolyte can be made from either a liquid or solid. For a high-efficiency cell (>10%), the redox electrolyte is made from acetonitrile containing

Organic, Inorganic, and Hybrid Solar Cells. By C.-F. Lin, W.-F. Su, C.-I Wu, and I-C. Cheng

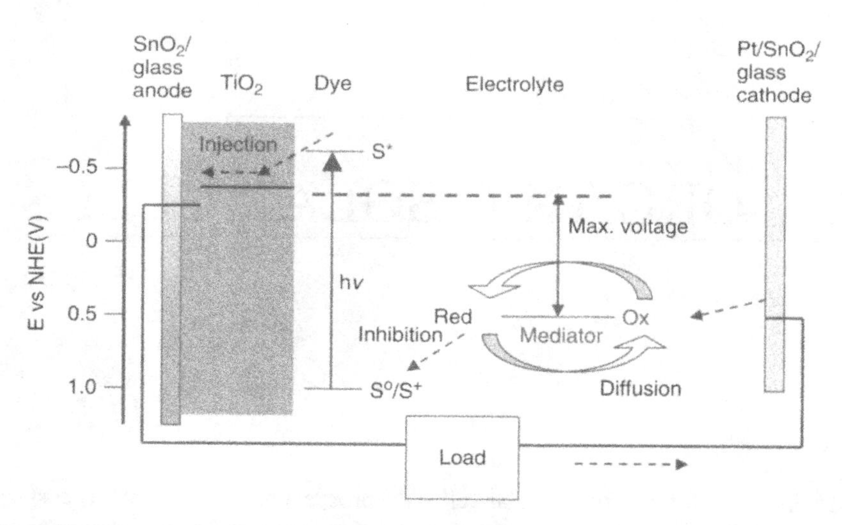

Figure 7.1. Energy band diagram and operation principle of a typical DSSC [3].

iodide/triiodide. The oxidized dye S+ is reduced by iodide while producing triiodide ions. This prevents any significant buildup of S+, which could recapture the conduction band electrons at the surface. The iodide is regenerated in turn by the reduction of the triiodide ions at the cathode, where the electrons are supplied via migration through the external load to complete the cycle. Thus, the cell is generating electricity from light without any permanent chemical transformation.

The voltage produced under sunlight corresponds to the difference between the chemical potential (Fermi level) of the electrons attained in the TiO_2 [μ(e−)], and the chemical potential of the holes in the hole conductor [μ(h+)]. For redox electrolytes, the latter corresponds to the Nernst potential. In the dark at equilibrium, $\mu(e-) = \mu(h+)$, that is, the Fermi level, is constant within the whole solar cell.

7.1.2 Principle of the DSSC and Development of the Dye

The absorption of dye is the determining factor for the power conversion efficiency of dye sensitized solar cells. Therefore, new dyes are constantly sought to absorb the whole solar spectrum, especially high absorption in the red and near-IR regions. The CYC-B1 dye contains extra thiophene moieties attached to the bipyridene ligand of N3 (Figure 4.17), which enhances the light harvesting. The alkyl-bis-thiophene moiety produces a strong new band centered at around 410 nm. The intensity of this transition maximizes the metal-to-ligand charge-transfer band in the visible region. The absorption region is shifted from 530 nm to 553 nm and the extinction coefficient is increased from 14,500 to 21,200 $M^{-1}cm^{-1}$ as compared with those of the N3 dye (Figure 7.2). The N719 is the di-tetra-butyl ammonium salt of N3. The thiocyanate donor ligand of the dye is believed to be the weakest part of the dye complex from a chemical stability point of view. A new

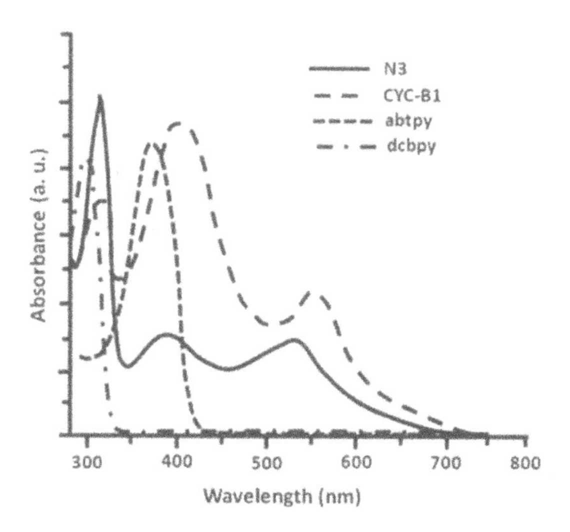

Figure 7.2. UV-Vis absorption spectra of N3, CYC-B1, dicarboxy-bipyridine (dcbpy), and alkylbisthiophene (abtp) ligands in DMF [2].

stable dye, YE05 (Figure 4.17), has been developed by replacing the two thiocyanate ligands of N719 with a stable cyclometalated ligand. There is a significant red shift in the absorption spectrum of YE05 as compared with N719. The cyclometalated ligand is a stronger donor than the thiocyanate ligand that results in destabilization of the $Ru(t_{2g})$ levels and narrowing of the gap between HOMO and LUMO of the dye.

The spectral response of the photocurrent generated by dye can be expressed by incident photon-to-current-conversion efficiency (IPCE) or external quantum efficiency (EQE). The IPCE can be expressed by the following equation:

$$\mathrm{IPCE}(\lambda) = \mathrm{LHE}(\lambda)\varphi_{inj}\eta_{coll} \tag{7-1A}$$

where LHE(λ) is the light-harvesting efficiency for photons of wavelength λ, ϕ_{inj} is the quantum yield for electron injection from the excited sensitizer; and η_{coll} is the electron collection efficiency. The EQE can be expressed by the following equation:

$$\mathrm{EQE} = \frac{J_{sc}(\mathrm{A})}{P(\mathrm{W})} \times \frac{1240}{\lambda(\mathrm{nm})} \times 100 \tag{7-1B}$$

where J_{sc} is the short-circuit current in amperes, and P is the input light intensity in watts.

Figure 7.3 shows the IPCE curves of three dyes—CYC-B1, C101 (Figure 4.17), and N719—when the cells were fabricated from the same nanocrystalline TiO_2 films and iodide/triiodide-based redox electrolyte. The thiophene moiety of CYC-B1 produces an onset shift of the photocurrent from 775 to 800 nm with a steep rise in the

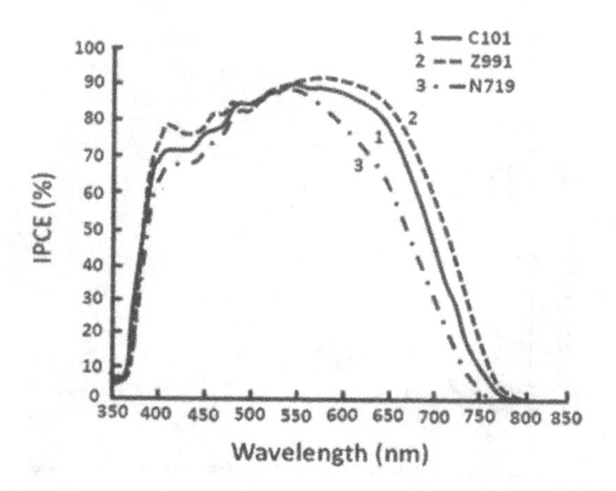

Figure 7.3. IPCE spectra of N719, C101, and CYC-B1 (Z991) [2].

IPCE curve with a high plateau value of 90% (Figure 7.3). That results in a highly efficient cell of CYC-B1 dye. The Ru complex dye is very expensive due to the use of the rare metal Ru. Thus, there are many efforts to replace it by pure organic dyes. Indoline dye (D205) achieved a power conversion efficiency of 9.5% in 2008. However, it is not stable enough for use outdoors. A new robust donor-pi-acceptor dye, C217 (Figure 4.17), with a cell efficiency of 9.8% has been obtained. The cyanoacrylate group of C217 can bound to the TiO_2 surface, which produces strong coupling of the excited-state wave function with the $Ti(3d,t_{2g})$ orbitals that form the conduction band of titanium dioxide. Thus, efficient and rapid electron injection from the excited state of the sensitizer into the conduction band of the oxide can be achieved.

The power conversion efficiency of the DSSC can be determined by the following equation:

$$\eta_{global} = \frac{J_{sc} \times V_{oc} \times FF}{I_s} \tag{7-2}$$

where the fill factor (FF) is the ratio of the maximum power of the solar cell (P_{max}) divided by the open-circuit voltage (V_{oc}) and the short-circuit current (J_{sc}) as follows:

$$FF = \frac{P_{max}}{J_{sc} \times V_{oc}} \tag{7-3}$$

where P_{max} is the product of the photocurrent and photovoltage at the maximum voltage of output power of the cell. The value of the fill factor reflects the extent of electrical (ohmic) and electrochemical (overvoltage) losses occurring during the operation of the cell. Increasing the shunt resistance and decreasing the series resistance, as well as reducing the overvoltage for diffusion and electron transfer, will lead to higher fill factors. Figure 7.4 shows the photocurrent voltage curves of YE05 under AM 1.5 standard

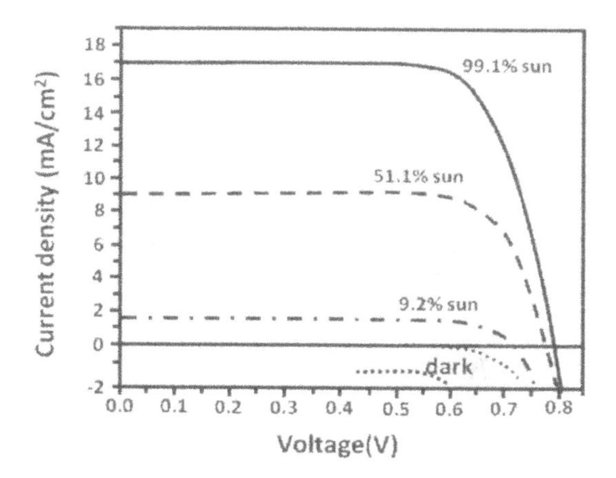

Figure 7.4. Photocurrent voltage curve of DSSC fabricated from YE05 dye [2].

sunlight irradiation, which produces a short-circuit photocurrent (J_{sc}) of 17 mA/cm^2, a V_{oc} of 800 mV and a fill factor (*FF*) of 0.74, corresponding to an overall conversion efficiency of 10.1%.

Table 7.1 summarizes the performance of different dyes used in DSSC. Figure 7.5 shows the progress of power conversion efficiency of DSSC over 30 years of development, from only about 2% in 1976 to 12.5% in 2009. The improvement is due to the development of highly light harvesting dyes.

7.1.3 Solid-State Dye-Sensitized Solar Cell

Intense research activity on solid-state dye-sensitized solar cells (SSDSCs) has solved the problems of corrosive and leakage of liquid electrolytes in DSSCs [4–9]. The operating mechanism [5] of the SSDSC involves two major steps: (1) dye absorbs light to generate excitons that dissociate into electrons and holes at the interface of the dye and the TiO_2 mesoporous layer; (2) the electrons are transported through the TiO_2 layer and the holes are transported through p-type materials (e.g., spiro-OMeTAD) to generate

TABLE 7.1. Summary of DSSC performance of different dyes

Dye	*PCE* (%)	J_{sc} (mA/cm^2)	V_{oc} (volt)	*FF* (%)
N3	8.4	18.4	0.69	66
N749	11.1	20.9	0.74	72
C101	11.0	17.9	0.78	77
CYC-B1	12.5	22.6	0.76	72
YE05	10.1	17.0	0.80	74
C217	9.8	16.4	0.80	76

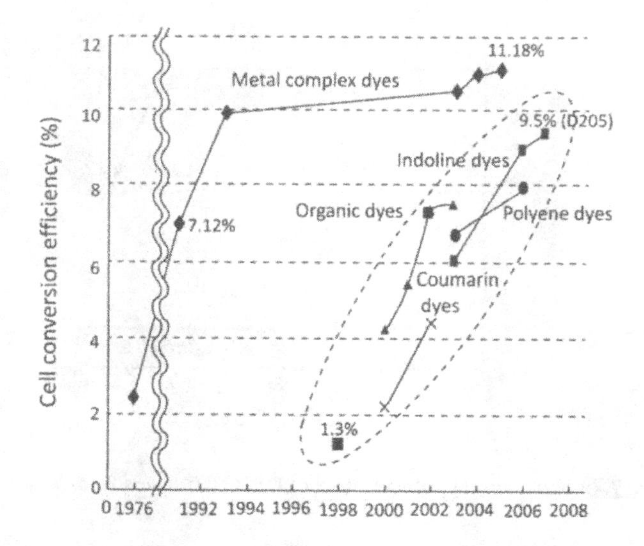

Figure 7.5. The progress of power conversion efficiency of DSSCs over 30 years at AM 1.5 [2].

current. In general, the power conversion efficiency of an SSDSC is lower than the conventional liquid DSSC due to the low optical harvesting capability from the thinner layer of TiO_2 (about one-fifth of the liquid cell, so that there is less dye on the TiO_2 layer) and lower charge mobility of the p-type materials as compared with liquid electrolytes. The diameter of the pores of the mesoporous TiO_2 layer is extremely small (in the nanometer range), which makes the infiltration of p-type material difficult. The diffusion length of the organic molecule is low (<20 nm). To facilitate the charge transport, the thickness of the TiO_2 layer has to be reduced, which results in less dye absorbed on the TiO_2 and reduces the light harvesting. Figure 7.6 shows the schematic structure, IPCE, and photocurrent curve of SSDSC fabricated from N3 dye with an efficiency of 0.74% [5]. However, 4% power conversion efficiency has been reached using organic dyes [6], ionic coordinating dyes [7], and hydrophobic dyes [8]. Recently, 5% efficiency has been achieved by using a strong optical reflector [10].

7.2 ORGANIC MOLECULE SOLAR CELL

The exciton binding energy (0.1–1 eV) of an organic molecule is higher than that of Si and the internal electric field of an organic device (10^6–10^7 V/m) cannot separate the exciton into charge carriers [11]. By using the structure of either a metal–semiconductor Schottky junction or a donor–acceptor p–n junction, the internal electric field will be lowered to allow the charge separation to occur [11]. The interface interactions between the all organic p–n junctions are expected to be better than those of heterojunction of organic and inorganic materials. The organic molecule solar cell has an advan-

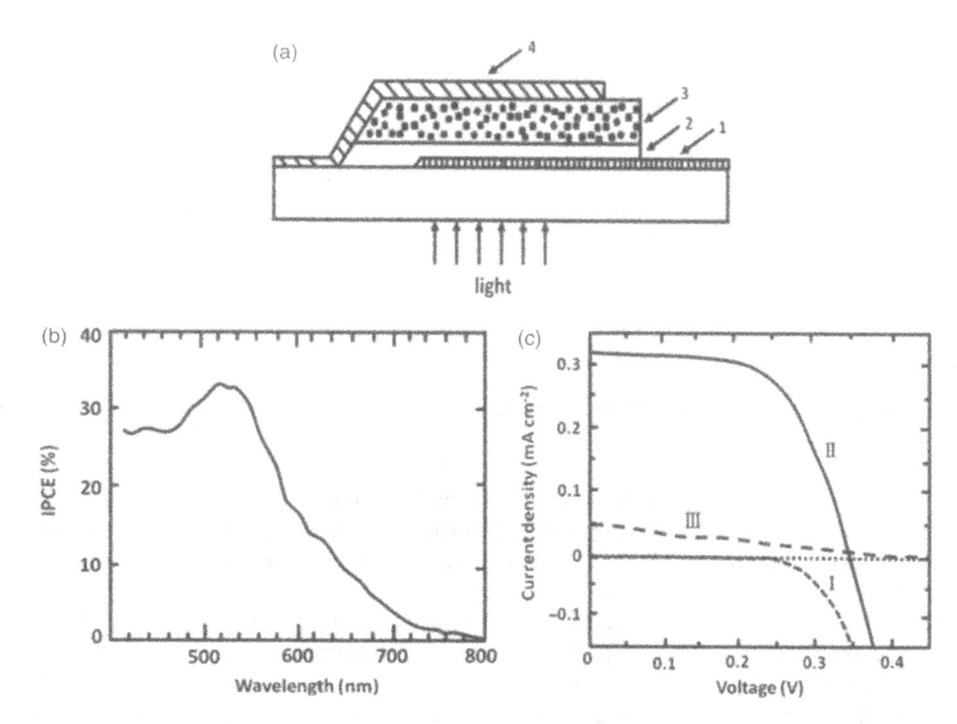

Figure 7.6. (a) Schematic illustration of a solid-state dye sensitized solar cell. Structure 1, F-doped SnO_2-coated glass; 2, compact TiO_2 layer; 3, dye-sensitized heterojunction; 4, gold electrode. (b) Photocurrent spectrum for a dye-sensitized heterojunction. (c) Device characteristics, obtained in the dark (I) and under white-light illumination at 9.4 mW cm^{-2} (II). The cell exhibits J_{sc} = 0.32 mA cm^{-2}, V_{oc} = 342 mV, and FF = 62%. Cell efficiency 0.74%. For comparison, the photocurrent-density/voltage characteristic of a cell containing no $N(PhBr)_3SbCl_6$ or $Li[(CF_3SO_2)_2N$ is also shown (III) [5].

tage over the SDDC due to its the easy low-temperature fabrication using a solution process. The DSSC process must produce the mesoporous TiO_2 electrode at 450°C, whereas the solution process can be carried out at room temperature.

The Schottky junction of an organic solar cell is fabricated from the thermal evaporation of organic molecule and placed in between two metal electrodes (one with a lower work function), as shown in Figure 7.7. At the junction, the energy level of the organic semiconductor is bent and forms a depleted region with a thickness of W. The exciton can only be separated at the depleted regions, so the device efficiency is limited by the diffusion of excitons. The exciton diffusion length of most organic molecules is less than 20 nm, so only an exciton located at a distance less than 20 nm to the metal electrode can generate photocurrent. This device usually exhibits high internal series resistance and low fill factor, so the efficiency of this solar cell is extremely low (<0.01%).

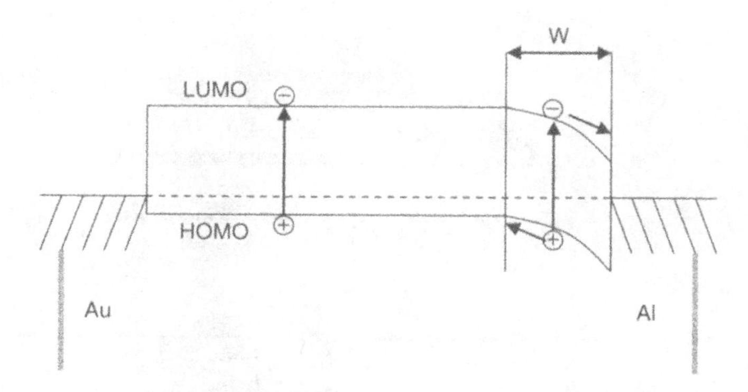

Figure 7.7. Schematic illustration of the Schotty junction of a single organic layer device.

The power conversion efficiency of the Schotty junction of a single organic layer solar cell can be improved by the p–n junction of the donor–acceptor double layer, as shown in Figure 7.8. The excitons are separated at the interface between the donor and acceptor. The ease of charge separation is determined by the band gap between the donor and acceptor. The HOMO of the donor and the LUMO of the acceptor are aligned with the respective electrode. The transport rate of electrons in the acceptor is much faster than the recombination rate of the holes in the donor, so the efficiency is greatly improved over the single layer solar cell. The cell exhibits a linear relationship of photocurrent with the irradiation strength of incident light and has a larger *FF* than that of single layer solar cell. When the semiconducting polymers are used as the donor and acceptor (donor:acceptor), the cell efficiency of the double layer solar cell is still relatively low because the charge mobility of semiconducting polymers is relatively low as compared with an inorganic semiconductor. Veenstra and coworkers fabricated a cell with a structure of ITO/PEDOT/MDMO-PPV:PC-NEPPV/LiF/Al that exhibits a cell efficiency of 0.75% [12]. Jenekhe and coworkers reported a cell efficiency of 1.5% with a structure of ITO/PPV:BBL/Al [13]. Friend

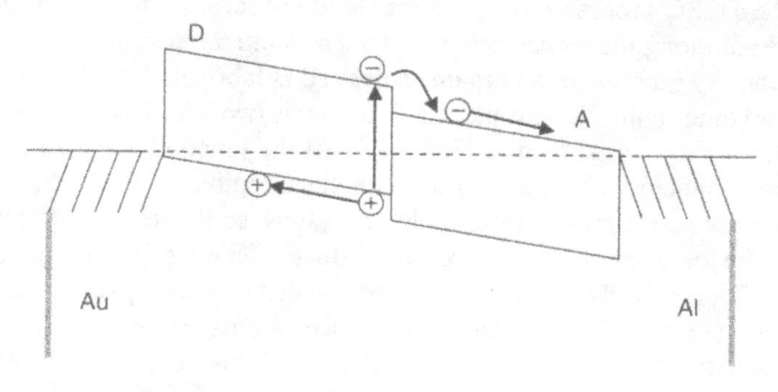

Figure 7.8. Schematic illustration of a p–n junction organic solar cell.

and coworkers showed a relatively high efficiency of 1.9% with a structure of ITO/POPT:MEH-CN-PPV/Al [14].

The small semiconducting organic molecule exhibits higher cell efficiency than that of the semiconducting polymer due to the higher charge mobility of the small molecule. A cell efficiency of 4,2% was reported by Forrest and coworkers [15] using a copper phthalocyanine:C_{60} bilayer structure. Forrest and coworkers also fabricated a tandem organic solar cell to harvest more sunlight, and the cell efficiency was increased to 5.7% [16], as shown in Figure 7.9. As compared with the polymer bilayer solar cell, the small organic molecule solar cell needs to be processed using vacuum evaporation and less stable for long-term durability. The efficiency of the p–n junction organic solar cell is highly dependent on the type of donor and acceptor. Further organic material research is needed to increase the cell efficiency to above 10%.

7.3 POLYMER SOLAR CELL

Crystalline silicon-based solar cells are heavy and expensive [17]. Thin-film amorphous Si solar cells have overcome the shortcomings of crystalline silicon, but lack long-term stability due to light-induced metastable changes in the amorphous silicon (Staebler–Wronski effect) [18].

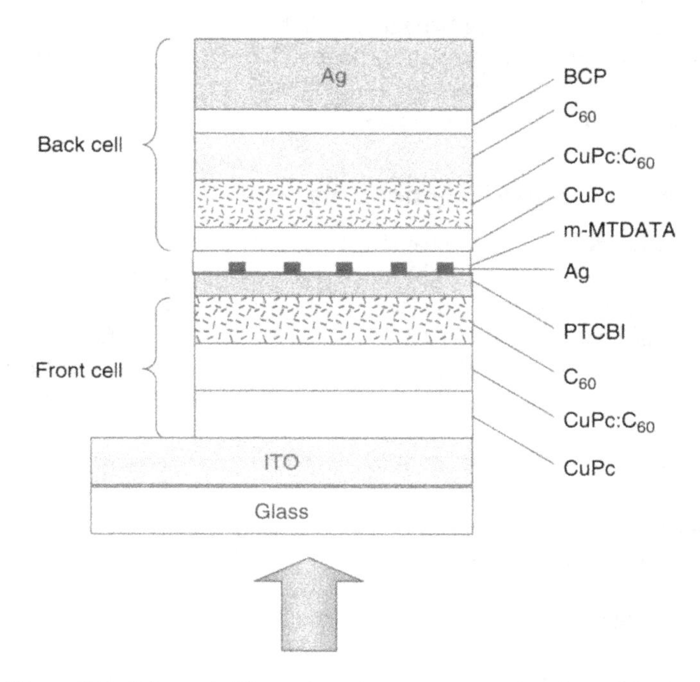

Figure 7.9. Schematic illustration of a tandem organic solar cell [16].

Nanocomposites made from conducting polymer–nanoparticle hybrids are attractive for photovoltaics because of the prospect of light weight, low cost, high throughput, and nonvacuum processes using reel-to-reel printing or spray deposition on flexible substrates. These materials also offer the potential for very thin and flexible photovoltaic devices with high energy densities [19–20]. In contrast to the Si solar cell, the conjugated conducting polymers are bounded electron–hole pairs (excitons) rather than free charge carriers; this is largely due to their low dielectric constant and the presence of significant electron correlation effects. In the absence of the exciton dissociation mechanism into free charge carriers, the exciton will undergo radiative and nonradiative decay with an exciton lifetime in the range of 100 ps to 1 ns [21]. Therefore, polymer solar cells based on Schottky diodes or donor/acceptor bilayer devices are relatively low in power conversion efficiency due to their extremely low diffusion length (~10 nm) and limiting film thickness (~100 nm, less than the optical absorption depth).

A key breakthrough in the utilization of conducting polymers for solar cells was the introduction of the bulk heterojunction [22] to overcome the issue of low diffusion length. A bulk heterojunction is made from a blend of polymers and nanoparticles that form a bicontinuous interpenetrating network. This enhances the donor:acceptor interfacial area available for exciton dissociation and thus reduces the distance the exciton needs to travel before reaching an interface. Charge photogeneration is thus increased. The conducting polymers have hole transport function and nanoparticles have electron transport function in addition to their shared functions of light harvesting and charge separation. At present, there are two kinds of hybrid materials being developed for solar cells: polymer–fullerene derivatives (e.g., PCBM) and polymer-semiconducting nanoparticles. An efficiency of 10–15% has been theoretically predicted for polymer–fullerene-based solar cells [19,23,24].

7.3.1 Principle of the Polymer Solar Cell

A schematic illustration of the fundamental principle of bulk heterojunction solar cells is shown in Figure 7.10. Solar light is absorbed by the hybrid material (polymer as donor and nanoparticle as acceptor) and excitons are generated. Then the excitons are diffused into the interface between polymer and nanoparticle and they are separated into electrons and holes. The electrons are transported through the acceptor phase to the metal electrode and the holes are transported through the donor phase to the transparent ITO electrode. Thus, current is generated.

The relative energy level diagram for light absorption and charge separation of the hybrid material is shown in Figure 7.11. After the donor polymer absorbs light, one of the electrons in the ground state (HOMO level) moves to a higher energy level (LUMO) and generates an exciton of an electron–hole pair [Figure 7.11(a)]. The electron at the LUMO level of the donor will transfer to the LUMO level of the acceptor to accomplish charge separation when the excited donor meets with the acceptor. Therefore the charges are separated at the interface between the donor and acceptor [Figure 7.11(b)]. Then the separated charges are transported through each domain of the bicon-

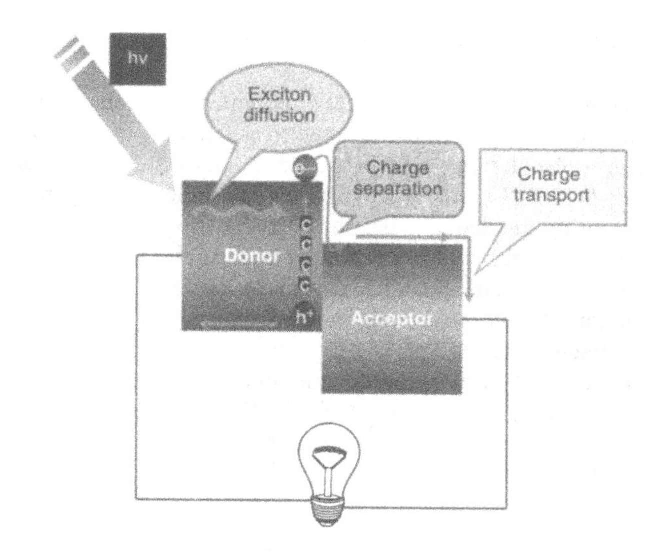

Figure 7.10. Schematic illustration of the fundamental principle of a bulk heterojunction solar cell.

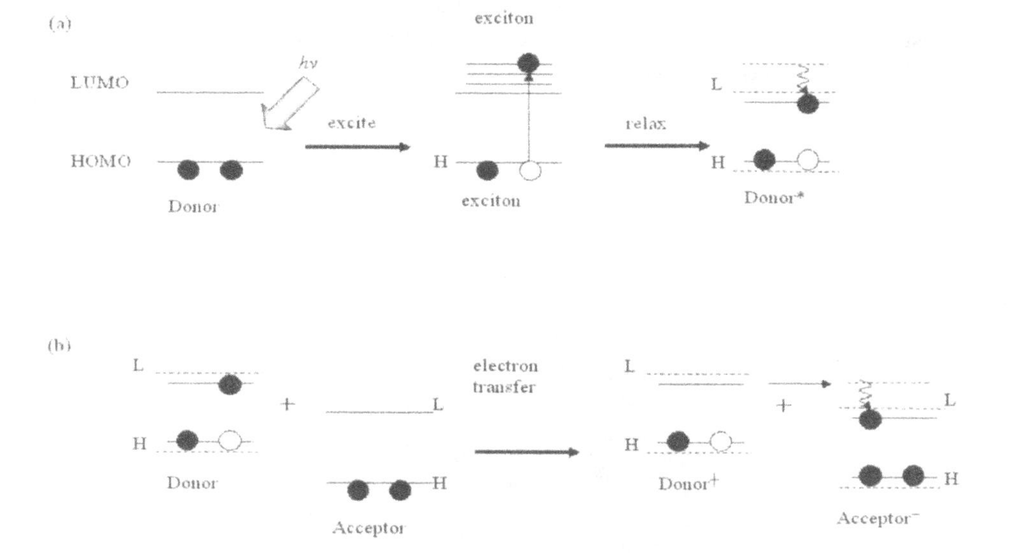

Figure 7.11. Energy level diagram of exciton formation (a) and charge separation (b) in hybrid materials.

tinuous phase of the hybrid to the opposite electrode and generates the current (Figure 7.12).

A key challenge for the development of polymer solar cells is to develop a predictive understanding of the relationship between molecular structure and solar cell performance. The molecular structure of polymers can influence the solar cell performance by their HOMO/LUMO energy gap for the optical band gap of the device and their charge mobility for charge collection efficiency.

The electron transfer step in the photocurrent generation of polymer solar cells does not necessarily directly generate free charge carriers. The hole is primarily localized on the donor HOMO orbital and the electron on the acceptor LUMO orbital, but the coulomb attraction (0.1–0.5 eV) between the electron and hole remains significantly greater than $k_B T$ (0.025 eV). These coulombically bound electron–hole pair states are comprised partially of charge-transfer (CT) states. A potential concern in this charge separation process is that the electron and hole must overcome their mutual coulomb attraction, V:

$$V = \frac{e^2}{4\pi\varepsilon_r\varepsilon_0 r} \tag{7-4}$$

where e is the charge of an electron, ε_r is the dielectric constant of the surrounding medium, ε_0 is the permittivity in vacuum, and r is the electron–hole separation distance. In conventional inorganic photovoltaic devices, such as those based on silicon p–n junctions, overcoming the coulomb attraction is easy due to the high dielectric constant of silicon ($\varepsilon_r \approx 12$) and because the electronic states involved are already highly delocalized (corresponding to a larger average r in Eq. 7-4). Similarly, dye-sen-

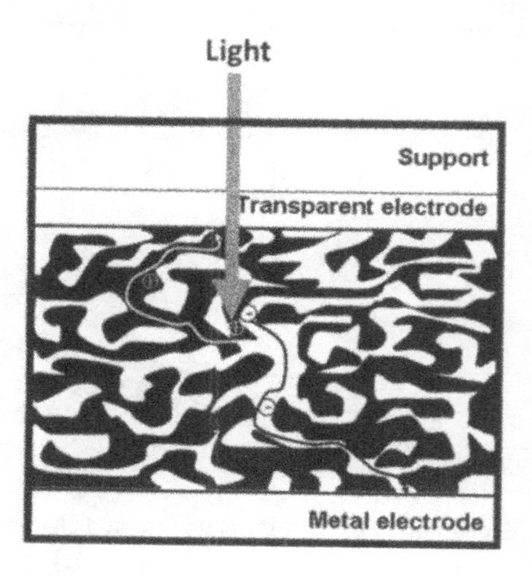

Figure 7.12. Schematic diagram shows transport paths of charge carriers in the bicontinuous phase of a hybrid material.

sitized photoelectrochemical solar cells exhibit a high dielectric constant for the electron acceptor nanoparticles (TiO_2, $\varepsilon_r \approx 80$) in addition to a high ionic strength electrolyte. As a result, after the initial electron injection step from the molecular excited state, the coulomb attraction of electrons and holes in dye-sensitized photoelectrochemical cells is effectively screened. However, for solar cells based on polymer materials, overcoming this coulomb attraction is significantly more demanding due to their smaller dielectric constant ($\varepsilon_r \approx 2$–4) and the more localized nature of the electronic states involved.

The device open-circuit voltage is primarily determined by the energy difference between the donor's ionization potential (HOMO) and acceptor's electron affinity (LUMO) as shown in Eq. 7-5. With real device efficiencies, after processing optimization, this typically is within 0.2 V of that predicted by this model. Furthermore, it has been clearly shown that the energy of the CT state correlates directly with V_{oc} for 26 different polymer:PCBM solar cells [23].

$$V_{oc} = \frac{\left|E_{\mathrm{HOMO(D)}} - E_{\mathrm{LUMO(A)}}\right|}{e} - 0.3 \qquad (7\text{-}5)$$

One of the approaches to increase the cell efficiency is reducing the optical band gap of the polymer so the light-harvesting efficiency can be enhanced and photocurrent can be increased. Many new polymers evaluated for their performance in solar cells have yielded much lower photocurrent densities and, consequently, poor device performance. In some cases, the low photocurrent densities can be attributed to poor charge collection due, for example, to low carrier mobility resulting in increased bimolecular recombination and space charge effects. In other cases, the low photocurrent can be attributed to poor exciton quenching due to insufficient mixing of the donor and acceptor components. However, in many cases, it appears that the low photocurrent derives from suboptimum dissociation efficiencies of the interfacial charge transfer states with the photo-generated CT states undergoing geminate recombination.

7.3.2 Polymer–Fullerene Solar Cell

Figure 7.13 shows the chemical structures of polymer and fullerene used in the solar cell. Their corresponding HOMO level and LUMO level are summarized in Table 7.2.

Brabec and coworkers have achieved a milestone polymer cell efficiency of 2.5% in 2001 using MDMOPPV:PCBM hybrid material [28]. In seven years, they doubled the efficiency to 5.2% using P3HT polymer and an annealing process [29]. The increase of efficiency is due to the P3HT having a higher charge mobility than that of MDMOPPV. The P3HT can be easily crystallized either by solvent annealing or thermal annealing. The type of polymer plays a major role in increasing the power conversion efficiency of solar cells. Recently, a new type of thieno[3,4-b]-thiophene and benzodithiophene alternating polymer (PTB7, Figure 4.19) has been developed by L. Yu's group [30]. These polymers have low band gaps and exhibit efficient absorption throughout the region of greatest photon flux in the solar spec-

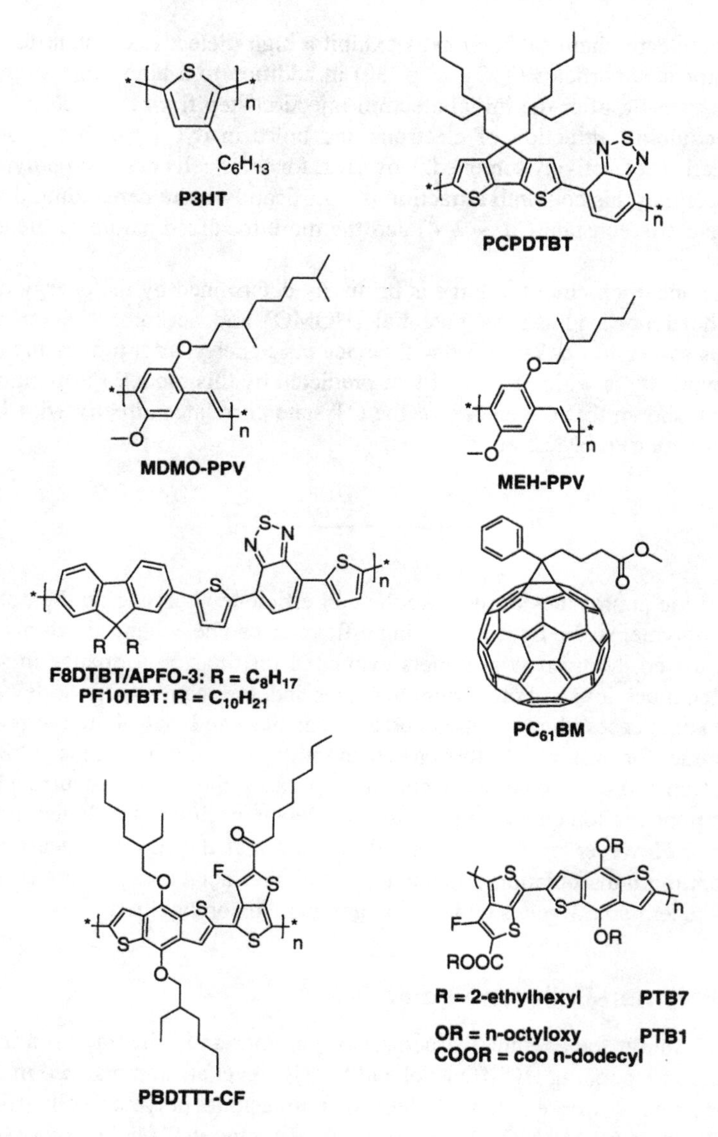

Figure 7.13. Chemical structures of polymer and fullerene used in organic solar cells.

trum (around 700 nm). They are the first kind of polymers with power conversion efficiency greater than 7%.

Symmetrical spheric C_{60}-based PCBM ($PC_{61}BM$) is the most commonly used fullerene in polymer–fullerene solar cells. However, $PC_{61}BM$ does not absorb solar light. The amount of $PC_{61}BM$ in the solar cell is usually equal to 50% by weight or higher. It is desirable to use solar-light-absorbing fullerene. Unsymmetrical oval C_{70}-

TABLE 7.2. HOMO and LUMO of materials used in polymer solar cells

Material	HOMO (eV)	LUMO (eV)	Reference
MDMOPPV	–5.40	–3.20	25
P3HT	–5.20	–3.20	24
PTB7	–5.15	–3.31	26
PBDTTT-CF	–5.22	–3.45	27
$PC_{61}BM$	–6.00	–4.20	24
$PC_{71}BM$	–6.10	–4.30	27

or C_{80}-based PCBM absorb solar light due to the ease of π—π^* absorption. Figure 7.14 shows that $PC_{71}BM$ harvests more light than $PC_{61}BM$ in the range of 350 nm—750 nm, which results in higher light absorption for the photoactive layer of P3HT:$PC_{71}BM$ in polymer solar cells [31]. Janssen and coworkers [32] used $PC_{71}BM$ instead of $PC_{61}BM$ for the MDMOPPV:fullerene solar cell, which increased the solar cell efficiency from 2.5% to 3.0%. Figure 7.15(a) shows that the absorption spectra of the PTB1/$PC_{61}BM$ film exhibited a weak absorption in the spectral range between 400 and 580 nm. The PTB1:$PC_{71}BM$ has a stronger absorption in that region, which led to a higher J_{sc} of 15.0 mA/cm^2, as indicated by the much higher EQE values [Figure 7.15(b)]. A PCE of 5.6% was obtained for PTB1/$PC_{71}BM$ [26].

The drawback of the $PC_{71}BM$ or PC81BM is that they are very expensive because the production yield of C_{70} and C_{80} is extremely low. Fullerenes are usually synthesized using an arc discharge between graphite electrodes (20 V, 60 A) in approximately 200 torr of He gas. This discharge produces carbon soot that can contain up to ~15% fullerenes: C_{60} (~13%) and C_{70} (~2%) [33]. Table 7.3 summarizes the performance of the state of the art of polymer:fullerene solar cells.

7.3.3 Effect of Active Layer Morphology on the Performance of Solar Cells

The morphology of the photoactive layer is also one of the key factors to control the performance of polymer solar cells. The morphology is highly dependent on the solvent used in the processing. The optimized morphology and bicontinuous phase of the photoactive layer of polymer solar cells can be controlled by using high-boiling-point solvent, mixed solvents, or solvent plus additive. Let us take the PTB7/$PC_{71}BM$ system as an example [26,30]. Figure 7.16 shows the I–V curves of PTB7/$PC_{71}BM$ devices using different solvents or using additional low-concentration additive 1,8-diiodooctane (DIO, m.p. 16–21°C, b.p. 167–169°C @ 6 mm Hg). Chlorobenzene (CB, b.p. 131°C) has a lower boiling point as compared with dichlorobenzene (1,4-DCB, b.p. 174°C), so it can be evaporated faster than the latter. Thus, there is more time for the formation of a bicontinuous phase and the domain size is smaller using DCB, which results in a higher efficiency solar cell. The DIO is a solid that cannot be evaporated. With a low melting point, it can act as a plasticizer for the ease of formation of the bicontinuous phase. Therefore, solvent CB plus DIO or 1,4-DCB plus DIO showed a large improvement in

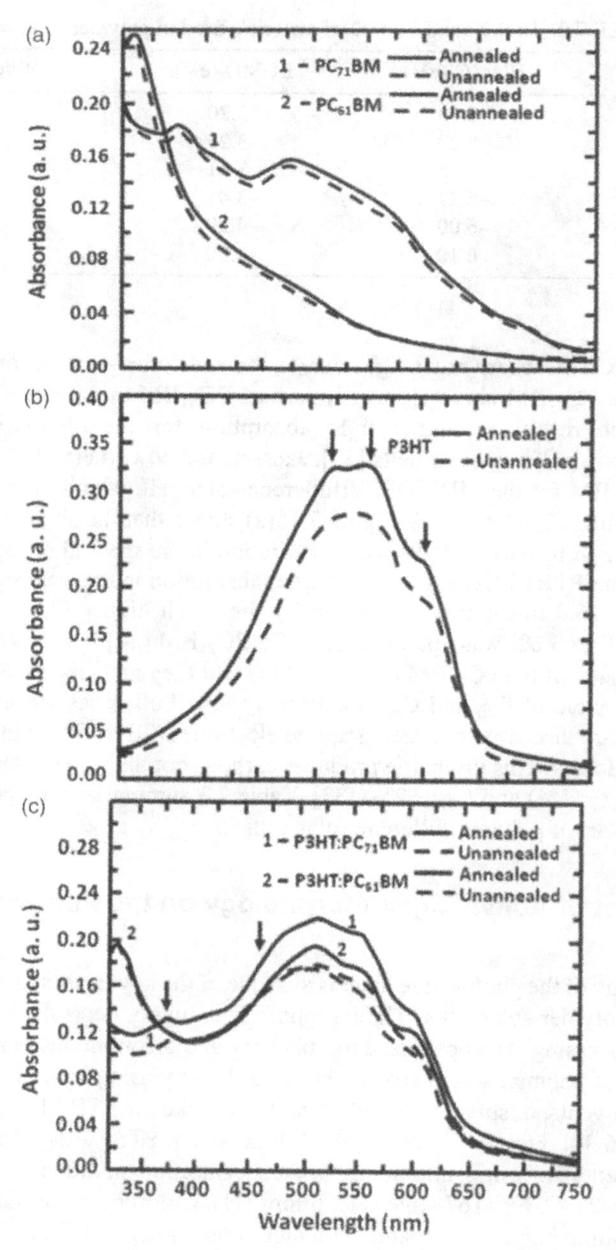

Figure 7.14. Absorption spectra of (a) $PC_{61}BM$ and $PC_{71}BM$, (b) P3HT, and (c) P3HT:$PC_{61}BM$ and P3HT:$PC_{71}BM$ nanocomposites before (dotted line) and after (solid line) thermal treatment at 150°C for 10 min [31].

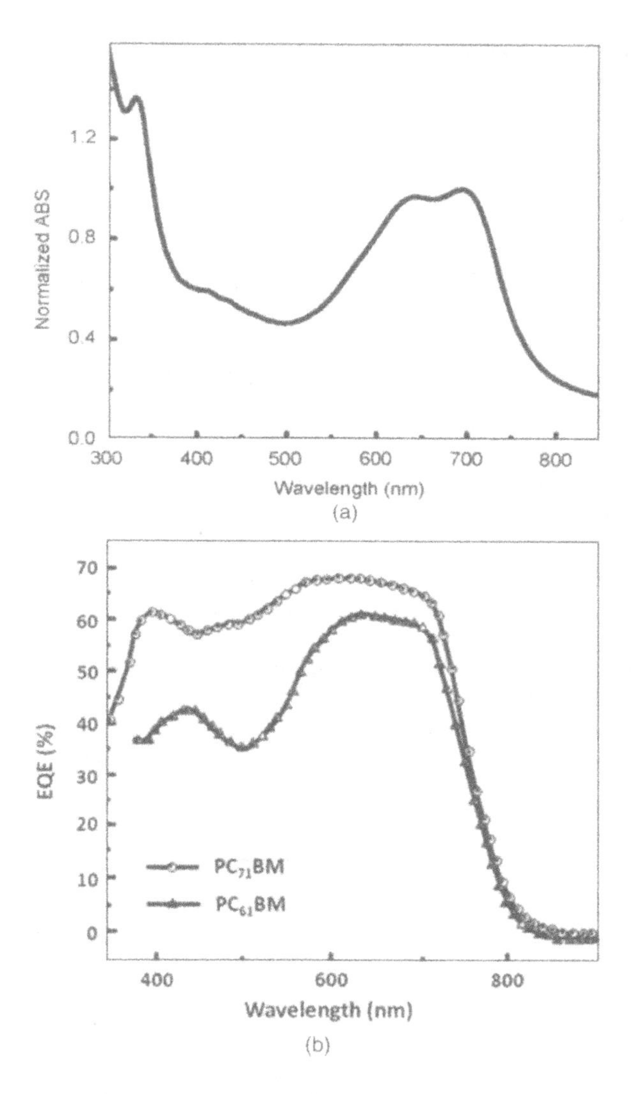

Figure 7.15. (a) UV-Vis absorption spectrum of PTB1:$PC_{61}BM$ and (b) external quantum efficiency of PTB1:$PC_{61}BM$ and PTB1:$PC_{71}BM$ devices [26].

solar cell efficiency. Table 7.4 summarizes the solar cell performance using different solvent systems. Figure 7.17 shows that the domain size of PTB7:$PC_{71}BM$ blend film prepared from CB (100–200 nm) is much larger than that from a mixture of 97 wt.% of CB and 3% DIO. The large-size domain is unfavored for an effective exciton separation. The use of a CB plus DIO mixture for processing promotes a good miscibility between PTB7 and $PC_{71}BM$ and the formation of interpenetrating networks.

TABLE 7.3. State of the art of the performance of polymer:fullerene solar cells

System	*PCE* (%)	V_{oc} (volt)	J_{sc} (mA/cm^2)	*FF* (%)	Reference
MDMOPPV:$PC_{61}BM$	2.5	0.82	5.25	61	28
MDMOPPV:$PC_{71}BM$	3.0	0.77	7.6	51	34
P3HT:$PC_{61}BM$	5.0	0.63	9.5	68	35
PTB7:$PC_{71}BM$*	7.4	0.74	14.50	69	30
PBDTTT-CF:$PC_{71}BM$*	7.7	0.76	15.20	67	27

*Mixed solvent containing 97% dichlorobenzene and 3% 1.8-diiodooctane was used.

7.3.4 Polymer:Semiconducting Nanoparticle Solar Cell

Polymer:semiconducting nanoparticle solar cells can solve the flammability problem of organic solar cells and can be used for building applications. The semiconducting nanoparticle is inorganic and nonflammable in nature, so it is expected to be fire retardant. The polymer:TiO_2 nanorod system is not only thermally stable but also low in cost (1/5000 of the cost of $PC_{61}BM$) and environmentally friendly. By using thermal atomic force microscopy, we have found that P3HT:TiO_2 is stable up to 150°C without aggregation but the P3HT:$PC_{61}BM$ system is not stable at 150°C with $PC_{61}BM$ aggregate size larger than one micron. The large-size aggregate disrupts the required bicontinuous phases for fast carrier transport, thus reducing the power conversion efficiency in the solar cell [36,37]. By blending TiO_2 nanorods with MEH-PPV, we obtained 0.5% cell efficiency [38] in 2006. A quadrupled efficiency of 2.2% was achieved recently [39,40] using the blend of surface-modified TiO_2 nanorods and P3HT. Although

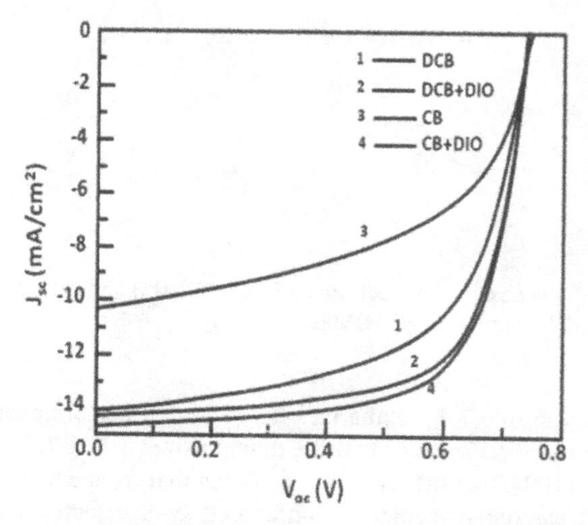

Figure 7.16. J–V curves of PTB7:PC71BM devices using (a) DCB only, (b) DCB with 3% DIO, (c) CB, and (d) CB with 3% DIO as solvents. The structure of PTB7 is shown in the inset [26].

TABLE 7.4. Device photovoltaic parameters of DCB only, DCB with 3% DIO, CB, and CB with 3% DIO as solvents and J_{sc} calculated from the EQE spectrum [26]

Solvent	V_{oc} (volt)	J_{sc} (mA/cm^2)	FF (%)	PCE (%)	J_{sc} (calcd) (mA/cm^2)
DCB	0.74	13.95	60.25	6.22	—
DCB/DIO	0.74	14.09	68.85	7.18	13.99
CB	0.76	10.20	50.52	3.92	—
CB/DIO	0.74	14.50	68.97	7.40	14.16

the polymer:nanoparticle system is more thermal stable than the polymer:fullerene system, the development of polymer:semiconducting nanoparticle solar cells is more difficult than that of polymer:fullerene solar cells due to the former containing two immiscible organic and inorganic components. The issues of compatibility and band alignment between donor and acceptor need to be addressed simultaneously.

The hybrid materials are generally prepared by blending the semiconducting nanoparticle with conducting polymer in the solvent. The semiconducting nanoparticle exhibits hydrophilic characteristics but the polymer is hydrophobic in nature; therefore, we need to modify the surface of the nanoparticle to be compatible with the polymer and solvent. The desired interface modifier should have the properties of (1) aligned band diagram with polymer and nanoparticle as shown in Figure 7.18, (2) electrical conductivity, and (3) amphiphilic characteristics.

A small molecule was used to modify the surface of TiO_2 nanoparticles and can reduce the electron–hole recombination effectively [39]. However, the small molecule exhibits high band gap and cannot align well with the hybrid blend. Ru dye has shown

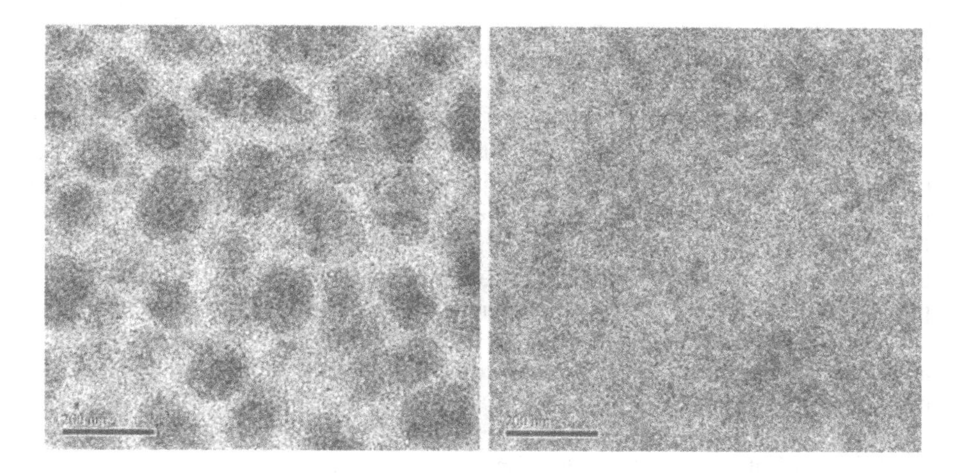

Figure 7.17. TEM images of PTB7:$PC_{71}BM$ blend film prepared from chlorobenzene without (left) and with (right) diiodooctane. The scale bar is 200 nm. (Reprinted from reference [26] with permission.)

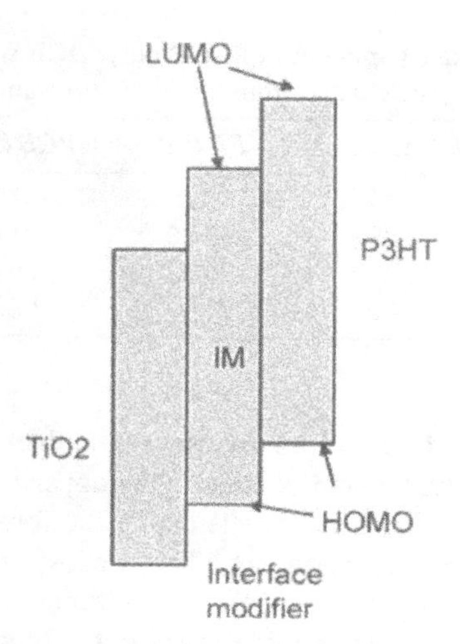

Figure 7.18. The band gap of the interface modifier is aligned with polymer and nanoparticle.

promising results to reduce charge recombination [40], but the dye is not electrically conducting and is very expensive due to the rare metal Ru. We have synthesized an Ru-free, carboxylic-terminated thiophene-conducting oligomer (mol.wt.~5,000). The oligomer can be adhered well on the surface of TiO_2 nanoparticles through the carboxylic functional group. The thiophene segment is compatible with P3HT, which will help the dispersion of TiO_2 nanoparticles in P3HT, as shown in Figure 7.19. The thiophene is electrically conducting, which will help the transport of charge carriers. The surface modification on the TiO_2 nanoparticle will reduce its surface defects, reducing the electron–hole recombination and facilitating the charge transport. Doubled cell efficiency has been achieved through this surface modification [40].

At present, the efficiency of the polymer:TiO_2 system is still low [40,41] because the TiO_2 does not absorb solar light and its electron transport rate is low (two orders lower than that of polymer). The electron transport through TiO_2 is via hopping the electrons along the TiO_2 network, which is at a lower rate than that of direct transport in highly ordered crystalline solids. The nanocrystalline TiO_2 also exhibits many surface and intrinsic defects in the crystal. Efforts are being sought to reduce the defects of nanocrystals such as surface modification of the TiO_2 and using ordered mesoporous TiO_2 thin films [42]. The small pore size of the mesoporous TiO_2 film still presents a problem in infiltrating the polymer properly, which results in a lower efficiency as compared with that of the surface-modified TiO_2 nanoparticle blended with a polymer system (0.5% vs. 2.2%).

The ZnO exhibits higher electron mobility than that of TiO_2. Janssen et al. [43,44] used ZnO nanoparticles to fabricate polymer:ZnO bulk heterojunction solar cells. However, the power conversion efficiency of the ZnO system is not high either, because of the compatibility and band alignment issues between polymer and ZnO. They have found that MDMOPPV:ZnO solar cells exhibited a higher efficiency

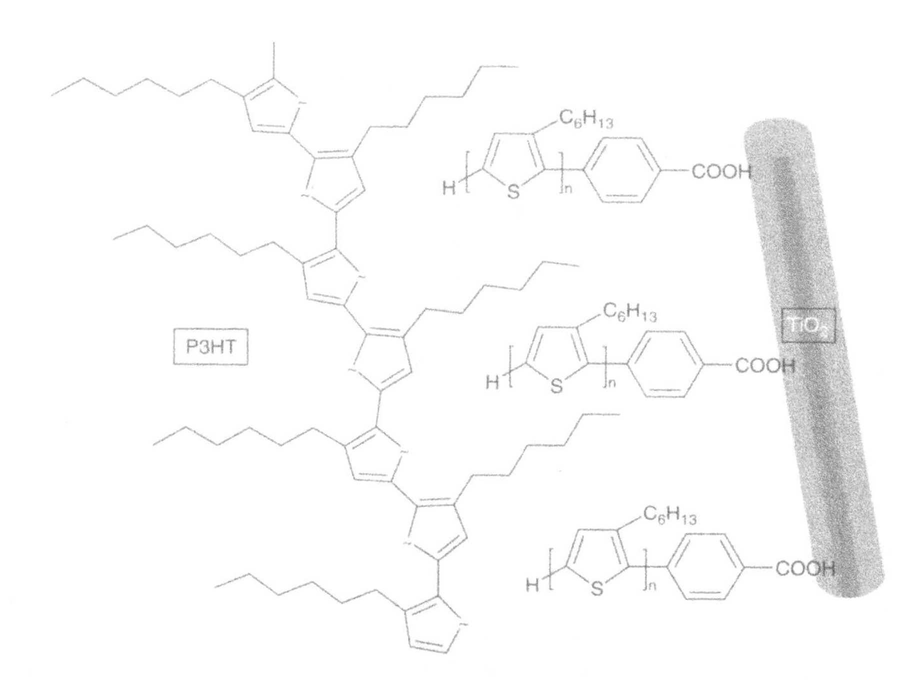

Figure 7.19. Schematic diagram of TiO_2/oligo-3HT-COOH/P3HT hybrid system.

(1.6%) than that of P3HT:ZnO (0.9%) due to the higher V_{oc} of the former (0.81 eV vs. 0.69 eV).

The CdSe nanoparticles absorb solar light so the polymer:CdSe system exhibits relatively higher power conversion efficiency as compared with that of polymer:metal oxide systems, as shown in Table 7.5. Alivisato's group reports an efficiency of 1.7% using P3HT:CdSe nanorod system [45]. Then the efficiency is increased to 2.6% by re-

TABLE 7.5. State of the art of polymer-semiconducting nanoparticle solar cells

System	*PCE* (%)	V_{oc} (volt)	J_{sc} (mA/cm^2)	*FF* (%)	Reference
P3HT:CdSe	1.70	0.70	5.70	40	45
APFO-3:CdSe (w:w = 1:6) (E_g = 2.0 eV)	2.60	0.95	7.23	38	46
PCPDTBT:CdSe (w:w = 1:9) (E_g = 1.4 eV)	3.19	0.68	10.1	51	47
P3HT:TiO_2	0.42	0.44	2.76	36	41
P3HT:TiO_2/Oligo-3HT-COOH	1.29	0.64	3.26	62	37
P3HT:TiO_2/N3 modifier	2.20	0.78	4.33	65	40
P3HT:ZnO	0.92	0.69	2.19	55	43
MDMOPPV:ZnO	1.60	0.81	2.40	59	44

placing P3HT with a high-V_{oc} polymer such as APFO-3 [46]. The efficiency is further increased to 3.2% by using low-band-gap polymer PCPDTBT and CdSe tetrapod nanoparticles [47]. The tetrapod structure of the nanoparticles is believed to afford a better electron transport rate than that of CdSe nanorod. It is interesting to note that the amount of CdSe has been increased from 50% to 90% by weight in the PCPDTBT:CdSe system. By increasing the nanoparticle concentration, it is possible to reduce the diffusion length of the exciton. The only problem associated with the CdSe nanoparticles is the toxicity and long-term stability of CdSe, which will not be a suitable material for environmentally friendly renewable solar cell systems.

Table 7.5 summaries the state of the art of polymer:semiconducting nanoparticle solar cells. In general, this type solar cell exhibits lower power-conversion efficiency than that of polymer:fullerene solar cells. The polymer:semiconducting nanoparticle system has three problems: (1) the size of semiconducting nanoparticles is five to ten times larger than fullerence (5–10 nm vs. 1 nm), which results in less bulk heterojunction and lower extent of charge separation; (2) surface defects of nanoparticles acting as traps for the recombination of electrons and holes; and (3) mismatched band alignment between polymer and nanoparticle. The surface defects can be decreased by surface modification, as discussed earlier. Table 7.6 shows the HOMO and LUMO of the semiconducting materials. The nanoparticle usually exhibits a higher band gap than its bulk band gap due to the quantum confinement. However, the value is still valid for comparison purposes. The low level of metal oxide HOMO presents a big gap between polymer and nanoparticle that results in high probability charge recombination and slow rate of charge transport. Therefore, the polymer used for fullerene is not optimized for semiconducting nanoparticles.

7.4 SCALE-UP, STABILITY, AND COMMERCIAL DEVELOPMENT OF ORGANIC SOLAR CELLS

Efficiency is the essential parameter for solar cells with respect to maximizing energy production and minimizing cost. Low device efficiency demands a larger photovoltaic active area in order to produce the same energy output. A device efficiency of ~10% and a module efficiency of ~5% are regarded as critical values for commercial products [51]. There are interconnection requirements in the module, so it has a photovoltaic active area with lower efficiency as compared with a device. Lifetime is the second most important parameter for the success of solar cell technology. A lifetime

TABLE 7.6. HOMO and LUMO of semiconducting materials

Material	HOMO (eV)	LUMO (eV)	Reference
TiO_2	–7.40	–4.20	48
ZnO	–7.39	–4.19	49
CdSe	–5.81	–3.71	50

TABLE 7.7. Performance and cost analysis of polymer-based solar cell modules

Year	*PCE* %	Life (yr)	Cost (USD/W_p)	Application
2010–2013	1–3	1–2	12	Consumer electronics
2012–2015	3–4	1–3	2–4	Small-size equipment
2015–2018	4–6	3–5	1	Residential buildings and large-size equipment
2018–2025	6–9	5–7	<0.5	Solar power plants

of 3–5 years (operation time of 3000–5000 hours) is usually required for the marketplace.

Organic solar cells can be processed by a solution process, so it can be easily scaled up by conventional coating processes such as printing, spraying, and dip coating. The development of dye sensitized solar cells (DSSCs) has been longer than that of the polymer-based solar cell. Therefore, the DSSC is closer to large-scale production at present.

A detailed procedure providing a guide to reproduce DSSC cell efficiency of 10% with a variation of 2–3% has been reported [52]. The manufacturing of DSSC modules on a semiautomated process [53] with a module efficiency of 8.2% has been published [54]. The DSSC has undergone stability evaluation of 85°C high-temperature stress, damp heat, and thermal cycling. The results have shown that the DSSC can sustain outdoor operation for 20 years and has potential for commercialization [55].

The polymer:fullerene solar cell has been analyzed to show a room-temperature lifetime of more than one year when the device is packaged between two glass plates. The accelerated 85°C test has shown the device life to be more than 2000 hours (defined by an efficiency drop of 50% from its initial value). The results are encouraging, but further experiments on the stability of the device under simulated solar light, and especially on the flexible plastic substrate, are necessary. Nielsen et al. [56] have done very detailed performance and cost analyses of polymer-based solar cells to determine the viability of the device as a commercial product over a period of 25 years, as shown in Table 7.7. The results show that polymer solar cells will become a widely used renewable power source after 2018.

REFERENCES

1. J. Desilvestro, M. Grätzel, L. Kavan, J. E. Moser, and J. Augustynski (1985), *J. Am. Chem. Soc., 107,* 2988–2990.
2. M. Grätzel (2009), *Accounts of Chemical Research, 42,* 1788–1798.
3. M. Grätzel (2001), *Nature, 414,* 338–344.
4. B. O'Regan and D. Schwartz, T. (1996), *J. Appl, Phys., 80,* 4749–4754.
5. U. Bach, D. Lupo, P. Comte, J. E. Moser, F. Weissortel, J. Salbeck, H. Spreitzer, and M. Grätzel (1998), *Nature, 395,* 583–585.

6. L. Schmidt-Mende, U. Bach, R. Humphry-Baker, T. Horiuchi, H. Miura, S. Ito, S. Uchida, and M. Grätzel (2005), *Adv. Mater., 17,* 813–815.
7. D. Kuang, C. Klein, H. J. Snaith, R. Humphry-Baker, S. M. Zakeeruddin, and M. Grätzel (2008), *Inorg. Chim. Acta, 361,* 699–706.
8. L. Schmidt-Mende, S. M. Zakeeruddin, and M. Grätzel (2005), *Appl. Phys. Lett., 86,* 013504.
9. P. Chen, J. H. Yum, F. DeAngelis, E. Mosconi, S. Fantacci, S. J. Moon, R. H. Baker, J. Ko, M. K. Nazeeruddin, and M. Grätzel (2009), *Nano Letters, 9,* 2487–2492.
10. H. J. Snaith, A. J. Moule, C. Klein, K. Meerholz, R. H. Friend, and M. Grätzel (2007), *Nano Letters, 7,* 3372–3376.
11. H. Hoppe and N. S. Sariciftci (2004), *J. Mater. Res., 19,* 1924–1945.
12. S. C. Veenstra, W. J. H. Verhees, J. M. Kroon, M. M. Koetse, J. Sweelssen, J. J. A. M. Bastiaansen, H. F. M. Schoo, X. Yang, A. Alexeev, J. Loos, U. S. Schubert, and M. M. Wienk (2004), *Chem. Mater., 16*(12), 2503–2508.
13. S. A. Jenekhe and S. Yi (2000), *Appl. Phys. Lett., 77*(17), 2635–2637.
14. M. Granström, K. Petritsch, A. C. Arias, A. Lux, M. R. Andersson, and R. H. Friend (1998), *Nature, 395,* 257–260.
15. J. Xue, S. Uchida, B. P. Rand, and S. R. Forrest (2004), *Appl. Phys. Lett., 84,* 3013.
16. S. R. Forrest (2005), *MRS Bulletin, 30,* 28–32.
17. S. E. Shaheen, D. S. Ginley, and G. E. Jabbour (2005), *MRS Bulletin, 30,* 10–15.
18. D. L. Staebler and C. R. Wronski (1980), *J. Appl. Physics,* June, *51*(6), 3262.
19. A. C. Mayer, S. R. Scully, B. E. Hardin, M. W. Rowell, and M. D. McGehee (2007), *Materials Today, 10*(11), 28–33.
20. C. J. Brabec and J. R. Durrant (2008), *MRS Bulletin, 33,* 671–675.
21. T. M. Clarke and J. R. Durrant (2010), *Chem. Rev.,* DOC:10.1021/cr90027/s.
22. G. Yu, J. Gao, J. C. Hummelen, F. Wudl, and A. J. Heeger (1995), *Science, 270,* 1789.
23. M. C. Scharber, D. Mühlbacher, M. Koppe, P. Denk, C. Waldauf, A. J. Heeger, and C. J. Barbec (2006), *Adv. Mater, 18,* 789.
24. J. Y. Kim, K. Lee, N. E. Coates, D. Moses, T. Q. Nguyen, M. Dante, and A. J. Heeger (2007), *Science, 317,* 222–225.
25. M. Al-Ibrahim, A. Konkin, H. K. Roth, D. A. M. Egbe, E. Klemm, U. Zhokhavets, G. Gobsch, and S. Senfuss (2005), *Thin Solid Films, 474,* 201–210.
26. Y. Liang and L. Yu (2010), *Acc. Chem. Res., 43*(9), 1227–1236.
27. H. Y. Chen, J. H. Hou, S. O. Zhang, Y. Y. Liang, G. W. Yang, Y. Yang, L. P. Yu, Y. Wu, and G. Li (2009), *Nat. Photonics, 3,* 649–653.
28. S. E. Shaheen, C. J. Brabec, N. S. Sariciftci, F. Padinger, T. Fromherz, and J.C. Hummelen (2001), *Appl. Phys. Lett., 78,* 841.
29. M. A. Green, K. Emery, Y. Hishikawa, and W. Warta (2008), *Progress in Photovoltaics: Research and Applications, 16,* 16–67.
30. Y. Y. Liang and L. Yu (2010), *Acct. Chem Res., 43,* 9, 1227–1236.
31. P. Boland, S. S. Sunkavalli, S. Chennuri, K. Foe, T. Abdel-Fattah, and G. Namkoong (2010), *Thin Solid Films, 518*(6), 1728–1731.
32. M. M. Wienk, J. M. Kroon, W. J. H. Verhees, J. Knol, J. C. Hummelen, P. A. van Hal, and R. A. J. Janssen (2003), *Angew. Chem. Int. Ed., 42,* 3371–3375.
33. G. Timp (Ed.) (1999), *Nanotechnology,* Chapter 5, Springer.

34. M. M. Wienk, J.M. Kroon, W. J. H. Verhees, J. Knol, J. C. Hummelen, P. A. van Hal, R. A. and J. Janssen (2003), *Angew. Chem. Int. Ed., 42,* 3371–3375.
35. W. Ma, C. Yang, X. Gong, K. Lee, and A. J. Heeger (2005), *Adv. Funct. Mater, 15,* 1617–1622.
36. Y.C. Huang, Y. C. Liao, S. S. Li, M. C. Wu, C. W. Chen, and W. F. Su (2009), *Solar Energy Materials and Solar Cells, 93,* 888–892.
37. Y. C. Huang, W. C. Yen, Y. C. Liao, Y. C. Yu, C. C. Hsu, M. L. Ho, P. T. Chou, and W. F. Su (2010), *Appl. Phys. Lett., 96,* 123501.
38. T. W. Zeng, Y. Y. Lin, H. H. Lo, C. W. Chen, D. H. Chen, S. C. Liou, H. Y. Huang, and W. F. Su (2006), *Nanotechnology, 15,* 5387.
39. Y. Y. Lin, T. H. Chu, C. W. Chen, and W. F. Su (2008), *Appl. Phys. Lett., 92,* 053312.
40. Y. Y. Lin, T. H. Chu, S. S. Li, C. H. Chuang, C. H. Chang, W. F. Su, C. P. Chang, M. W. Chu, and C. W. Chen (2009), *Journal of American Chemical Society, 131,* 3644.
41. C. Y. Kwong, W. C. H. Choy, A. B. Djurisic, P. C. Chui, K. W. Cheng, and W. K. Chan (2004), *Nanotechnology, 15,* 1156–1161.
42. C. Goh, S. R. Scully, and M. D. McGehee (2007), *J. Appl. Phys.*, 101, 114503.
43. W. J. E. Beek, M. M. Wienk, and R. A. J. Janssen (2006), *Adv. Funct. Mater., 16,* 1112–1116.
44. W. J. E. Beek, M. M. Wienk, and R. A. J. Janssen (2005), *J. Mater. Chem., 15,* 2985–2988.
45. W. U. Huynh, and Alivisatos (2002), *A. P. Science, 295,* 2425.
46. P. Wang and N. C. Greenham (2006), *Nano Lett., 6*(8), 1789–1793.
47. S. Dayal, N. Kopidakis, D. S. Ginley, and G. Rumbles (2010), *Nano Lett., 10*(1), 239–242.
48. L. Shen, G. Zhu, W. Guo, C. Tao, X. Zhang, C. Liu, W. Chen, S. Ruan, and Z. Zhong (2008), *Appl. Phy. Lett., 92,* 073307.
49. P. Ravirajan, A. M. Peiro,; M. K. Nazeeruddin, M. Graetzel, D. D. C. Bradley, J. R. Durrant, and J. J. Nelson (2006), *Phys. Chem. B, 110,* 7635–7639.
50. Y. Zhou, Y. Li, H. Zhong, J. Hou, Y. Ding, and C. Yang (2006), *Nanotechnology, 17,* 4041–4047.
51. C. J. Brabec, J. A. Hanch, P. Schilinsky, and C. Waldauf (2005), *MRS Bulletin, 30,* 50–52.
52. S. Ito, T. N. Murakami, P. Comte, P. Liska, Co. Grätzel, Md. K.Nazeeruddin, and M. Grätzel (2008), *Thin Solid Films, 516,* 4613–4619.
53. M. Späth, P. M. Sommeling, J. A. M. van Roosmalen, H. J. P. Smit, N. P. G. van der Burg, D. R. Mahieu, N. J. Bukker, and J. M. Kroon (2003), *Progr. Photovoltaics, 11,* 207–220.
54. L. Han, A. Fukui, Y. Chiba, A. Islam, R. Komiya, N. Fuke, N. Koide, R. Yamanaka, and M. Shimizu (2009), *Appl. Phys. Lett., 94,* 013305/1-013305/3.
55. Third International Conference on the Industrialization of Dye Sensitizer Solar Cells, Nara, Japan, April, 2009.
56. T. D. Nielsen, C. Cruickshank, S. Foged, J. Thorsen, and F. C. Krebs (2010), *Sol. Eng. Mater. Sol. Cells, 94*(10), 1553–1571.

EXERCISES

1. The liquid electrolyte of iodide/triiodide acetonitrile-based DSSC has demonstrated the highest power-conversion efficiency of all organic solar cells. However, this

type of DSSC is not safe for long-term usage due to the volatile nature of acetonitrile and corrosive nature of iodide/triiodide. Nonvolatile ionic liquid–solid electrolytes (e.g., spiro-OMeTAD, polyelectrolyte) have been proposed to replace liquid electrolytes but their power-conversion coefficients are always lower than that of liquid electrolytes. Explain the reasons for the low power-conversion coefficient of these two types of dye-sensitized solar cells and propose approaches to fabricate DSSCs that can achieve both safety and high power conversion efficiency at the same time. (Hint: use equation 7-2 as starting point.)

2. Compare the differences in operation principle between SSDSCs and polymer:nanoparticle solar cells. At present, polymer:fullerene solar cells have higher power-conversion efficiency than SSDSCs, but polymer:TiO_2 solar cells have lower power-conversion efficiency than SSDSCs. Explain the reasons for this.
3. The V_{oc} of the polymer:nanoparticle solar cell can be predicted well from the HOMO and LUMO of acceptor and donor according to equation 7-5. Calculate the V_{oc} of the active layers listed in the Table 7.3 and compare the calculated values with the experimental data. Show the results in a table. (Hint: use the Table 7.2 information.)
4. Draw the energy band diagram of P3HT:TiO_2 and P3HT:$PC_{61}BM$ systems, respectively. From the energy band diagram, show and explain which system exhibits better carrier transport. (Hint: use the LUMO and LUMO data of each material)
5. Compare the differences between polymer:fullerene solar cells and polymer:semiconducting nanoparticle solar cells. Up to now, the efficiency of the latter has been lower than that of the former. Explain why. Propose approaches that can improve the efficiency of polymer:semiconducting nanoparticle solar cells, so they can have efficiencies comparable to those of polymer:fullerene systems.

8

ORGANIC–INORGANIC HYBRID SOLAR CELLS

Ching-Fuh Lin

8.1 FUNDAMENTAL CONCEPTS OF ORGANIC–INORGANIC HYBRID SOLAR CELLS

Conventional organic solar cells with a device architecture of ITO/PEDOT:PSS/P3HT:PCBM/Al easily deteriorate because several device components are highly sensitive to moisture and air [1–3]. For example, the PEDOT:PSS layer is acidic and has side effects on the device performance due to its corrosion to ITO and electrical inhomogeneities [2,4]. In addition, the hygroscopic nature of PEDOT:PSS and the easy oxidation of the back contact (Al) are harmful to the device characteristics [1,3]. To avoid those adverse effects, they are usually fabricated and tested in inert atmosphere and then encapsulated, so they are not directly exposed to the usual atmosphere that contains oxygen and moisture.

An alternative architecture, called the inverted structure, is used to overcome the above limitations of Al and PEDOT:PSS [1,3]. The inverted configuration utilizes an air-stable high-work-function electrode as the back contact anode and the ITO as the cathode [5–9]. The reason that it is called the inverted structure is because the electrons and holes transport in the direction opposite to that in the conventional organic solar cells. In conventional organic solar cells, electrons transport toward the metal electrode, Al, and holes transport toward the ITO side. For inverted structures, electrons transport toward the ITO electrode, while holes transport toward the metal electrode.

To further improve the performance of organic solar cells of the inverted structure, another transition metal oxide is introduced as anodic modification between the organic layer and the top metal electrode. This configuration has a sandwiched structure, as shown in Figure 8.1. The organic layer is sandwiched between two oxide layers, oxide 1 and oxide 2. The entire solar cell consists of both the organic layer and the inorganic oxide layers and is called an organic–inorganic hybrid solar cell.

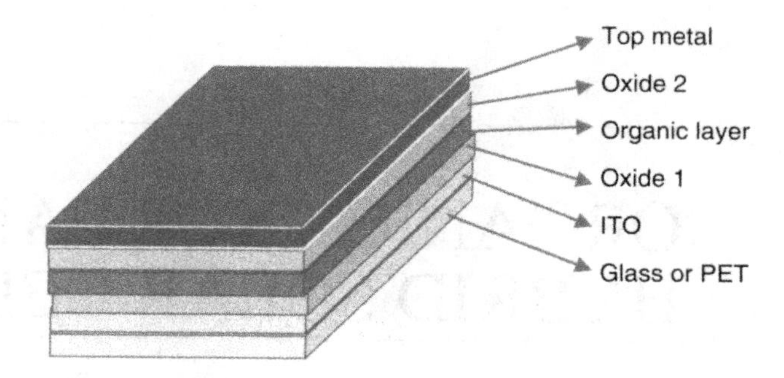

Figure 8.1. A schematic of the inverted structure for organic solar cells.

The oxide layers have several functions. First, they protect the organic layer from direct contact with air and moisture, so the organic layer can last a longer time before degradation. Second, with proper thickness of the oxide layer, in particular the oxide 2, the standing wave, formed by the incident wave and the reflected wave from the top metal, can have a maximum intensity inside the organic layer. In this way, the light absorption can be enhanced. Third, the two oxide layers serve as electron or hole blockers if they have proper conduction and valence bands. A detailed explaination follows.

Figure 8.2 shows a band diagram of the inverted structure of organic solar cells. There are two important points in this diagram. First, oxide 1 has a valence band lower than the highest occupied molecule orbital level (HOMO) of the organic materials, so holes are blocked by this oxide. As a result, holes are forced to move toward the top-metal electrode. Similarly, oxide 2 has a conduction band higher than the lowest unoc-

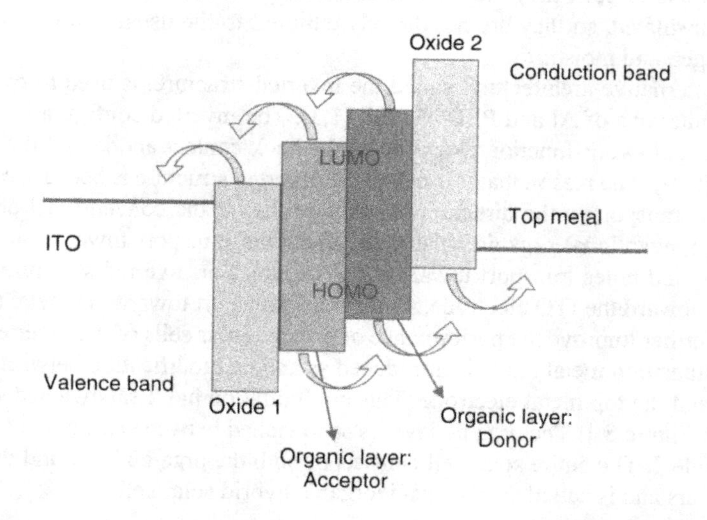

Figure 8.2. Band diagram of the inverted structure for organic solar cells.

cupied molecule orbital level (LUMO) of the organic materials, so electrons are blocked and forced to transport only toward the ITO electrode, which serves as the cathode for the inverted structure. In this way, electrons and holes are forced to separate and transport in opposite directions.

With the above three functions provided by the oxides, inverted organic solar cells with the sandwiched structure are expected to have higher carrier collection efficiency and longer lifetime, compared to conventional solar cells, if the same organic layer is used.

On the other hand, the oxides cannot be arbitrarily chosen. In addition to the requirements of the conduction band and valence band for electron or hole blocking, they have to be conducting and transparent. Therefore, semiconductors with a wide band gap around 3 eV are commonly used.

8.2 SANDWICHED STRUCTURES OF ORGANIC–INORGANIC HYBRID SOLAR CELLS

8.2.1 Fabrication of Sandwiched Structures

As shown in Figure 8.1, the solar cell has oxide 1 deposited on glass that is coated with the transparent and conducting material indium tin oxide (ITO). Oxide 1 could be any oxide that is conducting, transparent, and has a valence band much lower than the HOMO level of organic materials. ZnO is commonly used due to its easy preparation. The ZnO layer is deposited as follows.

First, 0.5 M ZnO sol–gel is prepared by dissolving zinc acetate dihydrate [Zn(CH3COOH)2·2H2O] in 2-methoxyethanol (2MOE) and stirred at 80°C for 2 hours. This ZnO sol–gel solution is coated on the ITO-coated glass substrate and baked in oven at 200°C to form a ZnO film. The thickness of this ZnO layer is around 30–40 nm. After baking at 200°C, the solvent is removed and the ZnO becomes a film of small grainy nanoparticles.

The conduction-band minimum of ZnO is –4.4 eV, which is lower than the LUMO level (–3.7 eV) of the electron acceptor in the organic layer, so electrons can transfer from the organic layer to the ZnO film. On the other hand, the valence band maximum of ZnO is –7.8 eV, which is much lower than the HOMO level (–6.1 eV) of the donor in the organic layer. Therefore, holes are almost completely blocked by the ZnO layer and cannot transport toward the ITO electrode. In this way, holes are forced to move only toward the metal electrode on the other side.

Another material, $CsCO_3$, has also been used for oxide 1 layer [10]. As long as it exhibits the conducting and transparent properties and has a valence band much lower than the HOMO level of organic materials, it can function well for this layer.

After the deposition of oxide layer 1, the organic blend that consists of the acceptor and the donor materials is spin-coated onto the oxide 1 layer. Because the oxide 1 layer has already solidified after being baked at a high temperature, the subsequent deposition of the organic layer causes no damage to it. The deposition of the organic layer is similar to that for conventional organic solar cells and, hence, is not elaborated here. However, the postdeposition treatment is not same. The reason is possibly because the interface condition between the organic and ZnO layers is not the same as that between the organic and the PEDOT:PSS layers. The posttreatment is mainly for the purpose of improving self-organization of the P3HT or other organic acceptors, so holes can easily move out. For conventional organic solar cells using P3HT, the postdeposition treatment usually involves annealing at 150°C. However, for inverted structures, the annealing temperature is only 120°C or lower, and 2–4 days are required for the active organic layer to self-organize. On the other hand, the posttreatment also varies with the organic acceptor. If organic materials different from P3HT are used, the posttreatment can be very different. This is still an area that needs significant amounts of research.

On top of the organic layer is the oxide 2 layer. Several transition metal oxides such as V_2O_5 [7,11], MoO_3 [9], and WO_3 [8] can be used as anodic modifications in the inverted OSCs. They are usually deposited by the vacuum evaporation, which has the advantage of affecting the organic layer less as long as the temperature during the evaporation is controlled to be much less than 150°C. However, the vacuum evaporation of these oxides detracts from the advantage of the ease of large-area fabrication and meanwhile increases the fabricating cost. A very good advantage in the fabrication of organic solar cells is the possibility of using a solution process which is also highly desirable for the oxide layers. The solution process for the oxide layers has been developed but it is the most difficult step because the organic layer underneath can be easily damaged.

To avoid mixing or damaging the photoactive organic layer during the coating of the metal oxides, the solvent for the metal oxides has to be properly selected. An important concept to achieve this purpose is that the solvent must be orthogonal to dichloromethane (DCB), which is the solvent used for the P3HT/PCBM blend. The polar characteristics of these solvents are critical. The polarity of a solvent is related to its dielectric constant [12]. Because the relative dielectric constant (ε_r) of isopropanol is ~20, which is much larger than that of the P3HT:PCBM film (ε_r ~3.5) [13], it has negligible influences on the deposited photoactive organic layer. Table 8.1 summarizes the relative dielectric constants of several solvents and the photoactive organic layer of the P3HT/PCBM blend. From this table, the dielectric constant of isopropanol is the largest, so it is a good orthogonal solvent [14]. Other solvents such as toluene, chlorobenzene, 2-methoxyethanol, and dichloroethane had also been experimented with, but they did not function as well as isopropanol did.

TABLE 8.1. Relative dielectric constants of common organic solvents and P3HT:PCBM layer [14]

Organic solvents or P3HT:PCBM	Relative dielectric constant (ε_r) at 300 K
Toluene	2
Chlorobenzene (CB)	5.6
2-methoxyethanol	16
Dichloromethane (DCB)	7.5
Isopropanol	20
P3HT:PCBM film	3.5

After the careful selection of the solvent for the oxides, an oxide ink is prepared by dispersing and suspending some metal oxide powder homogeneously in isopropanol by using ultrasonic agitation, forming a colloidal solution with different concentrations. Usually, a concentration of 0.1 mg/ml is a good choice to give an appropriate thickness for oxides such as V_2O_5, WO_3, NiO, or CuO. The prepared oxide ink is then spin-coated on the photoactive organic layer at a rate of 4000 rpm. Figure 8.3 schematically shows the process of spin-casting the solution that contains the metal oxides. Finally, the top electrode is a silver film of ~200 nm, which is deposited in a vacuum of 2×10^{-6} torr. Figure 8.4 shows the device structure of ITO/ZnO thin film/P3HT:PCBM/metal oxide/Ag.

8.2.2 Performance of Organic–Inorganic Hybrid Solar Cells with Sandwiched Structures

The cells are characterized after the fabrication. A solar simulator with an intensity of 100 mW/cm^2 and AM 1.5G filters is used for the characterization. The J–V curves are measured with a Keithley 2400 source measurement unit. The illumination intensity is calibrated by a NREL calibrated Si solar cell with KG-5 color filter. A shadow mask is used to block the light outside of the active area, which is 10 mm^2, to avoid the collection of extra current outside of the defined region.

Figure 8.5 shows the J–V characteristics of inverted organic solar cells with different metal oxides as anodic modification. The control device is the one without any oxide for the anodic modification. It exhibits a J_{SC} of 10.67 mA/cm^2, a V_{oc} of 0.53 V, an *FF* of 54.82%, and a *PCE* of 3.1%. The performance is similar to other reports on the

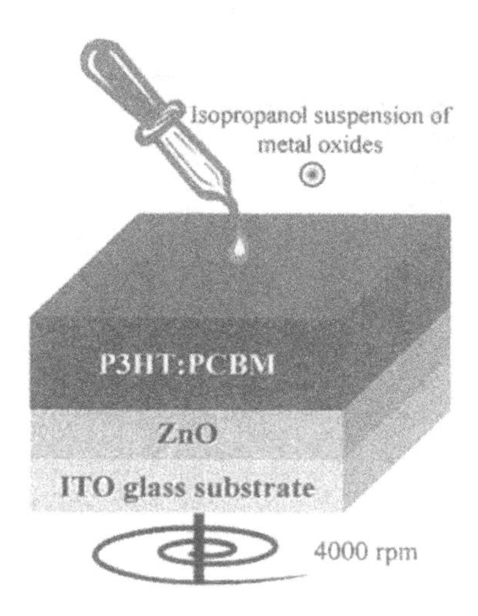

Figure 8.3. Schematic diagram of depositing the metal oxides [14].

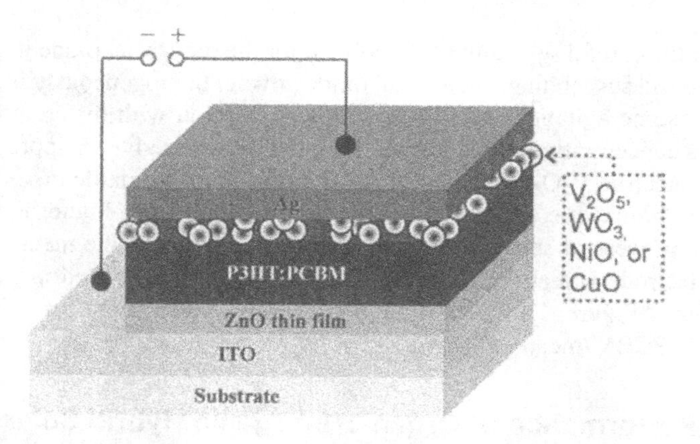

Figure 8.4. Schematic device structure of ITO/ZnO thin film/ P3HT:PCBM/metal oxides/Ag [14].

same device structure [6,15]. Because the control device does not have the oxide on top of the organic layer, it is not a sandwiched structure.

Introducing one of the metal oxides mentioned previously (NiO, V_2O_5, WO_3, or CuO) to form a sandwiched structure, the device performance is notably improved. With NiO, the organic solar cell exhibits a J_{SC} of 11.92 mA/cm^2, a V_{OC} of 0.54 V, and an FF of 58.41%, resulting in an improved PCE of 3.76%. The device with V_2O_5 exhibits a PCE of 3.75% with J_{SC} of 11.51 mA/cm^2, V_{OC} of 0.54 V, and FF of 60.33%. As WO_3 is used, the PCE is improved to 3.73% with J_{SC} of 12.34 mA/cm^2, V_{OC} of

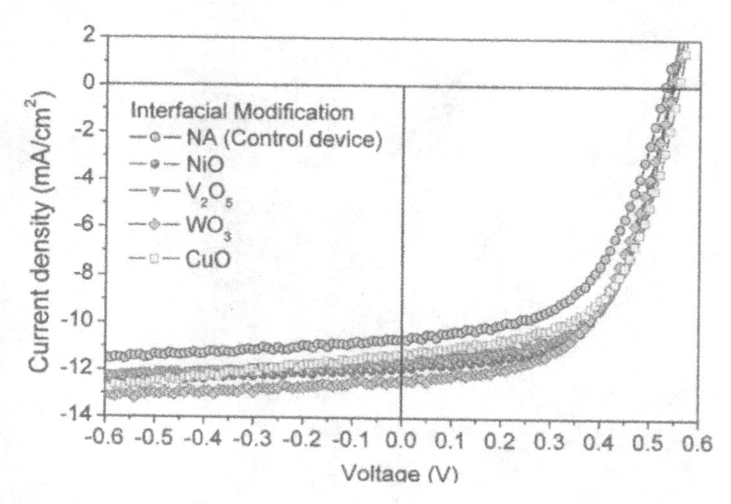

Figure 8.5. J–V characteristics of inverted organic solar cells with different metal oxides as anodic modification.

0.545 V, and *FF* of 55.46%. The use of CuO enhances the V_{OC} to 0.56 V, J_{SC} to 11.33 mA/cm², *FF* to 57.21%, and *PCE* to 3.62%. Therefore, the organic solar cells have 14%, 21%, 20%, and 17% enhancements in the PCE after inserting NiO, V_2O_5, WO_3, and CuO as the anodic modification, respectively. The device parameters are summarized in Table 8.2, which gives the device parameters with each of the above four metal oxides, which can all enhance the photovoltaic performance.

Table 8.2 shows that each metal oxide leads to different degree of improvement in the device performance (*FF*, J_{SC}, or V_{OC}), implying that different mechanisms might be involved. To explain those mechanisms, energy levels of those materials are shown in Figure 8.6. The valence band maximum (VBM) of NiO is 5.4 eV [16]. The VBM of V_2O_5 is 4.7 eV [17]. Both are close to the HOMO level P3HT, 4.8 eV [17]. Therefore, the holes may transport from the HOMO level of P3HT through the VBM of NiO or V_2O_5, then to Ag with work function of ~5 eV (due to its oxidation) [18]. The good match of the energy levels gives rise to a smooth transport of holes, leading to small se-

TABLE 8.2. Photovoltaic parameters and efficiencies of inverted organic solar cells with different metal oxides as anodic modification [14]

Oxide Modification	J_{SC} (mA/cm²)	V_{OC} (V)	*FF* (%)	*PCE* (%)	R_S (Ω-cm²)	R_{SH} (Ω-cm²)
NA (control)	10.67	0.53	54.82	3.1	4.35	524
NiO	11.92	0.54	58.41	3.76	1.78	629
V_2O_5	11.51	0.54	60.33	3.75	2.13	683
WO_3	12.34	0.545	55.46	3.73	1.33	604
CuO	11.33	0.56	57.21	3.62	2.35	421

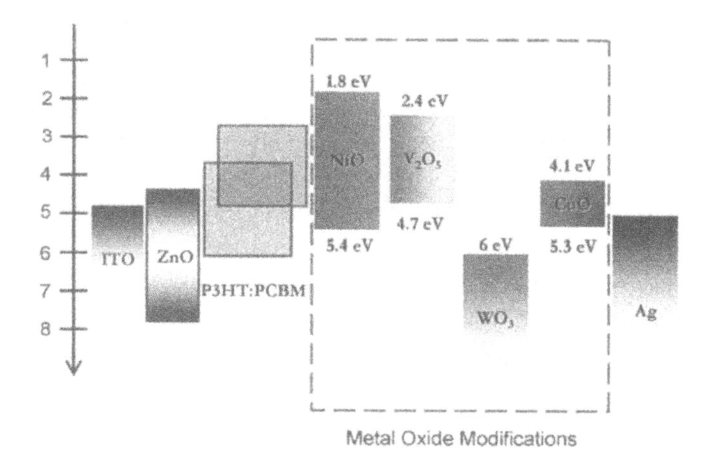

Figure 8.6. Energy level diagrams of the device components, including cathode, photoactive layer, anode, and metal oxides [14].

ries resistance (R_S), 1.78 Ω-cm^2 and 2.13 Ω-cm^2, respectively, for devices with NiO and V2O5. Both are smaller than the R_S of the control device (4.35 Ω-cm^2).

On the other hand, the conduction band minimum (CBM) of NiO is 1.8 eV [16]. The CBM of V_2O_5 is 2.4 eV [17]. They are higher than the LUMO level of PCBM, which is 3.7 eV [17]. Thus, electrons cannot transport from the PCBM to Ag. As a result, the reverse flow of electron is blocked, leading to the increase of shunt resistance (R_{SH}). The devices with NiO and V_2O_5 have R_{SH} of 629 Ω-cm^2 and 683 Ω-cm^2, respectively. Both are larger than the R_{SH} of the control device (524 Ω-cm^2). The increased R_{SH} indicates that the device has reduced leakage current and enhanced *FF* [19]. Therefore, the *FF*s of the devices with NiO and V_2O_5 are 58.41% and 60.33%, respectively, which are better than the *FF* (54.82%) of the control device.

The function of tungsten trioxide (WO_3) is different from the previous two metal oxides, NiO and V_2O_5. It is a naturally n-type semiconductor but possesses an efficient hole-injection property, so it has been used in organic light-emitting diodes (OLEDs) [20–23]. WO_3 has a work function of about 6 eV, as shown in Figure 8.6. Hence, holes can be transported from the HOMO level of P3HT to Ag through the work function of WO_3 and the R_S of the device with WO_3 is reduced to 1.33 Ω-cm^2.

For copper oxide (CuO), the band gap is only 1.2–1.5 eV [24–26]. Its ionization energy is 5.3 eV [27]. The VBM of CuO is 5.3 eV, which is close to the HOMO level of P3HT. Therefore, the mechanism of the hole transport from P3HT to Ag is similar to those of NiO and V_2O_5, from the HOMO level of P3HT through the VBM of CuO, then to Ag, so CuO is beneficial for hole transport and collection. Hence, the R_S of the devices with CuO is 2.35 Ω-cm^2, smaller than the R_S of the control device. However, the narrow band gap makes the CBM of CuO lower than the LUMO of PCBM. As a result, electron flow from PCBM to Ag is not forbidden, leading to leakage current. Thus, the corresponding R_{SH} is only 421 Ω-cm^2, even smaller than the R_{SH} of the control device.

The above results clearly show that the metal oxides are helpful for carrier transport and collection. The reason can be explained using their energy band levels. In addition, the metal oxide is inserted between the photoactive layer and the metal electrode. It may give rise to better match of spatial redistribution of the light intensity inside the photoactive layer if the metal oxide is uniformly distributed in this layer. It may also result in scattering of reflected light into the photoactive layer, effectively increasing the optical path if the metal oxide is not uniformly distributed in this layer.

8.2.3 Crystal Phase of Metal Oxides Used for Organic–Inorganic Hybrid Solar Cells

The layer of the metal oxide consists of many nano-size particles. Their quality is examined using X-ray diffraction (XRD) with Cu Kα radiation (λ = 1.54178 angstrom). The measured XRD spectra of NiO, V_2O_5, WO_3, and CuO are shown in Figure 8.7. The diffraction peaks in the 2θ range correspond to the crystalline structures of the cubic NiO, orthorhombic V_2O_5, monoclinic WO_3, and monoclinic CuO, respectively.

The crystal quality of the V_2O_5, WO_3, and CuO particles is further examined using the high-resolution transmission electron microscopy (HRTEM). The field-emission

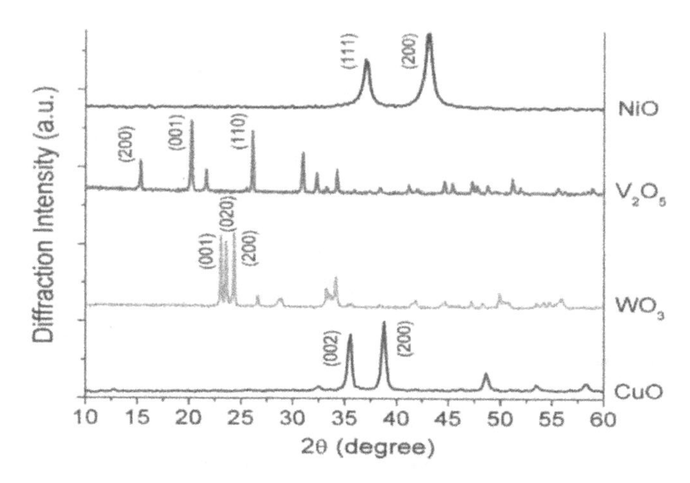

Figure 8.7. XRD spectra of NiO, V_2O_5, WO_3, and CuO used for organic–inorganic hybrid solar cells [14].

TEM (Philips Tecnai G2 F20 field-emission, gun-transmission microscope operated at 200 kV) is used for such investigation. The samples for HRTEM are prepared by dripping a few drops of the metal oxide solutions onto carbon-coated copper grids. The HRTEM images of the V_2O_5, WO_3, and CuO are shown in Figure 8.8. The HRTEM image of V_2O_5 shows a lattice spacing of 0.267 nm, corresponding to the (020) planes (JCPDS Card No. 41–1426). The HRTEM image of WO_3 exhibits a well-defined lattice fringe separation with 0.38 nm, corresponding to the (001) planes of a monoclinic WO_3 crystal (JCPDS Card No. 75–2072). The lower portion of Figure 8.8 shows the HRTEM image of CuO. The spacing between the clear lattice planes is 0.23 nm, which corresponds well to the (200) planes of monoclinic-phase CuO (JCPDS Card No. 05–0661). The crystal characteristics of HRTEM images agree with their XRD spectra. Both the XRD spectra and HRTEM images reveal that the nanosize particles of those metal oxides have high-quality crystal phases, so they function well for transporting holes.

8.3 EFFECT OF MIXED-OXIDE MODIFICATION ON ORGANIC–INORGANIC HYBRID SOLAR CELLS

In the previous section, an individual metal oxide deposited using the solution process was shown to enhance the performance of solar cells with organic–inorganic sandwiched structure. We may further speculate that the solar-cell performance may be further improved by suitable combination of different oxides that are complementary. In particular, the solution process provides the convenience of easily mixing various metal oxides with a desired ratio in the same solution. This cannot be easily achieved using

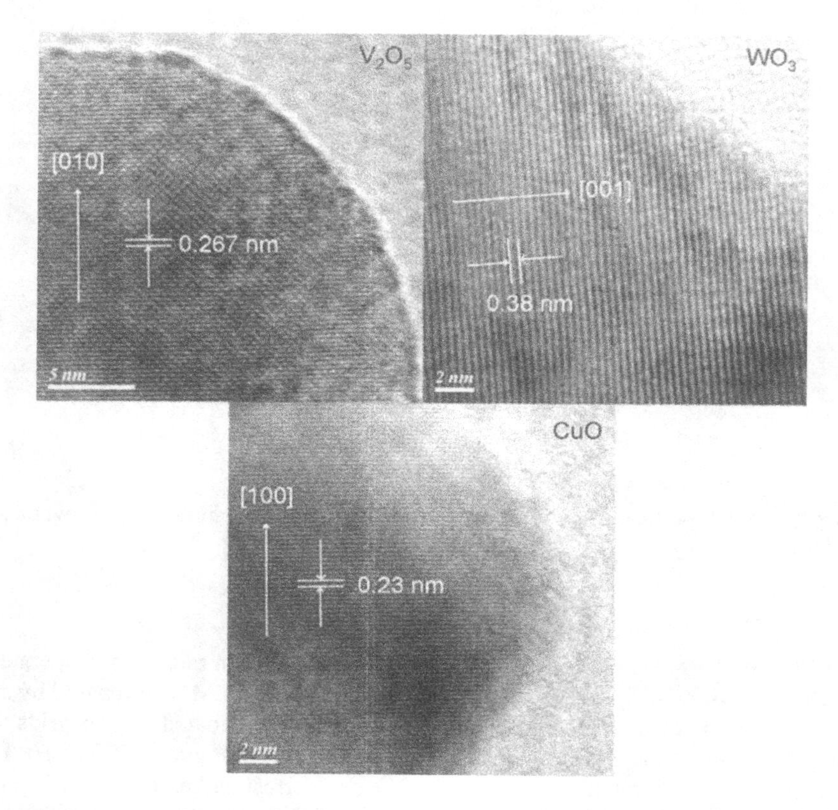

Figure 8.8. HRTEM images of V_2O_5, WO_3, and CuO used for organic–inorganic hybrid solar cells [14].

the coevaporation method because different metal oxides have different boiling points. Therefore, the effect of mixing different metal oxides, called mixed oxides, on the performance organic–inorganic hybrid solar cells is investigated in this section. WO_3 and V_2O_5 were chosen to form the mixed oxides [28]. The schematic diagram of the process for depositing the mixed oxides is shown in Figure 8.9. Two different photoactive layers, the blend of P3HT:PCBM and Plexcore PV2000, are used in the device structure of ITO/ZnO thin film/photoactive layer/mixed oxides/Ag. The Plexcore PV 2000 photoactive ink is a blend of OPV-grade regioregular P3HT and a proprietary n-type acceptor. It is obtained from Plextronics, Inc.

The mixed oxide ink is prepared with the following procedure. 1-mg V_2O_5 powder (Riedel-de Haën, 99%) and 1-mg WO_3 nanopowder (<100 nm, Aldrich, 99.9%) are homogeneously dispersed in 10-ml isopropanol by using ultrasonic agitation, so those metal oxides of particles are suspended in the solution. Field-emission transmission electron microscopy (TEM) (Philips Tecnai G2 F20 field-emission, gun-transmission microscope operated at 200 kV) is used to characterize the crystal quality. Similar to

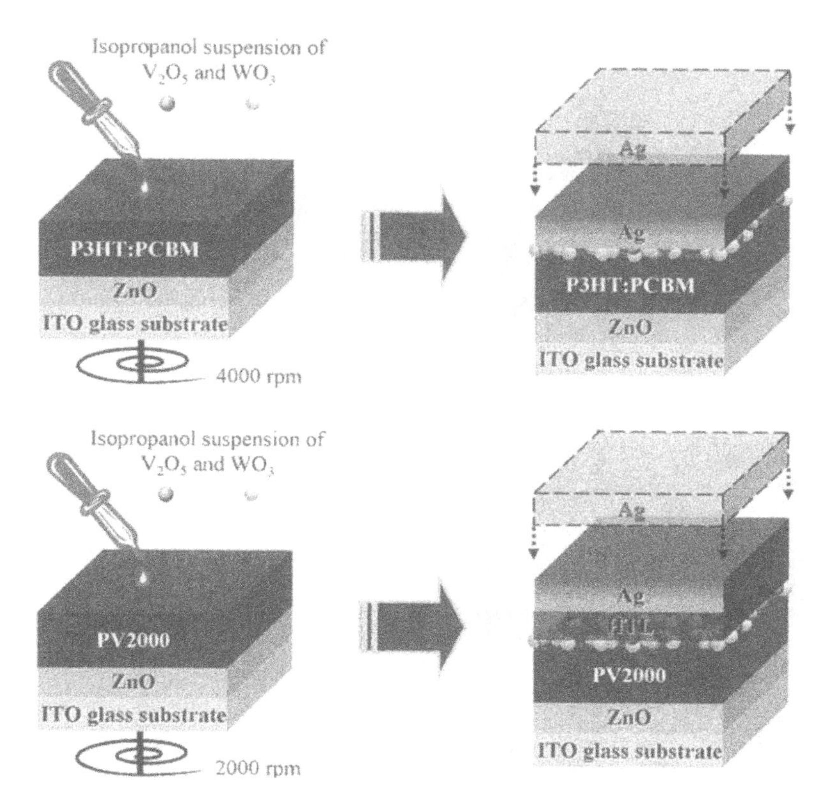

Figure 8.9. Schematic diagram of depositing the mixed oxides as well as the device structure of ITO/ZnO thin film/photoactive layer/mixed metal oxides/Ag, where the photoactive layer is PV2000 or the P3HT:PCBM system [14].

the previous investigation, the samples for TEM are prepared by dripping a few drops of the WO_3–V_2O_5 mixed-oxide solution onto a carbon-coated copper grid.

Both the bright-field TEM and high-resolution TEM (HRTEM) lattice imaging are applied to investigate the morphology and structural characteristics of the WO_3–V_2O_5 mixed oxide. A low-magnification TEM image of the mixed oxides is shown in Figure 8.10(a). Two protrusions in the upper-right portion of Figure 8.10(a) are further investigated using HRTEM. The upper one is identified as WO_3, whereas the lower one is V_2O_5. Figure 8.10(a) shows that the particles of WO_3 and V_2O_5 are very close spatially. Their tip-to-tip spacing is ~50 nm. This indicates that the WO_3 and V_2O_5 particles are well mixed in the solvent and in the film. Although the film of the mixed oxides on the active layer cannot be directly observed, we expect that they are well mixed, too. The HRTEM image taken from the tip of the protruding WO_3 particle is shown in Figure 8.10(b). A well-defined lattice fringe separation of 0.38 nm is observed. It corresponds to (001) planes of a monoclinic WO_3 crystal (JCPDS Card No. 75-2072).

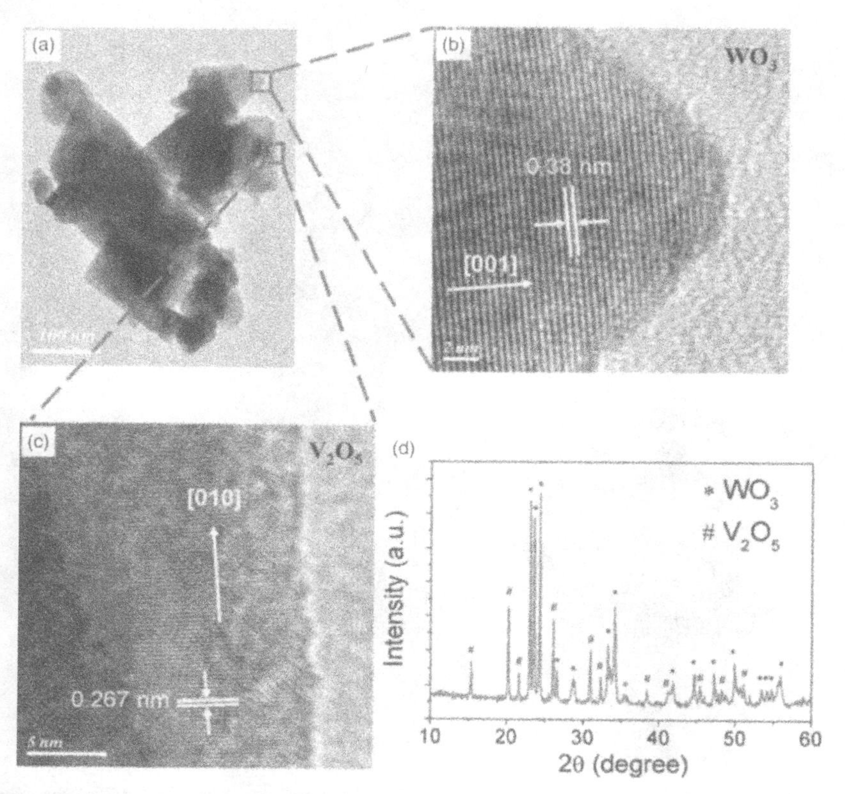

Figure 8.10. (a) TEM images of the WO_3–V_2O_5 mixed oxides. HRTEM images of (b) WO_3 and (c) V_2O_5 from the mixed oxides. (d) X-ray diffraction spectrum of the mixed oxides [14].

Shown in Figure 8.10(c) is the HRTEM image of the V_2O_5 particle. It has a single-crystalline nature with a lattice spacing of 0.267 nm, corresponding to the (020) planes of V_2O_5 (JCPDS Card No. 41–1426). X-ray diffraction (XRD) with Cu Kα radiation (λ = 1.54178 angstrom) is further used to analyze the crystal phase. The diffraction peaks in the XRD spectra well characterize the orthorhombic crystalline structure of V_2O_5 and monoclinic crystalline structure of WO_3. These observations show that the two oxides still maintain their individual characteristics although they are mixed. No chemical reaction occurs as they are placed in the same solvent.

8.3.1 Effect of Mixed Oxide on P3HT:PCBM–Inorganic Hybrid Solar Cells

The prepared mixed oxide ink is spin-cast onto the organic photoactive layer, which is a P3HT:PCBM blend, with the procedure schematically shown in Figure 8.9. Silver film is then deposited on the layer of the mixed oxides in a vacuum of 2×10^{-6} torr. After the devices are fabricated, their J–V curves are measured with a Keithley 2400 source measurement unit with dark condition and light-illuminated condition. Again, a solar simulator with an intensity of 100 mW/cm^2 and AM 1.5G filters is used for the characterization. The illumination intensity is calibrated by a NREL calibrated Si solar cell with KG-5 color filter. A shadow mask is also used to block the light outside of the active area to avoid the collection of extra current.

The dark J–V characteristics of the devices without the top oxide (control device), with the single oxide, and with mixed oxides are shown together in Figure 8.11 for easy comparison. Rectification ratio (RR), defined as the current ratio at ±1.5 V (forward and reverse bias, respectively) from the J–V curves measured in the dark condition, is usually used to characterize the diode property. For the control device, the RR is 4.69×10^2. As either WO_3 or V_2O_5 is used for the anode modification, the RR is increased to over 10 times. With the mixed oxides, the RR is further enhanced to 3.65×10^4, nearly two orders of magnitude improvement. It clearly shows that the mixed oxides suppress the leakage current more significantly than the individual oxide.

The J–V characteristics of these devices under AM 1.5G irradiation of 100 mW/cm^2 are shown in Figure 8.12. Again, devices of various anode modification are shown together for comparison. The J–V curves show that the performance of the control device is the worst. It exhibits J_{SC} of 10.67 mA/cm^2, V_{OC} of 0.53 V, and FF of 54.82%, resulting in a *PCE* of 3.1%. With the anode modification using single oxide, the device performance is improved, as discussed before. As the layer of the mixed ox-

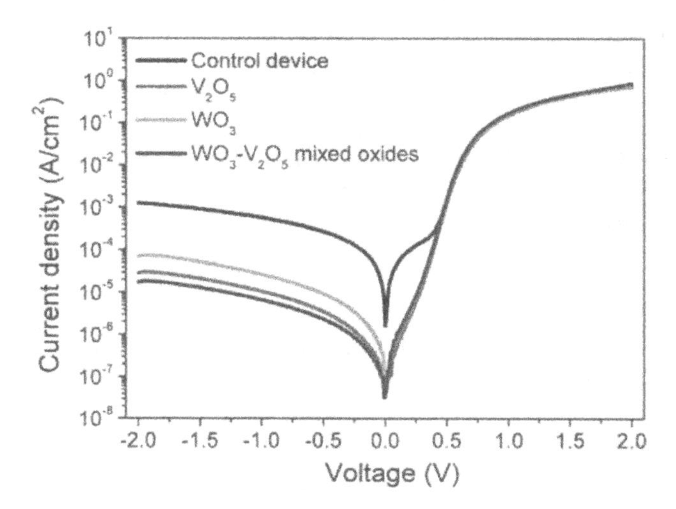

Figure 8.11. The J–V curves of the photovoltaic devices with the different oxide modifications measured in the dark [28].

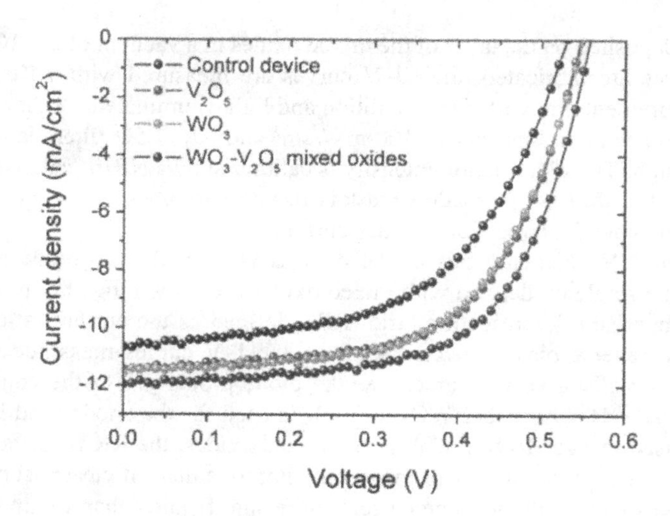

Figure 8.12. The J–V curves of the photovoltaic devices with the different oxide modification measured under 100 mW/cm^2 AM 1.5G irradiation [28].

ides is used, the device has the best performance. J_{SC} increases to 11.97 mA/cm^2, V_{OC} to 0.56 V, and FF to 62.06%. The PCE is also notably improved to 4.16%, a 34% enhancement compared to that of the control device.

The device parameters are shown in Table 8.3. The mixed oxides obviously improve the device performance better than the pristine oxides. The FF, R_{SH}, and R_S are all improved. In particular, R_{SH} significantly increases from 524 to 1122 Ω-cm^2 and R_S decreases from 4.35 to 1.19 Ω-cm^2, so FF is greatly enhanced from 54.82% to 62.06%. The high R_{SH} also indicates less leakage current across the cell. The decreased R_S indicates the good hole transport from P3HT to Ag, which may result from two paths, as shown in Figure 8.13. One is the path from the HOMO level of P3HT through the valence band of V_2O_5 (4.7 eV) and then to Ag. The other is through the work function of WO_3 (6 eV). The two paths are complementary, so the two oxides improve the device performance with different mechanisms and give rise to the lowest series resistance.

TABLE 8.3. Photovoltaic parameters and efficiencies of inverted OSCs with different metal oxides as anodic modification [28]

Anodic modification	J_{SC} (mA/cm)	V_{OC} (V)	FF (%)	PCE (%)	R_S (Ω-cm^2)	R_{SH} (Ω-cm^2)
NA (control)	10.67	0.53	54.82	3.1	4.35	524
V_2O_5	11.45	0.55	60.5	3.81	2.13	776
WO_3	11.57	0.55	58.77	3.74	1.93	716
Mixed oxides	11.97	0.56	62.06	4.16	1.19	1122

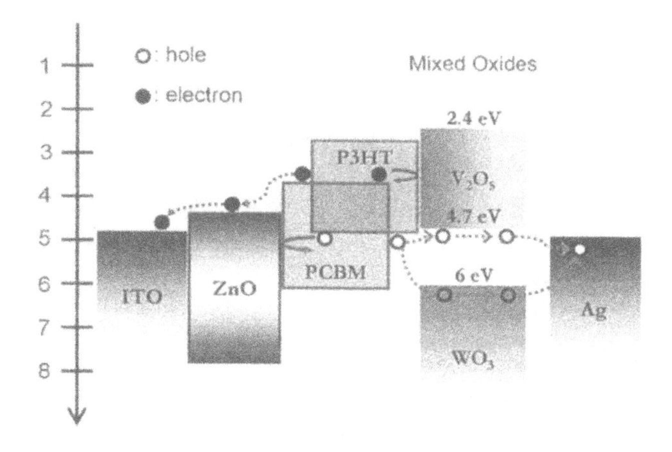

Figure 8.13. Energy level diagram of the device with mixed oxides [14].

8.3.2 Effect of Mixed Oxides on PV2000–Inorganic Hybrid Solar Cells

The effect of mixed oxides had also been investigated in another blend—Plexcore PV2000. The Plexcore PV 2000 photoactive ink is a blend of OPV-grade regioregular P3HT and a proprietary n-type acceptor, bis-indene[C60]. Because the bis-indene[C60] has a higher LUMO level than that of the PCBM, it could give the solar cells a larger open-circuit voltage.

As shown in Figure 8.9, the fabrication procedure is similar to that for devices using P3HT:PCBM blend. Here the PV2000 is dried slowly over a period of 40 minutes in air. The thickness of the photoactive layer is about 250 nm. The prepared ink of mixed oxides is spin-cast onto the photoactive layer at a spin rate of 2000 rpm. Different from the previous devices with P3HT:PCBM blend, here another hole transport layer (HTL), also provided by Plextronics, Inc. is spin-cast on the layer of metal oxides at a spin rate of 4000 rpm. Then silver electrode is deposited on top of the HTL layer in a vacuum of 2×10^{-6} torr. Finally, the devices are annealed at 160° in an N_2 filled glove box for 10 min. The measurement conditions are the same as before.

For comparison, two other types of solar cells using the Plexcore PV2000 as the photoactive material were also fabricated, as schematically shown in Figure 8.14. Device A denotes the control device that has neither metal oxides nor the HTL layer, so its layer structure is ITO/ZnO/photoactive layer/Ag. Device B has the HTL layer on PV2000, but no metal oxides, so the layer structure is ITO/ZnO/photoactive layer/HTL/Ag. Device C is the one described in the previous paragraph. The layer structure is ITO/ZnO/photoactive layer/mixed oxides/HTL/Ag.

The J–V curves for the above three types of devices measured under AM 1.5G irradiation of 100 mW/cm^2 are shown in Figure 8.15. Obviously, Device C has the best performance. It has J_{SC} of 11.23 mA/cm^2, V_{OC} of 0.77 V, FF of 59.33%, and PCE of 5.13%. In comparison, Device A, which has neither mixed oxides nor HTL,

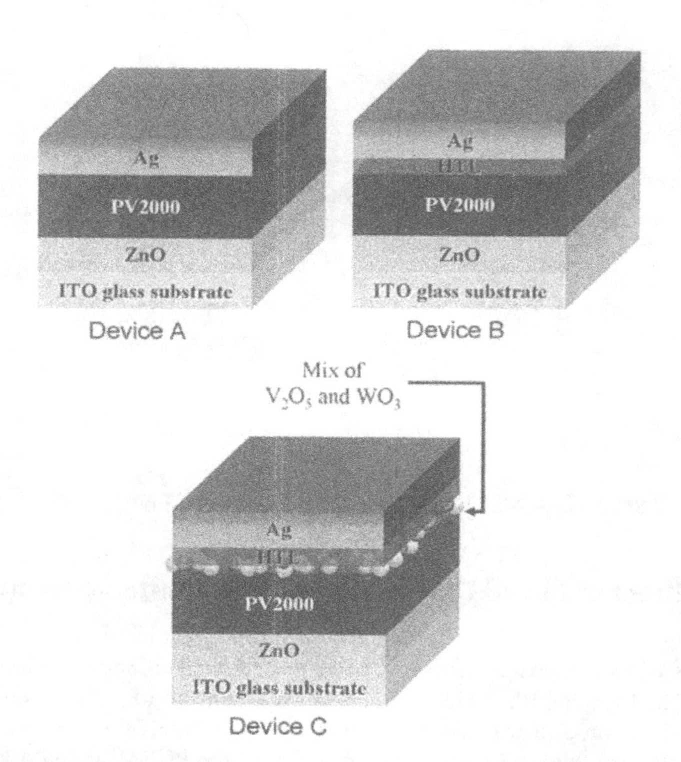

Figure 8.14. Layer structures of three different types of devices. Device A: ITO/ZnO/PV2000/Ag; Device B: ITO/ZnO/PV2000/HTL/Ag; Device C: ITO/ZnO/PV2000/ mixed oxides/HTL/Ag [14].

performs the worst. It has J_{SC} of 9.86 mA/cm^2, V_{OC} of 0.68 V, FF of 47.28%, and PCE of 3.17%. Device B, which has just HTL but no mixed oxides, has performance between Device A and Device C. It has J_{SC} of 10.24 mA/cm^2, V_{OC} of 0.67 V, FF of 50.28%, and PCE of 3.45%. Compared to Device A and Device B, Device C has 62% and 49% enhancement in PCE, respectively. This shows that the mixed oxides do help. Table 8.4 shows the device parameters. V_{OC} from the Plexcore PV2000 devices is larger than that from the standard P3HT:PCBM system because the acceptor of PV2000 is bis-indene[C60]. It has a higher LUMO level than that of the PCBM. The increased V_{OC} makes the PCE much larger than that with the standard P3HT:PCBM system.

Table 8.4 also shows that R_S and R_{SH} are improved as the mixed oxides are used. Device C has R_S of 3.1 Ω-cm^2, which is four times lower than those of Devices A and B. Its R_{SH} of 636 Ω-cm^2 is 1.5 times and three times larger than that of Device B and Device A, respectively. The increased R_{SH} is an evidence of suppression of the leakage current across the cell due to the mixed oxides. It also further contributes to the increased V_{OC} and the improved FF [19]. Further evidence for the suppression in the

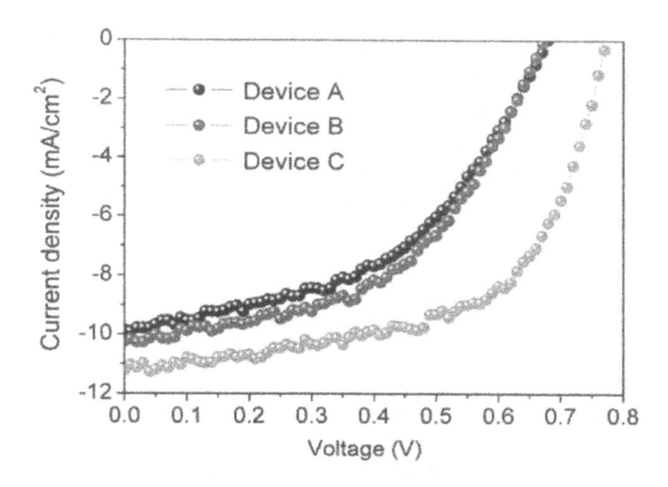

Figure 8.15. J–V characteristics under 100 mW/cm^2 AM 1.5G irradiation of three devices [14].

leakage current is the rectification ratio (*RR*), which was also measured. The *RR* of 7.69×10^4 from Device C is two orders of magnitude larger than these of Device A (2.65×10^2) and Device B (5.16×10^2). The mixed oxides in the sandwiched structure greatly improve R_S, R_{SH}, and *RR*, and so the device performance.

8.3.3 Enhancement of Optical Absorption and Incident Photon-to-Electron Conversion Efficiency

As shown in Table 8.4, the J_{SC} of Device C is 11.23 mA/cm^2, which is much larger than 9.86 mA/cm^2 of Device A, indicating that the combination of mixed oxides and HTL greatly enhances short-circuit current. To further understand the role of this combination in the sandwiched structure, incident photon-to-electron conversion efficiency (IPCE) is measured. The IPCE spectra of Devices A and C are shown in Figure 8.16. Over the entire spectral range, Device C has a higher IPCE than Device A. In particular, the maximum value at 520 nm for Device C and Device A is 65% and 55%, respectively, indicating an 18% enhancement for Device C. The higher IPCE for Device C agrees with the increased short-circuit current. The absorption

TABLE 8.4. Photovoltaic parameters and efficiencies of the three OSCs [14]

Devices	PCE (%)	V_{OC} (V)	J_{SC} (mA/cm^2)	FF (%)	R_S (Ω-cm^2)	R_{SH} (Ω-cm^2)
A	3.17	0.68	9.86	47.28	12.43	256
B	3.45	0.67	10.24	50.28	14.76	429
C	5.13	0.77	11.23	60	3.12	636

spectra of the two devices are also shown in the inset of Figure 8.16. A substantial absorption enhancement is observed in Device C. Two possible reasons account for this. One is that the metal oxides possibly give rise to scattering of reflected light back to photoactive layer and, hence, effectively increase the light path in the photoactive layer. The other reason is that the mixed oxides and HTL together may induce a better match between the spatial distribution of the light intensity and the photoactive organic layer [29].

The internal quantum efficiency (IQE) is also plotted in Figure 8.17 for Devices A and C. The IQE spectrum can be derived from the total absorption spectrum and the IPCE spectrum. As shown in Figure 8.17, the IQE of Device C is larger than that of Device A for the entire spectral range. Also, the IQE of Device C remains near 90% over the absorption spectrum, 420–700 nm. The high IQE of Device C means that almost every absorbed photon gives rise to a separated electron–hole pair. In addition, the separated carriers are collected at the electrodes. In comparison, Device A does not have a high IQE. The advantage of using both mixed oxides and HTL is proved from the above data.

The same advantage of mixed oxides is also demonstrated in the P3HT:PCBM–inorganic hybrid solar cells. Figure 8.18 shows the IPCE spectra of the devices with and without mixed oxides. The control device has the structure of ITO/ZnO thin film/P3HT:PCBM /Ag. The other one is with mixed oxides and has the structure of ITO/ZnO thin film/P3HT:PCBM/mixed oxides/Ag. Again, the device with mixed oxides has a higher IPCE than the control device over the entire spectral range. The max-

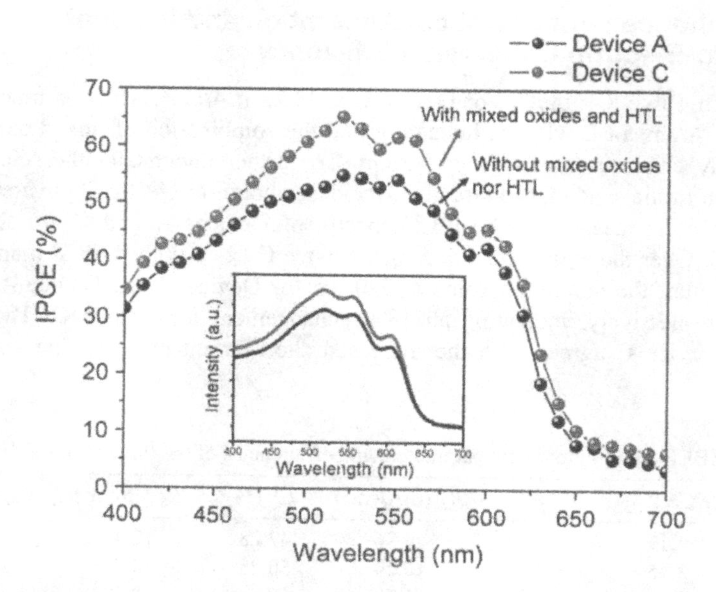

Figure 8.16. IPCE spectra of Device A and C. Inset: absorption spectra of the both devices [14].

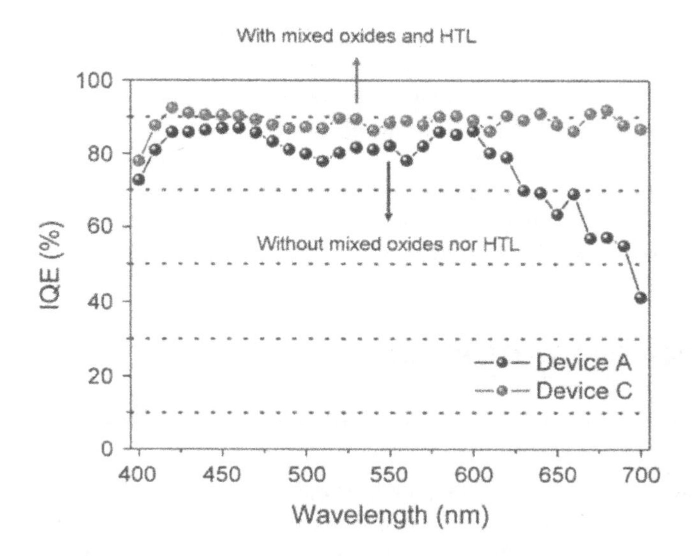

Figure 8.17. IQE spectra of Devices A and C [14].

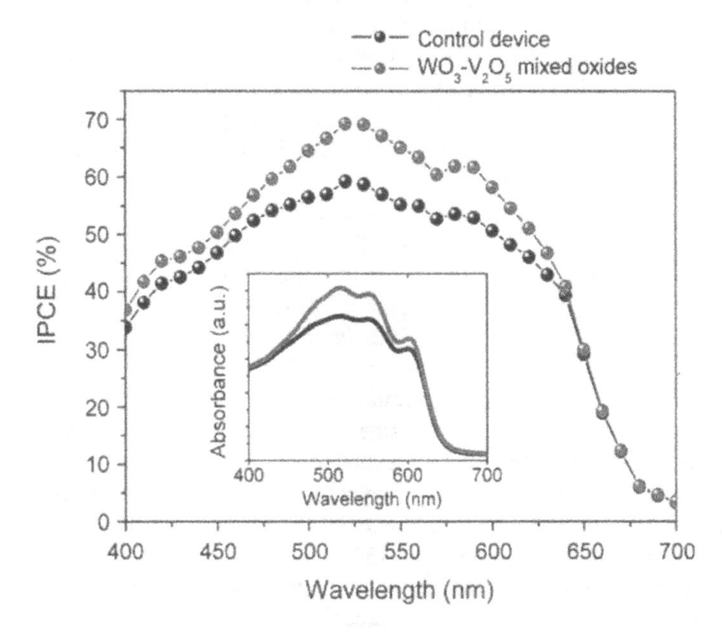

Figure 8.18. IPCE spectra for the devices with and without mixed oxides. Inset: absorption spectra of both devices [28].

imum value also occurs at 520 nm because both have P3HT as the light absorption material. The maximum IPCE for the device with mixed oxides is 69%, whereas the control device has a maximum IPCE of 59%. The absorption spectra of the two devices are also shown in the inset of Figure 8.18. The absorption is substantially enhanced for the device with mixed oxides due to the same reasons. One is the metal-oxide induced scattering of reflected light back to photoactive layer. Another is a better match between the spatial distribution of the light intensity and the photoactive organic layer [29] due to the insertion of the oxide layer.

8.4 IMPROVEMENT OF STABILITY

8.4.1 Improvement of Stability Using Mixed Oxides of WO_3 and V_2O_5

Because conventional organic solar cells (OSCs) have bad stability, many people doubt their practical applications in electricity generation. The reason for bad stability is mainly due to the use of PEDOT:PSS and the Al electrode, so most conventional OSCs have their efficiency reduced to nearly zero only a few days after fabrication. However, the sandwiched structure with two oxides protecting the organic layer is expected to greatly improve stability, as we discussed in the beginning of this chapter. Here, we will discuss the details of device performance in terms of stability. The reliability test has been performed on three kinds of devices, all with sandwiched structures: the P3HT:PCBM–inorganic hybrid solar cells using mixed oxides of WO_3 and V_2O_5 as the protection layer, the PV2000–inorganic hybrid solar cells using mixed oxides of WO_3 and V_2O_5 as the protection layer, and P3HT:PCBM–inorganic hybrid solar cells using CuO_x as the protection layer.

The P3HT:PCBM–inorganic hybrid solar cells fabricated on the ITO glass are periodically tested over 1200 hours to gauge the device stability. Between the measurements, the devices are placed under ambient conditions without encapsulation. The PCE is normalized with its maximum value for easy comparison of the variation. Figure 8.19 shows the normalized PCEs of the devices with and without the mixed oxide interlayer as a function of time. The maximum *PCE* usually occurs at around 3-5 days after the fabrication for the inverted structure of OSCs [30,31] and conventional OSCs with NiO interlayer [16]. Two mechanisms account for such phenomena. One is the morphological evolution of the donor/acceptor blend over the time. Tsai and coworkers found that the photoactive organic layer will evolve and the grain size of the donor/acceptor domain increase over the first 120 hours. The time scale is in good agreement with the device performance [32]. The morphological evolution involves phase separation and self-organization, which approaches the stable condition when P3HT and PCBM have a large interfacial area. The situation will result in efficient carrier separation and attain a crystalline order that improves charge transport [33,34]. The other is the oxidation of Ag. The surface of silver gradually oxidizes in air. Its work function increases from 4.3 to 5.0 eV [18], thus increasing the built-in electro-

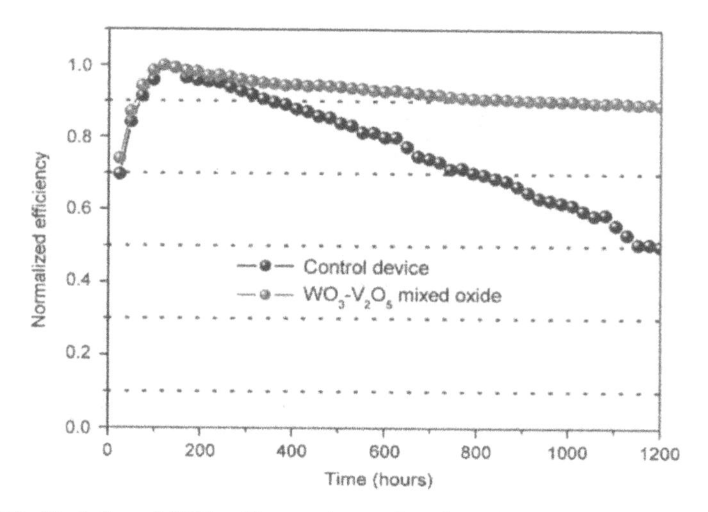

Figure 8.19. Variation of PCEs of inverted organic solar cells with and without mixed oxides over 1200 hour exposure to air [28].

chemical potential asymmetry and V_{OC}. In addition, the barrier for hole transport to Ag is reduced, so J_{SC} increases.

It takes about five days for the above two mechanisms to reach stable conditions, so *PCE* increases over the first 120 hours after fabrication, as shown in Figure 8.19. The J–V curves in Figure 8.12 and Table 8.3 are taken after 120 hours of fabrication, when the devices have the highest *PCE*. The control device without the oxide interlayer reaches a maximum *PCE* of ~3.1%. The *PCE* then gradually reduces to ~1.6% after 1200 hours. It has a ~50% degradation. In contrast, the device with the WO_3–V_2O_5 mixed oxide interlayer has its *PCE* reaching a maximum of ~4.16%. The *PCE* also gradually decreases, but at a much lower rate. It declines to ~3.75% after 1200 hours, only 10% degradation. The decrease of *PCE* is mainly due to the decrease of J_{SC} because both the *FF* and V_{OC} remain relatively constant over the period of 1200 hours. Therefore, the mixed oxides do give rise to improved stability.

The stability for PV2000 systems with mixed oxides and HTL in PV2000 systems is also examined. The J–V curves are periodically measured over 500 hours to check the device stability. Between the measurements, the devices are also placed under ambient conditions without encapsulation. Device A and Device C are those that are discussed in Section 8.3.2 and Section 8.3.3. Figure 8.20 shows the *PCE*s of Device A and Device C as a function of time. As shown in Figure 8.20, the devices with PV2000 have their *PCE* increase for the first 48 hours after fabrication. The shorter time to reach maximum *PCE* is probably due to the annealing process at 160°C. The annealing process can accelerate the silver oxidation and the phase separation of the photoactive layer.

After 48 hours, the *PCE* of both devices starts to decrease. For Device C, the *PCE* declines from 5.13% to ~3.94% after 500 hours of fabrication, with 23% degradation.

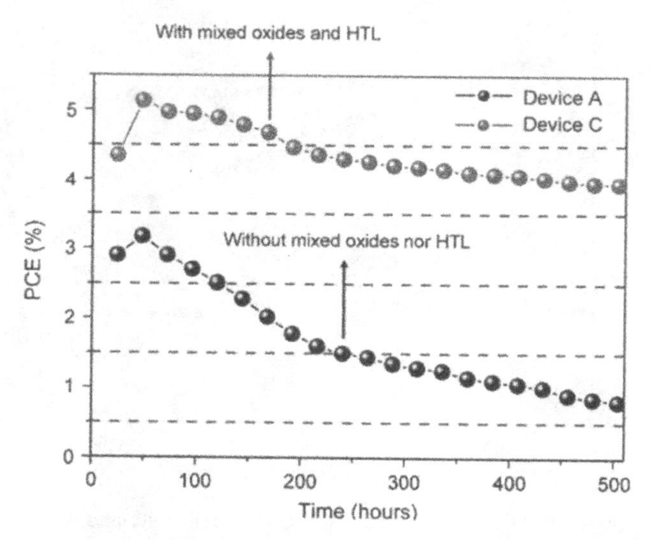

Figure 8.20. Variation of PCEs of Devices A and C over 500 hour exposure to air [14].

The declined *PCE* is attributed to the decrease of J_{SC} and *FF*. In contrast, Device A has its *PCE* degrade for more than 70% over the same period of time. This investigation shows again that the metal oxides can act as a barrier to prevent oxygen or water from degrading the photoactive layer. Because the oxides are relatively insensitive to water and stable in air, they are able to give good protection to the organic layer and improve the stability of devices with the sandwiched structures.

8.4.2 Improvement of Stability Using Sol–Gel Processed CuO_x

The above results show that the mixed oxides of WO_3 and V_2O_5 are able to significantly improve the stability of the devices. The other oxide, CuO_x, has also been investigated and proved to improve the stability of P3HT:PCBM–inorganic hybrid solar cells. Here the CuO_x is deposited from a sol–gel solution. As mentioned before, using solution-processed metal oxide as an anode interlayer is challenging because of the hydrophobic property of the organic active layer [35]. Therefore, the selection of solvent for the sol–gel is an important issue. 2-methoxyethanol (2MOE) is usually used as the solvent in sol–gel methods [36]. However, to be compatible with the photoactive organic layer, iso-propanol (IPA) instead of 2MOE is used as the solvent because it is orthogonal to dichloromethane (DCB), the solvent of the P3HT/PCBM blend [37]. After the photoactive organic layer is deposited, the CuO_x layer is spin-coated from a 0.25 mol solution of copper (II) acetate monohydrate in IPA mixed with monoethanolamine and deionized water. For comparison, another sample without the CuO_x layer is also prepared.

The sol–gel derived thin film of CuO_x is examined using scanning electron microscopy (SEM) and Fourier transform infrared spectroscopy (FTIR). Figure 8.21

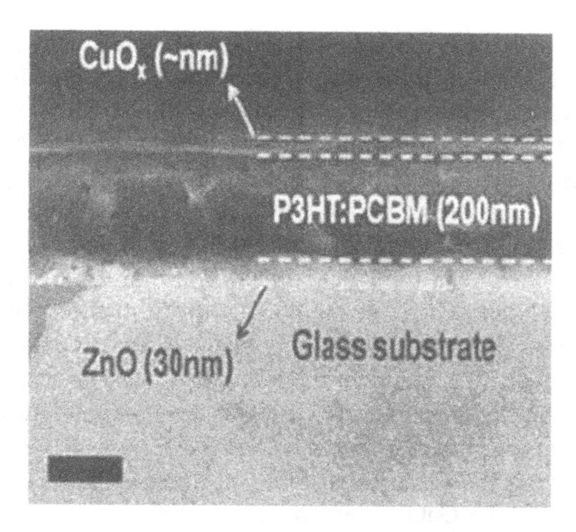

Figure 8.21. SEM cross-sectional image of the glass/ZnO/P3HT:PC61BM/CuO_x [37] (scar bar: 100 nm).

shows the SEM cross-sectional image of the glass/ZnO/P3HT:$PC_{61}BM$/CuO_x. The thicknesses of the ZnO film and active layer are about 30 nm and 200 nm, respectively. The sol–gel-derived CuO_x thin film (~1 nm) can cover the active organic layer well to prevent the organic layer from interaction with oxygen and moisture.

The FTIR spectra of the polymer thin films with and without CuO_x prepared and placed in air for 9 days are shown in Figure 8.22(a) and 8.22(b). In Figure 8.22(a), the C=O peak at 1740 cm^{-1} [38] of the sample with oxide layer appears after 3 days and then becomes stable, revealing that the CuO_x can protect the polymer layer from oxygen. In Figure 8.22(b), the C=O peak at 1740 cm^{-1} of the sample without oxide layer is found to

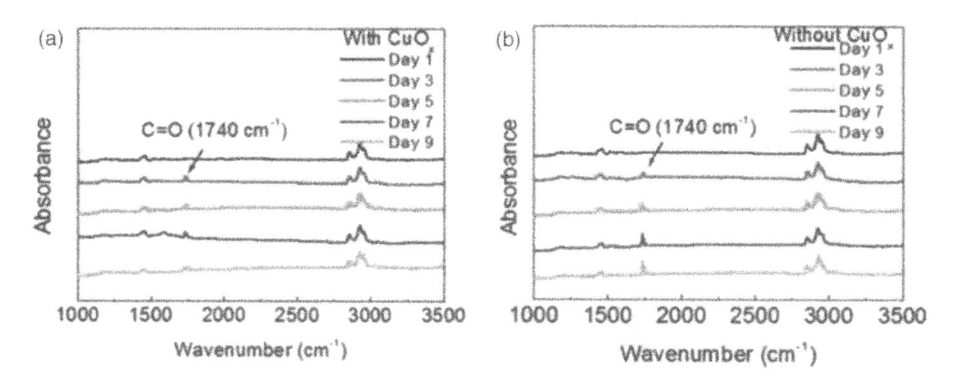

Figure 8.22. (a)The FTIR spectrum of the polymer thin film with CuO_x in the air for 9 days. (b) The FTIR spectrum of the polymer thin film without CuO_x in the air for 9 days [37].

gradually increase with time, indicating that the thin film is continually oxidized. It implies that the CuO_x can protect the photoactive organic layer from interaction with oxygen.

The thin-film composition is identified from the comparison of the FTIR spectra with and without CuO_x, as shown in Figure 8.23(a). The CuO_x film consists of Cu_2O (630 cm^{-1}, 606 cm^{-1}, and 588 cm^{-1}) [39,40], Cu_2S (617 cm^{-1}) [32], and CuO_2 (646 cm^{-1}) [41]. The CuO_x may react with the S atom of the P3HT at the interface, resulting in a peak at 617 cm^{-1} [32]. The weak signals at 646 cm^{-1} and 617 cm^{-1} indicate that the CuO_2 and Cu_2S contained in the CuO_x thin film are both rare. To further clarify the containment of CuO_2 and Cu_2O in the CuO_x thin film, the spectra are simulated by Gauss functions with 588 cm^{-1}, 630 cm^{-1}, and 646 cm^{-1} peaks, as shown in Figure 8.23(b). By calculating the area under the individual peak, we can roughly estimate the containment of a variety structures [42]. Here, the containment of CuO_2 in CuO_x is about 1%, which is much smaller than the Cu_2O.

Figure 8.24 shows the variation of PCEs of unencapsulated inverted organic solar cells with and without the CuO_x interlayer over 1000 hours in air. The device without CuO_x reaches a maximum PCE of ~3.1% and then decreases to ~1.5% after 1000 hours, a ~51% degradation. In comparison, the PCE of the device with the CuO_x interlayer reaches a maximum of ~4.0% and then declines to ~3.5% after 1000 hours, only a 13% degradation. The improved stability is mainly attributed to the CuO_x interlayer, similar to the protection provided by the mixed oxides of WO_3 and V_2O_5.

8.5 ORGANIC-NANOSTRUCTURED–INORGANIC HYBRID SOLAR CELLS

In Chapter 7, we had discussed the necessity of forming nanomorphology in the P3HT:PCBM blend because the exciton diffusion length of polymers is very short, typically less than 10 nm. The short diffusion length causes the electron and the hole to

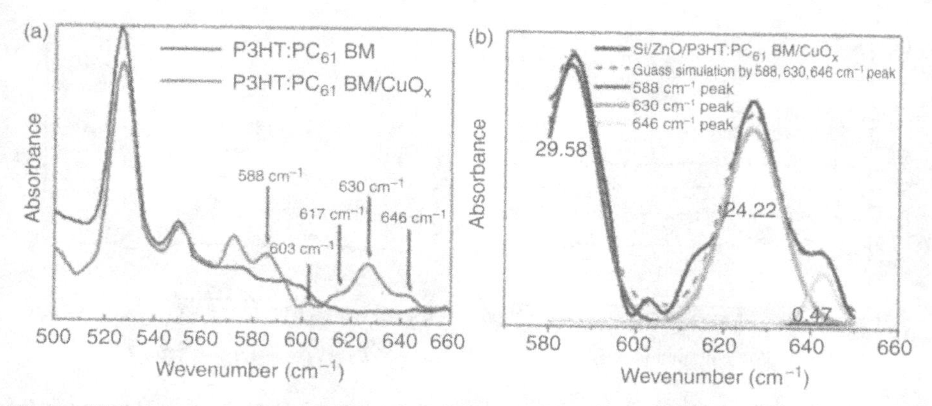

Figure 8.23. (a) The FTIR spectrum of the polymer thin film with and without CuO_x. (b) The FTIR spectra of the CuO_x thin film and the spectra simulated by variety peaks [37].

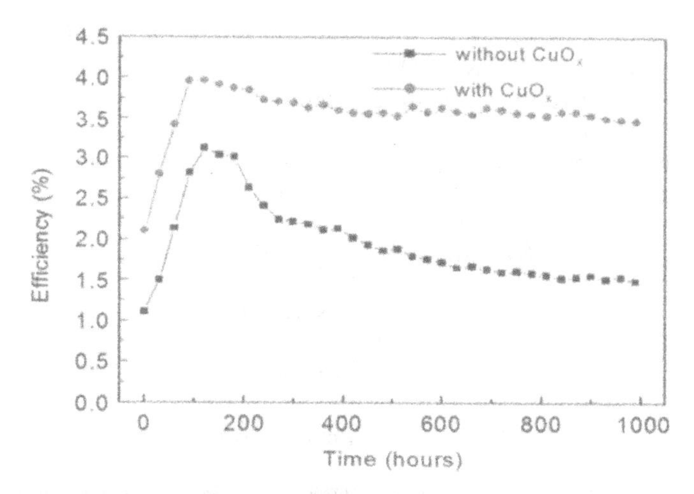

Figure 8.24. Variation of PCEs of inverted organic solar cells with and without the CuO_x interlayer over 1000 hour exposure to air [37].

easily recombine before they reach the junction. Currently, the most successful way to form the nanomorphology is through the self-organization of organic polymers under certain conditions. However, this process is not well controlled, so quite a few researchers try to accomplish the nanomorphology through nanostructures of inorganic semiconductors such as ZnO [31,43], TiO_2 [44,45], and CdSe nanorods [46]. Here we will describe the two most common ways of using nanostructured inorganic semiconductors for organic–inorganic hybrid solar cells. One is to incorporate ZnO nanorods and the other is to use TiO_2 nanorods.

8.5.1 Organic–ZnO Nanorod Hybrid Solar Cells

The fabrication process is similar to previous organic–inorganic hybrid solar cells except that ZnO nanorods are grown on the ZnO thin film before the deposition of the photoactive organic layer. The procedure is described briefly as follows:

1. Clean the ITO-coated glass.
2. Deposit ZnO thin film as the seed layer using the sol–gel method, as described previously.
3. Grow ZnO nanorods on the ZnO seed layer.
4. Spin-coat the photoactive organic layer on the ZnO nanords.
5. Deposit the metal electrode of Ag.

8.5.1.1 Growth of ZnO Nanorods. Because the growth of ZnO nanorods was not described previously, here we will elaborate a little bit more. The growth occurs in a solution that consists of a mixture of zinc nitrate hexahydrate ($Zn(NO_3)_2 6H_2O$) and hexamethylenetetramine ($C_6H_{12}N_4$, HMT) dissolved in deionized water. Zinc nitrate is a water-soluble salt. It is dissociated in the form of Zn^{2+} ions in water. At a growth temperature of around 90°C, HMT is thermally decomposed to release hydroxyl ions (OH^-), which further react with Zn^{2+} ions to form ZnO [47]. The above chemical interaction is summarized in the following equations:

$$(CH_2)_6N_4 + 6H_2O \longleftrightarrow 6HCHO + 4NH_3 \tag{8-1}$$

$$NH_3 + H_2O \longleftrightarrow NH_4{}^+ + OH^- \tag{8-2}$$

$$2OH^- + Zn^{2+} \longleftrightarrow Zn(OH)_2 \tag{8-3}$$

$$Zn(OH)_2 \xrightarrow{\Delta} ZnO(s) + H_2O \tag{8-4}$$

Equation (8-1) describes the reaction of HMT and water to form aldehyde (HCHO) and ammonia (NH_3). Then the NH_3 dissociates in water and releases NH^{4+} ions and the desired OH^- ions. In addition, the HMT also acts as a kinetic pH buffer to regulate the pH value of the solution because the rate of its hydrolysis decreases with increasing pH and vice versa [48]. As the concentration of these Zn^{2+} and OH^- ions exceeds a critical value, the ZnO nuclei start to precipitate, as expressed in Eq. (8-3). Later on, the zinc–hydroxyl complexes [$Zn(OH)_2$] transform into the solid phases and Zn—O—Zn bonds are formed as a result of the dehydration reaction [43]. The ZnO crystal is gradually constructed by the continuous dehydration reaction between the OH^- ions on the surface of the growing crystals and the OH^- ligands of the zinc–hydroxyl complexes.

The crystal habits of ZnO make the nuclei form a hexagonal shape, as shown in Figure 8.25. The ZnO has polar and nonpolar faces in the wurtzite hexagonal phase, which is the typical crystal structure formed using the hydrothermal method. In the polar face, which is along the c-axis, the zinc atom and the oxygen atom are arranged alternately and the top surface is usually Zn-terminated (0001) and catalytically active,

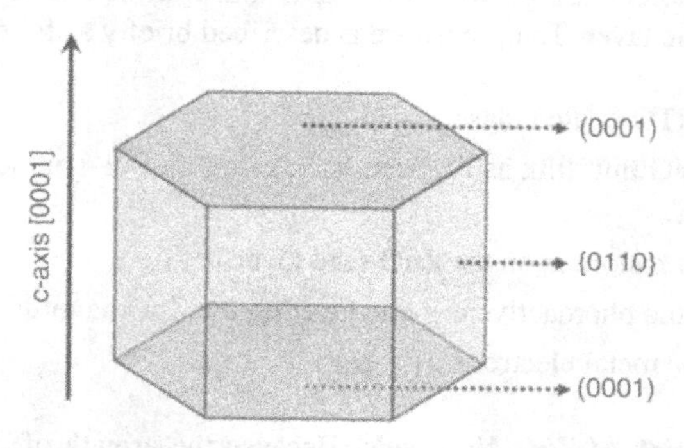

Figure 8.25. Schematic illustration of hexagonal wurtzite structure of ZnO crystal [14].

whereas the bottom surface is O-terminated $(000\bar{1})$ and chemically inert. The polar face with surface dipoles is thermodynamically unstable, compared to nonpolar faces. Consequently, the polar face often undergoes rearrangement to minimize the surface energy and tends to grow rapidly. Therefore, the [0001] direction usually has the largest growth speed under hydrothermal conditions [50] and the nanorods are grown mainly along the c-axis direction. The nanorods exhibit polar Zn-terminated (0001) top and O-terminated $(000\bar{1})$ bottom surfaces and are bounded with the six crystallographic nonpolar $\{01\bar{1}0\}$ planes [47]. The freedom of motion for those ions in the solution makes the hydrothermally synthesized nanorods fully consistent with the typical growth habit of the ZnO crystals.

The grown ZnO rods usually have different rod sizes, depending on the growth conditions and the seed-layer crystals. For applications in organic solar cells, the rod size and the rod spacing should be as small as possible. The minimum achievable rod size is about 30 nm. A typical top view of an SEM image is shown in Figure 8.26. The grown length is between 120 nm and 140 nm to match the thickness of the photoactive organic layer to be deposited. The density is as high as 767.9 rods/μm^2. The rod size is between 30 nm and 40 nm.

8.5.1.2 Influence of Drying Time. Afterward, the photoactive organic blend of P3HT and PCBM is spin-coated on top of the ZnO nanorods. However, the organic blend cannot easily infiltrate into the hollow spaces of the ZnO nanords, making the device performance very poor. Several techniques can be used to overcome this difficulty, including slow drying by either reducing the spinning rate or adding a high-boiling-point solvent like 1-chloronaphthalene (Cl-nap) or 1-bromonaphthalene (Br-

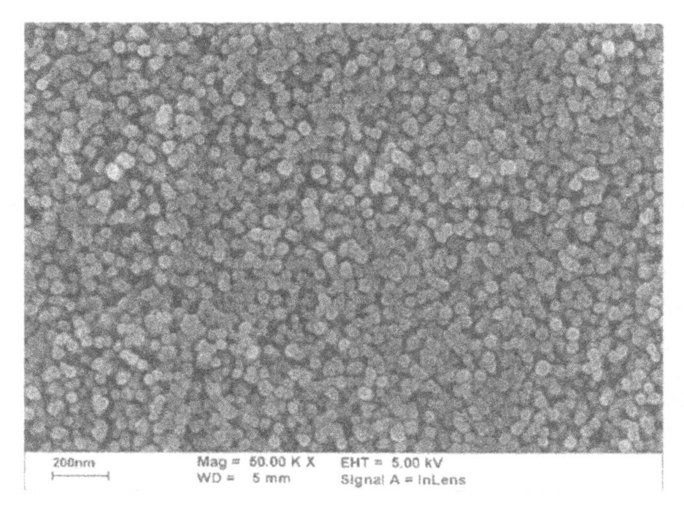

Figure 8.26. A top view of the grown ZnO nanorods with most of them orienting upward [14].

naph), and adding PCBM clusters between the ZnO nanorods and the photoactive organic layer.

Adding a solvent with a high boiling point could enable the polymers to have sufficient time to self-organize. Atomic-force microscopy (AFM) is used to investigate the solvent influence on the surface morphology [51]. Figure 8.27(a) and 8.27(b) show the AFM images of the surface of the P3HT/PCBM film on the ZnO nanorods, with and without high-boiling-point solvent, respectively. In (a), the surface is relatively smooth, with a root-mean-square (rms) roughness of 10.6 nm. In (b), the added high-boiling-point solvent gives the polymers more time to self-organize and form hill-like structures. The rms roughness increases to 13.0 nm, indicating the self-organization and crystallization of polymers. The PCE also increases from 2.10% to 3.39% as the drying time changes from 22 minutes to 46 minutes by adding 3% of Br-naph to the DCB solvent.

Although the addition of high-boiling-point solvent could increase the drying time for improved self-organization, such additional solvent may create undesired recombination centers. Thus, it is also desirable to increase drying time without additions. Fortunately, this can also be achieved by slowing down the spin rate [51]. Figure 8.28 shows the variation of drying time with spin rate. As the spin rate is changed from 1000 rpm to 400 rpm, the drying time increases from 5 minutes to 56 minutes. Meanwhile, the film thickness also increases from 240 nm to 400 nm. In conventional organic solar cells, the increase of organic-film thickness usually leads to a decrease in carrier transport. However, with the help of ZnO nanorods, the thick film of organic layer does not decrease the carrier transport because electrons can easily reach ZnO nanorods.

The slow spin rate has two advantages. First, the polymers have more time to self-organize, as described previously. Second, the film thickness increases, so the optical absorption also increases. Therefore, as the spin rate changes from 400 rpm to 1000 rpm, the *PCE* of the devices increases from 1.58% to 3.58%. The *FF* also increases from 39% to 58% and the short-circuit current density increases from 8.4 mA/cm^2 to 11.7 mA/cm^2. The improvement is very significant. Figure 8.29 shows the absorption

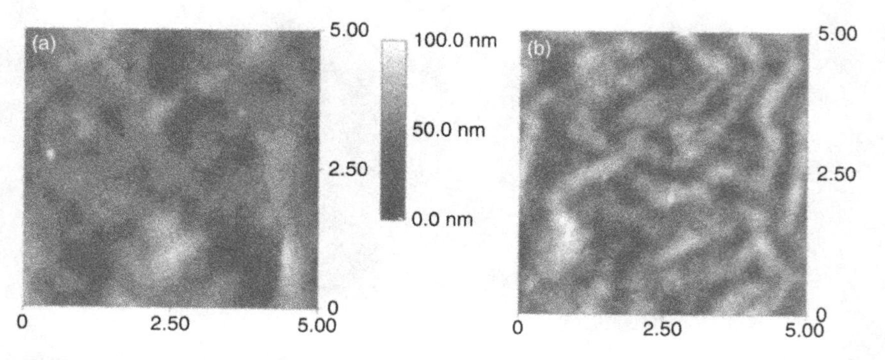

Figure 8.27. AFM images: (a) no addition of high-boiling-point solvent; (b) with high-boiling-point solvent [51].

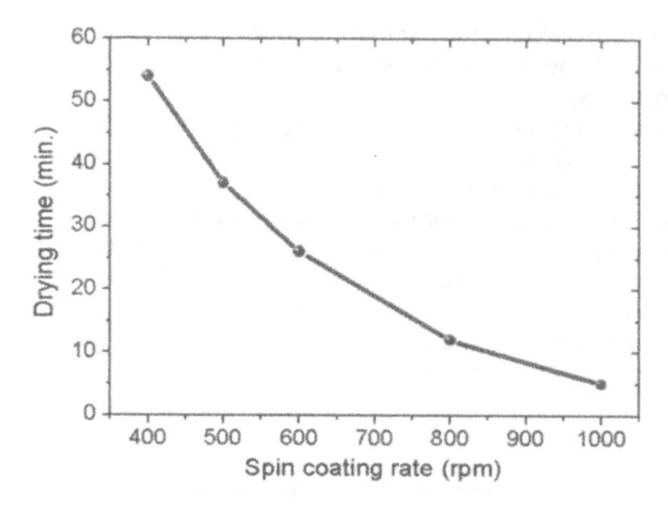

Figure 8.28. Variation of drying time with spin rate [51].

spectra of the photoactive organic layers spin-coated at different spin rates. The variation of the absorption with the spin rate is very significant. In addition to the increase in the absorption, the thick film also has its absorption red-shifted from 500 to 513 nm, indicating the improved crystallization of polymers and the increased length of the molecule with the π-conjugate bonding. The AFM images also show significant differences in surface morphology between the photoactive organic layers deposited at a fast

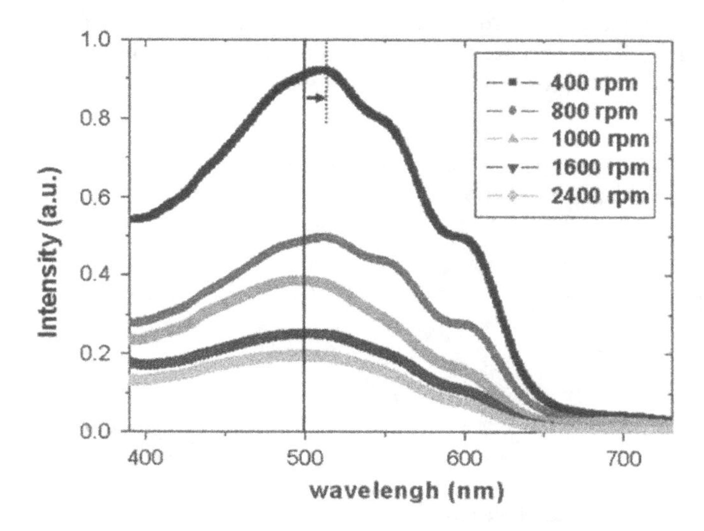

Figure 8.29. Absorption spectra of the photoactive organic layers spin-coated at different spin rates [51].

spin rate and a slow spin rate. Similar to the slow drying technique of adding Br-naph to the DCB solvent, the rms roughness increases as the spin rate is slowed down, so drying time increases.

In conventional organic solar cells, the annealing process improves the self-organization of polymers. However, in the organic–ZnO nanorod hybrid solar cells, annealing is detrimental to the devices. The annealing process causes the P3HT polymer to aggregate and induces a stress that pulls the ZnO nanorods, causing the nanorods to break. The broken ZnO nanrorods can be clearly observed from the comparison of the field-emission scanning electron microscopy (FESEM) images shown in Figure 8.30(a) and 8.30(b) before and after the annealing process. X-ray diffraction (XRD) spectra also confirm this. Without annealing, the XRD spectrrum shows only one strong peak at $2\theta = 34.4°$, corresponding to the (0002) direction of the ZnO crystal. After annealing, the XRD peak value at $2\theta = 34.4°$ is reduced by two orders of magnitude and other peaks also appear.

8.5.1.3 Effect of Additional PCBM Clusters Deposited on ZnO Nanorod Arrays. Jing-Shun Huang and coworkers had investigated the effect of an additional layer of PCBM between the ZnO nanorod arrays and the donor–acceptor blend [52]. Figure 8.31 shows the device structure. Before the P3HT:PCBM blend was deposited, the PCBM clusters were spin-coated in air. The PCBM materials were first dissolved in a dichloromethane solution. The FESEM images in Figures 8.32(a) and (b) are the ZnO nanorod arrays before and after the PCBM clusters were deposited. The ZnO nanorod arrays are well formed, vertically aligned, and densely packed on ITO glass substrates. The middle curve in Figure 8.32(c) is the XRD spectrum of the ZnO nanorod arrays. The curve consists of a peak at $2\theta = 30.3°$ from ITO and a strong peak at $2\theta = 34.4°$ from ZnO [53]. As the ZnO nanorod arrays are covered with the PCBM clusters, we cannot easily see the individual ZnO nanorods in the FESEM image, as shown in Figure 8.32(b) because the PCBM molecules fill in the hollow spaces among the ZnO nanorods. The corresponding XRD spectrum (top curve) in Figure 8.32(c)

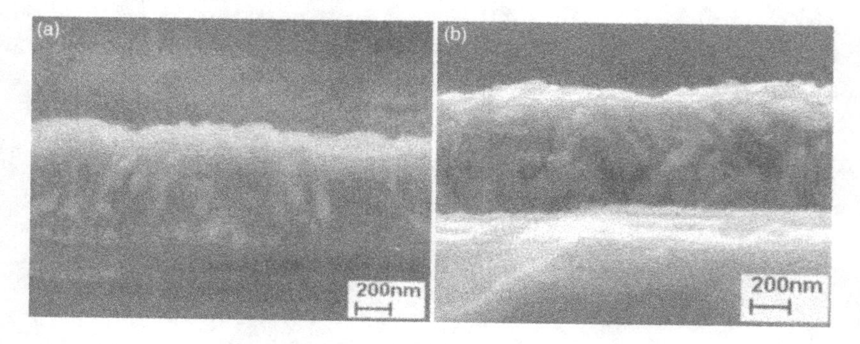

Figure 8.30. FESEM images of the cross sections of the devices with ZnO nanorods: (a) before annealing; (b) after annealing [51].

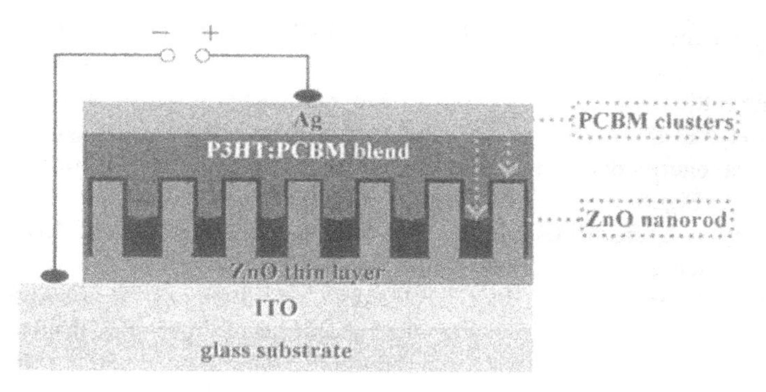

Figure 8.31. Hybrid device structure of ITO/ZnO nanorod arrays/PCBM clusters/P3HT:PCBM/Ag.

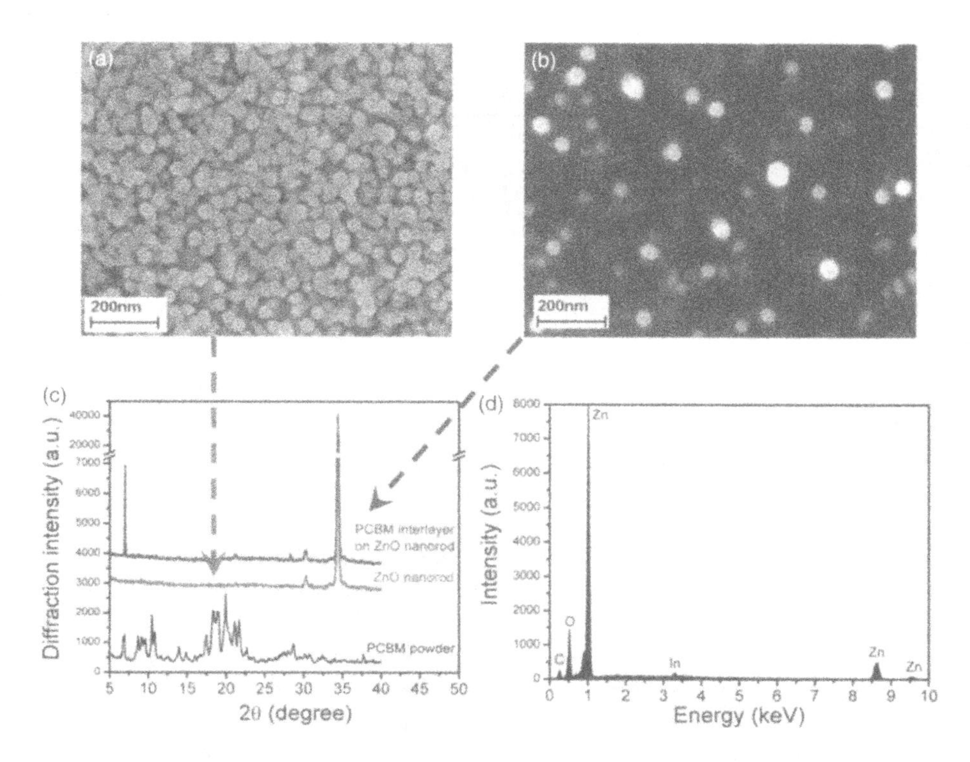

Figure 8.32. (a) FESEM image of ZnO nanorod arrays on the ITO glass substrate. (b) FESEM image of PCBM clusters on the ZnO nanorod arrays. (c) XRD spectra of pure PCBM powder (bottom curve), ZnO nanorod arrays on ITO (middle curve), and PCBM clusters on the ZnO nanorod arrays (top curve). (d) EDX spectrum of the PCBM clusters on the ZnO nanorod arrays [52].

shows a narrow peak at $2\theta = 7.03°$ from the PCBM molecules. The PCBM crystallites present a layered structure due to the van der Waals interaction [54] from the corresponding lattice constant of 1.26 nm. This indicates that the PCBM molecules aggregate orderly in the ZnO nanorod spacing and so are beneficial for electron transport. The typical energy dispersive X-ray spectroscopy (EDX) obtained from Figure 8.32(b) is shown in Figure 8.32(d). The peak at 0.26 keV results from the element of carbon, which originates from PCBM. The FESEM images and XRD and EDX spectra confirm the presence of the PCBM clusters on the ZnO nanorods.

The additional layer of PCBM clusters also influences the absorption spectra of the P3HT:PCBM film. As shown by the two curves in Figure 8.33(a), for the film with the PCBM layer, the peak at 515 nm and the two shoulders at 550 and 600 nm become more pronounced due to the enhanced ordering of P3HT. This indicates that the PCBM clusters help the self-organization of P3HT chains during the solidification of the blend films. To confirm this, AFM was used to measure the surface topography. Figure 8.33(b) and Figure 8.33(c) show the corresponding AFM images of the photoactive

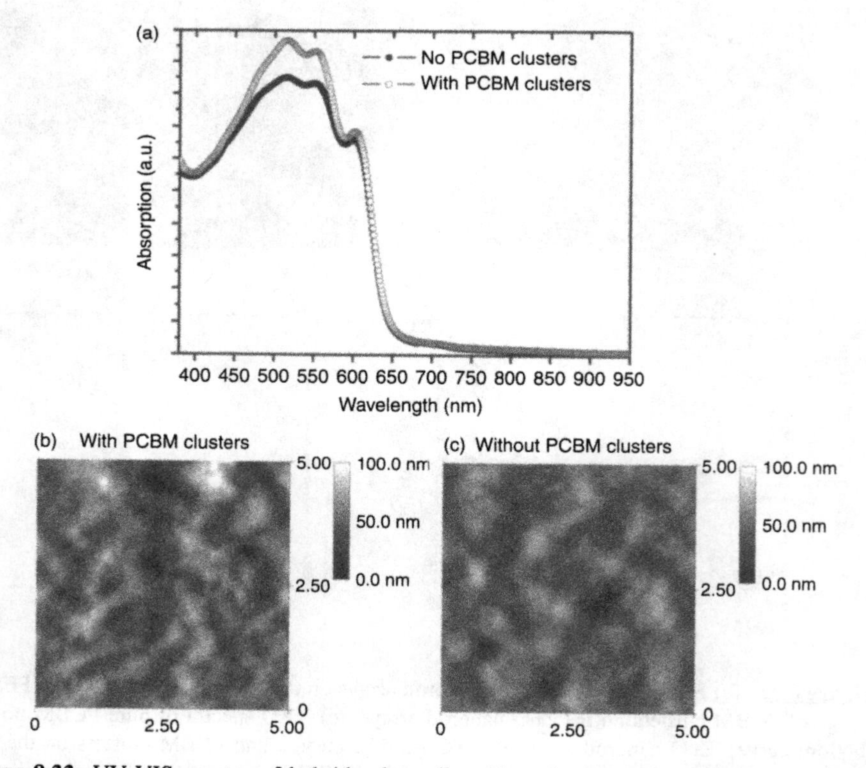

Figure 8.33. UV-VIS spectra of hybrid solar cells with and without the PCBM clusters (a). AFM images of photoactive layers with (b) and without (c) PCBM clusters. AFM image scans are 5 × 5 μm.

layers with and without the additional layer of PCBM clusters, respectively. The PCBM clusters enhance the surface roughness. The rms roughness (σ_{RMS}) is 15 nm, as shown in Figure 8.33(b). In comparison, the surface of the photoactive film without the PCBM clusters underneath is relatively smooth, with σ_{RMS} ~7 nm, as shown in Figure 8.33(c). The enhanced surface roughness is a signature of polymer self-organization [33] and, thus, a higher degree of ordering of P3HT. Thus, the internal series resistance of the device is reduced.

The PCBM layer also helps wet the interface, so the P3HT:PCBM blend can better fill in the spaces between ZnO nanorods. The cross-sectional FESEM images of the devices with and without the PCBM clusters are shown in Figure 8.34(a) and Figure 8.34(b), respectively. Figure 8.34(a) shows that the organic materials in the device infiltrate well into the spaces among ZnO nanorods due to the PCBM modification of the interface between the ZnO nanorods and the P3HT:PCBM blend. In contrast, some void spaces are observed in the region of the ZnO nanorod arrays for the device without the fullerene modification, as shown in Figure 8.34(b). This causes two bad effects. First, the effective thickness of the light absorption layer is reduced, so its absorption is not as much as that of the device with the fullerene modification, as shown in Figure 8.33(a). Second, the interfacial area between the ZnO nanorod arrays and the organic

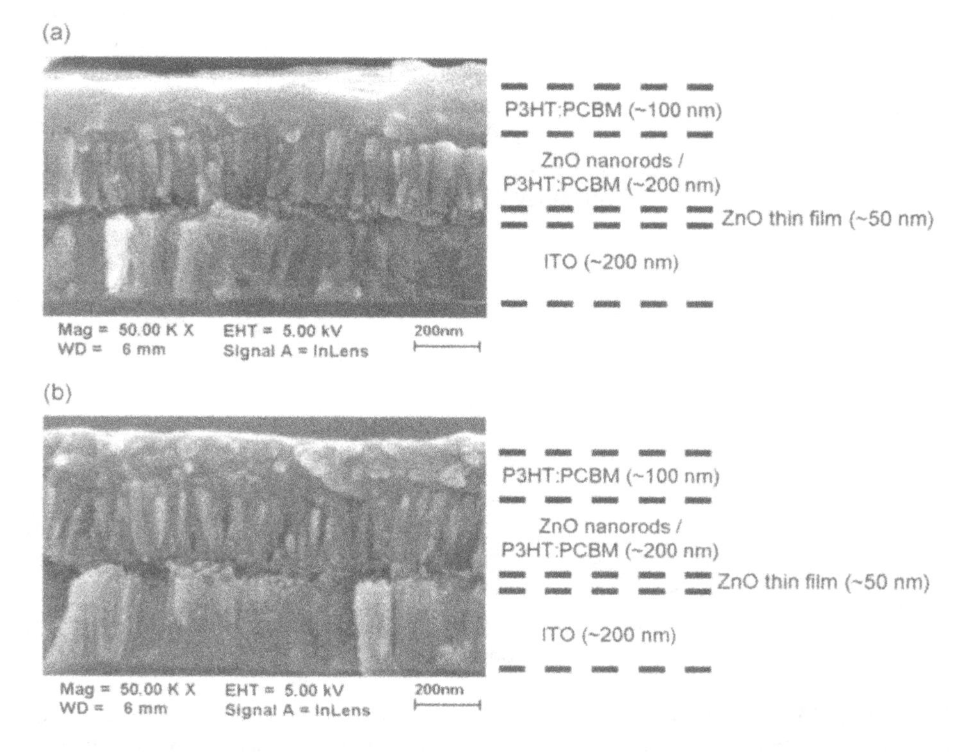

Figure 8.34. Cross-sectional SEM images of devices (a) with and (b) without PCBM clusters.

materials is also reduced, leading to the increased series resistance (R_S, defined from the J–V curves near 1.5 V under light illumination). The device without the PCBM clusters has R_S of 5.83 Ωcm^2, while the device with the PCBM clusters has the R_S significantly decreased to 1.67 Ωcm^2. The reduction of R_S could further enhance the *FF*.

From the above discussions, we expect that the device with the additional layer of PCBM clusters should perform better than that without such a layer. Figure 8.35 shows the J–V characteristics for both types of photovoltaic devices. Device A and Device B are the solar cells with and without the PCBM clusters, respectively. The comparison between the two devices clearly reveals that the insertion of the PCBM clusters between the P3HT:PCBM blend and the ZnO nanorod arrays provides significant improvement in the device performance. Device B has a J_{SC} of 10.13 mA/cm^2, V_{OC} of 0.51 V, *FF* of 45.5%, and *PCE* of 2.35%, whereas Device A exhibits an improved J_{SC} of 11.67 mA/cm^2, V_{OC} of 0.55 V, and *FF* of 49.9%. Thus, the *PCE* of Device A is enhanced to 3.2%.

In the report by Jing-Shun Huang and coworkers [52], eight devices for each type (with PCBM and without PCBM modification) were tested. The mean V_{OC} is 0.51 V with a standard deviation of 0.0014 V for those without the PCBM clusters. In comparison, the mean V_{OC} increases to 0.55 V with a standard deviation of 0.0015 V for those with the PCBM clusters. The small standard deviations indicate that those measured V_{OC}s are reliable. To explain the reason for the increased V_{OC}, the relative energy levels of those materials are plotted in Figure 3.36. It has been reported that the V_{OC} of the BHJ solar cells is directly related to the energy difference between the HOMO level of the donor, which is P3HT here (HOMOP3HT), and the LUMO level of the acceptor [56]. If the donor of P3HT directly contacts ZnO, then ZnO can also function as the acceptor. However, the energy difference between the conduction bands (CB) of ZnO and the HOMOP3HT is only 0.4 eV. It is smaller than the energy difference ($\Delta E_{PCBM,P3HT} = 1.1$ eV)

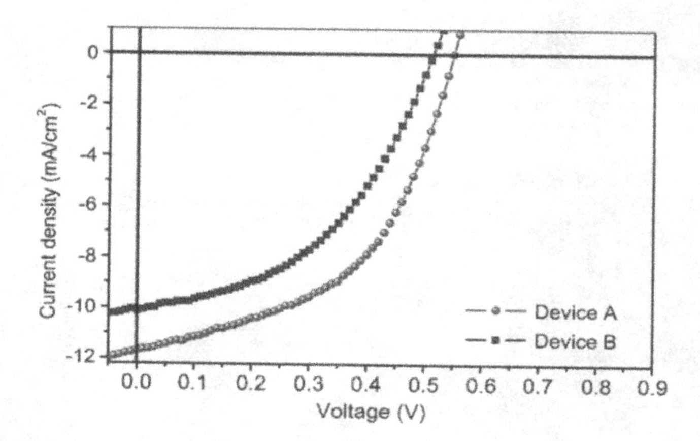

Figure 8.35. The J–V characteristics of the photovoltaic devices under 100 mW/cm^2 AM 1.5G irradiation. Device A represents hybrid solar cells with solution-processed PCBM clusters. Device B denotes hybrid solar cells without PCBM clusters [14].

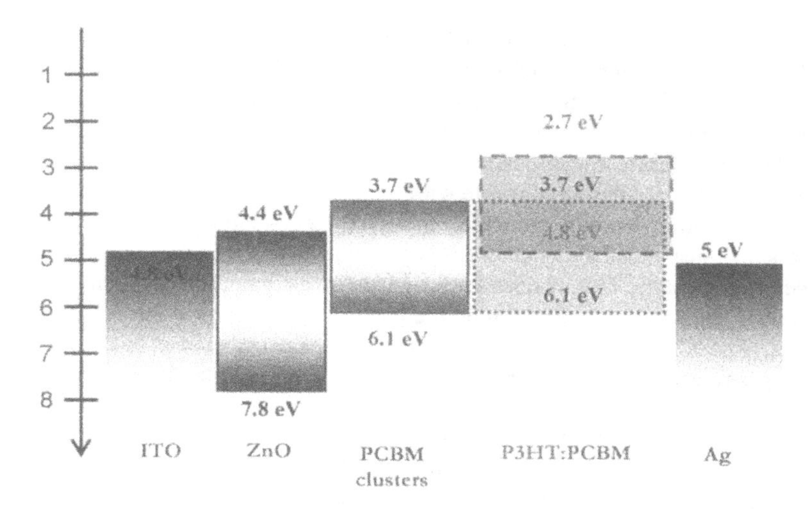

Figure 8.36. Energy band diagram for hybrid solar cells with a structure of ITO/ZnO nanorod arrays/PCBM clusters/P3HT:PCBM/Ag [14].

between the LUMO level of PCBM and the $HOMO_{P3HT}$. This leads to the reduction of V_{OC}. The additional layer of PCBM clusters prevents P3HT from directly contacting ZnO, so the devices with this layer have increased V_{OC}. Some other experimental works on the P3HT/ZnO nanorod array solar cells also demonstrated smaller V_{OC} (0.4–0.45 V) [57–59] than P3HT:PCBM BHJ solar cells (0.5–0.6 V).

Devices without ZnO nanorod arrays have also been tested for the effect of the additional layer of fullerene clusters. It is observed that devices with and without the fullerene clusters have similar photovoltaic performance, indicating that the additional layer of PCBM clusters does not help. The reason is because the layer of P3HT:PCBM blend actually has more PCBM aggregate near the ZnO film, whereas P3HT is more distributed near the top electrode side of Ag [60]. Therefore, the additional coating of the PCBM layer between the P3HT:PCBM blend and the ZnO film does not help much. When ZnO nanorods are used, the rough surface makes the P3HT have a greater probability of directly contacting ZnO, so the additional layer of PCBM clusters inserted between the ZnO nanorods and the P3HT:PCBM blend could prevent this from happening.

8.5.2 Effect of an Additional Layer of TiO_2 Rods Deposited on ZnO Film

The above discussion concludes that ZnO nanorods do not help device performance much compared to the device without ZnO nanorods, although an interpenetrating nano-network may be better formed with the nanorods. The main reason for the minimal effect is that the ZnO nanorods grown by the hydrothermal method are too big in diameter. If the nanorods have a much smaller diameter, the interpenetrating nano-network should help improve the device performance. Here, TiO_2 nanorods are used to study such an effect. Small-scale TiO_2 nanorods with diameter of 5 nm and length of 30–40 nm are inserted between the ZnO layer and photoactive layer [61].

The device fabrication procedure is as follows. ZnO sol–gel film is spin-coated onto a cleaned ITO-coated glass substrate with a sheet resistance of 7 Ω/□ from a 0.5 M zinc acetate solution in 2-methoxyethanol, followed by annealing at 200°C. Then the solution that has suspended TiO_2 nanorods is spin-coated at a spin rate of 2000 rpm. Several concentrations of the TiO_2 nanorod solution are investigated. Subsequently, the PV2000 solution is spin-coated at 600 rpm on the layer of TiO_2 nanorods and dried slowly over the course of 30 min in nitrogen atmosphere. Afterward, a layer of NiO contained in the solution is spin-coated at 2000 rpm, and dried in air. An additional hole transport ink, developed by Plextronics Inc., is used as the hole transporting layer (HTL) at the anode side and coated on the NiO layer. Finally, the Ag metal (250 nm) is thermally evaporated on top of the device with a shadow mask to define the active area of 6.00 mm^2. After the device structure is fabricated, the devices undergo a postannealing process at 170°C for 10 minutes in a nitrogen environment. Current density–voltage (J–V) characteristics of the devices are measured with a Keithley 2400 source measurement unit.

8.5.2.1 Effect of NiO Layer. The NiO layer functions as the electron-blocking layer, as described previously. In addition, the interfacial layer of NiO helps the deposition of the HTL layer. The reason is as follows. In the current device of inverted structure, the HTL is placed on the photoactive layer, in contrast to the conventional structures in which the HTL is deposited under the photoactive layer. However, the hydrophobic characteristics of the photoactive layer make the subsequent deposition of the hydrophilic HTL very difficult, leading to bad contact of the interface. The use of an interfacial layer of NiO here helps solve this problem due to the enhanced adhesion between the HTL and the NiO layer.

A schematic of the finished device structure is shown in Figure 8.37(a). The difference between the devices with and without the NiO layer can be clearly shown by the current density–voltage (J–V) characteristics shown in part (b). The measurement is performed under 100 mW/cm^2 light illumination. With the NiO layer, the *PCE* is effectively improved from 5.07% to 5.61%, *FF* from 56.6% to 63.5%, and the series resistance is reduced from 3.3 Ω-cm^2 to 2.17 Ω-cm^2, indicating that this NiO layer can suppress leakage current, as expected for the electron-blocking layer, and play the role in improving interface contact.

8.5.2.2 Effect of TiO_2 Nanorods. The influence of the TiO_2 nanorods can be seen from the density–voltage (J–V) characteristics shown in Figure 8.38. Under 100 mW/cm^2 light illumination, the device without TiO_2 nanorods shows a short-circuit current density (*Jsc*) of 10.16 mA/cm^2, an open-circuit voltage (V_{OC}) of 0.74 V, and a fill factor (*FF*) of 58.5%, resulting in a PCE of 4.4%. With the array of small-scale TiO_2 nanorods, the *PCE* was improved to 5.61% under the same conditions, as a result of the increased value for J_{SC} of 11.31 mA/cm^2, V_{OC} of 0.78 V, and *FF* of 63.5%.

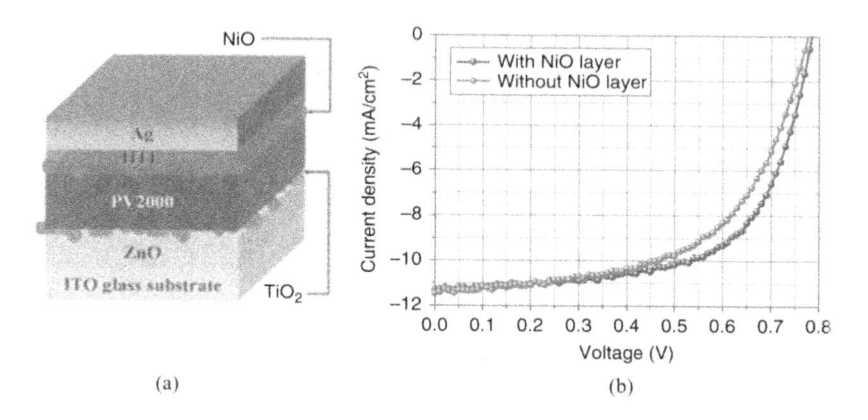

Figure 8.37. (a) A schematic of the finished device structure; (b) the current density–voltage (J–V) characteristics [61].

These results show that the TiO_2 nanorods can act as a functional layer to enhance device performance.

The influence of the TiO_2 nanorods is further investigated from the absorption spectra of the devices, as shown in Figure 8.39, for the device with and without the TiO_2 nanorods interlayer. The situation without the photoactive layer but with nanorods is shown in structure C [ITO/ZnO/TiO_2 nanorods] (lower line), indicating that the nanorod layer barely absorbs light. The device without the nanorod layer (mid-

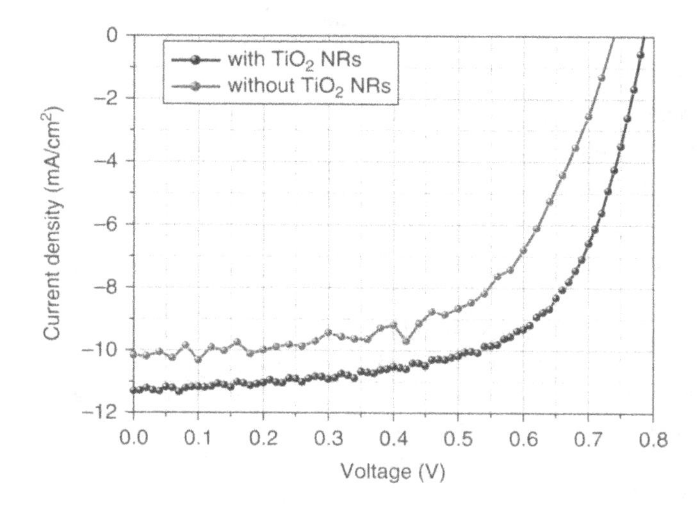

Figure 8.38. The current density–voltage (J–V) characteristics of the devices with and without TiO_2 nanorods [61].

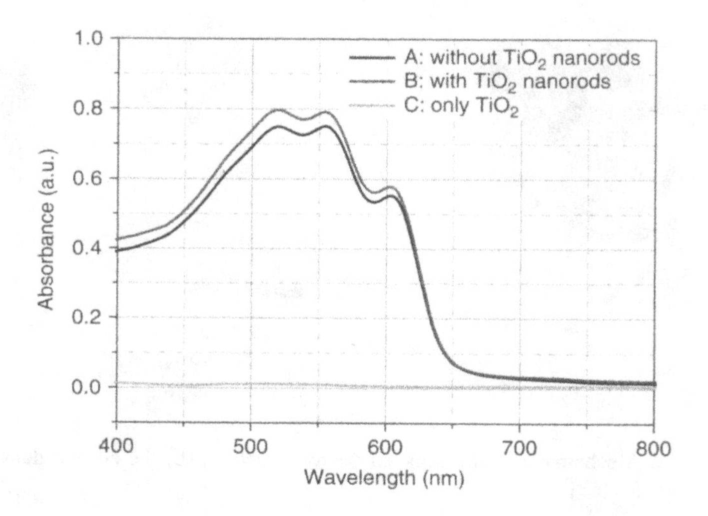

Figure 8.39. Absorption spectra of the devices [61].

dle line) shows the absorbance of 0.75 at 515 nm, whereas the device with the nanorod layer (top line), the main absorption band from 450 nm to 600 nm shows a significant increase, and the peak at 515 nm reaches 0.8. This indicates that the device with nanorods could enhance the optical absorption of the photoactive layer, resulting in improvement of device performance.

FESEM cross-section images are also taken to closely examine the photoactive organic layers of the devices fabricated with the photoactive organic layer coated at the same spin rate of 600 rpm but with and without TiO_2 nanorods. The FE-SEM images are shown in Figure 8.40. The thickness of the photoactive layer without TiO_2 nanorods is ~200 nm, whereas for the device with TiO_2 nanorods, the thickness reaches ~280 nm, which demonstrates an obvious enhancement of 40%. Although the small-scale nanorods are vague in the figure, the thickness of PV2000 provides evident enhancement, which leads to an improved light harvest. It is clear that this increase in the absorption and the J_{SC} (from 10.16 mA/cm^2 to 11.31 mA/cm^2) is a result of the thickened photoactive layer. However, in the absence of TiO_2 nanorods, when we use a photoactive layer spin-coated at a lower speed of 500 rpm as authentication, the thickness is effectively increased and J_{SC} is raised to 10.99 mA/cm^2, but the *FF* decreased to 53.4%, resulting in an efficiency of only 4.11%. In this investigation, it is considered that because of the poor carrier mobility in organic film, part of the charges have the possibility of recombination before reaching the electrode and, therefore, the thickness of the photoactive layer has its limitations. As the thickness exceeds the limitation, it causes low *PCE* due to the enhancement of series resistance, which derives from the increased charge recombination [43]. Consequently, the TiO_2 interlayer in this device can lead to a thicker photoactive organic layer, but does not induce additional charge recombination.

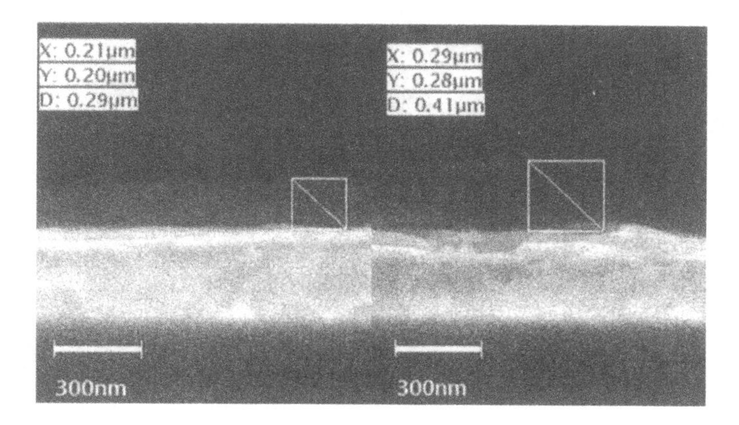

Figure 8.40. Field emission scanning electron microscopy (FE-SEM) cross section images of the devices: without TiO_2 nanorods (left) and with TiO_2 nanorods (right) [61].

8.5.2.3 Influence of TiO_2 Nanorods on the Surface Morphology. The FE-SEM top image of TiO_2 nanorods arrays deposited by solution processing is shown in Figure 8.41(a). The nanorods are spin-coated unevenly onto the ZnO thin film. In some places, large numbers of nanorods pile up to form islands, whereas other places have only a small number of nanorods. Such morphology on the ZnO surface results in relatively large ups and downs. The atomic force microscope (AFM) is used to closely investigate the surface morphology of the ZnO layer surface with and without the TiO_2 nanorods, as shown in Figure 8.41(b). In the structure ITO/ZnO case (the left image), the rms roughness is 7.0 nm and the ups and downs of morphology are approximately 10 nm. In the ITO/ZnO/TiO_2 case (the right image), the rms roughness is 12.6 nm, which is nearly twice as large as the device without TiO_2 nanorods, and the dramatic ups and downs are upgraded to 40 nm. These nanostructures, with thickness ranging from 10 nm to 50 nm, are formed by stacked nanorods. Such morphology provides the photoactive organic layer and the inorganic layer a larger contact area with which to collect the charges.

8.5.2.4 Overall Effect of TiO_2 Nanorods on the Device Characteristics. By introducing inorganic material of high carrier mobility in the form of the nanostructure, the charges in the organic film can quickly pass through this inorganic channel to the ITO electrode. Therefore, the characteristics of devices with inverted structures are improved efficiently [61]. The photoactive layer is thickened due to the introduction of TiO_2 nanorods, indicating improved light harvesting and photocurrent. In addition, the roughened interface between the TiO_2 nanorods and the photoactive organic layer enhances light scattering due to their large difference in refractive index, so the effective light path in the active layer is increased. This also results in more light absorption. Moreover, the charges in the photoactive layer are collected and transported via the

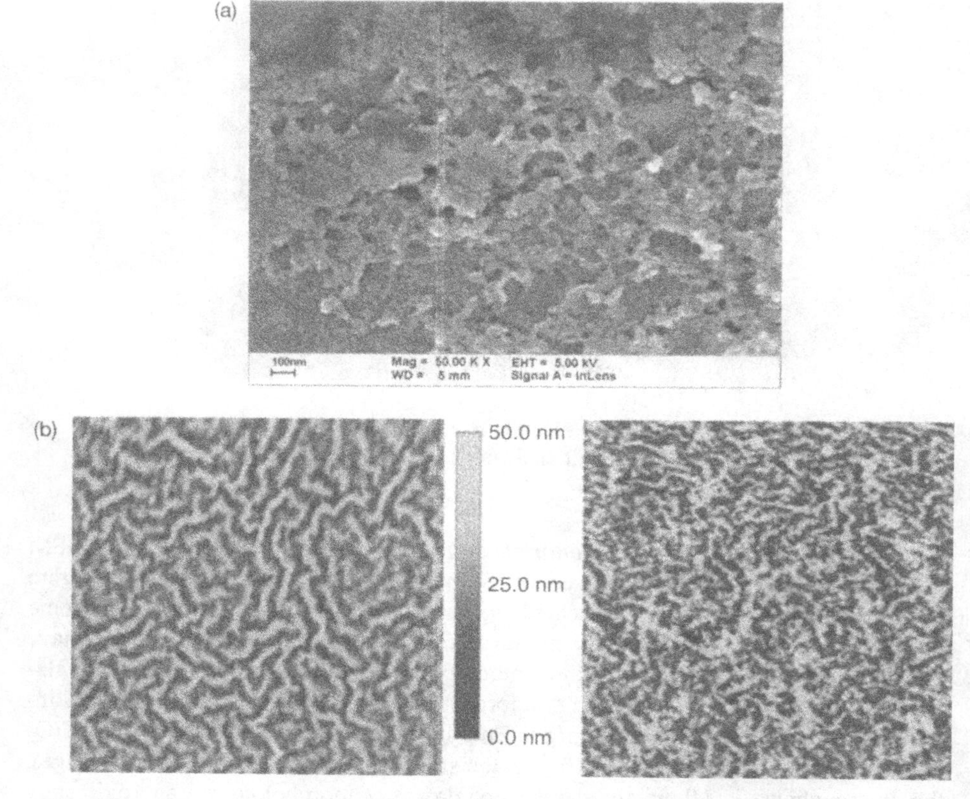

Figure 8.41. (a) FE-SEM top image of TiO_2 nanorods arrays; (b) atomic force microscope images without TiO_2 nanorods (left) and with TiO_2 nanorods (right) [61].

nanorods with relatively higher carrier mobility, thereby improving the series resistance. As a result, it is possible to achieve both improved optical absorption and charge transportation.

Furthermore, with the TiO_2 nanorods interlayer, the device shows significant improvements in V_{OC} (from 0.74 to 0.78 V). This indicates that the morphology effectively suppresses the leakage currents. Even though the ZnO layer acts as a hole blocking layer, this sol–gel layer is still not sufficiently dense to prevent holes from passing through the gaps between ZnO grains, and leads to charge recombination at the ITO surface. Therefore, this TiO_2 layer can be a functional layer to make up for the ZnO deficiency. The valence band of TiO_2 (7.7 eV) is lower than the HOMO of P3HT (5.1 eV), showing that the TiO_2 layer can block the reverse hole flow from P3HT to ITO, effectively preventing current leakage at the organic/ITO interface. In addition, the conduction band of TiO_2 (4.3 eV) is close to that of ZnO (4.4 eV), making it easy to transport electrons from TiO_2 to ZnO.

Photoluminescence (PL) spectra of the devices with and without TiO_2 nanorods are also measured at room temperature and shown in Figure 8.42. The optical emission from 625 nm to 750 nm is due to the recombination of electron–hole pairs (excitons). The main peak occurs at 644 nm, corresponding to the bandgap at 1.9 eV, which is consistent with the description of the previous study with the P3HT. The comparison of the PL spectra of these two structures reveals that the light intensity of the device with TiO_2 nanorods is significantly weaker, indicating that nanostructures can indeed improve electron–hole dissociation, leading to so-called PL quenching. The morphology (TiO_2 nanorods) provides the photoactive layer and the inorganic layer a larger contact area, meaning that the excitons near the nanorods can be separated not only by n-type material, but also by an addition of n-type TiO_2, thereby effectively reducing the occurrence of exciton recombination and thus improving performance.

On the other hand, the islands formed by the irregular arrangement of nanorods lead to increased defects in the interface, further enhancing the contact resistance. Consequently, it is important to improve the device performance by adjusting the concentration of the TiO_2 nanorod solution. The effect of the TiO_2 nanorod concentration on device performance is investigated and shown in Figure 8.43.

The V_{OC} and J_{SC} are evidently increased by the introduction of the nanorod interlayer, and the power conversion efficiency is improved from 4.40% to 5.61% with J_{SC} of 11.31 mA/ cm^2, V_{OC} of 0.78 V, and *FF* of 63.5%. The detailed results are summarized in Table 8.5. These results showed that adding small-scale TiO_2 nanorods is an effective method to improve efficiency. The optimum concentration of the TiO_2 solution is 3 mg/ml. At a low concentration of TiO_2 solution (1.5 mg/ml or less), the ZnO surface is only lightly covered by nanorods, but the performance is improved. It is evi-

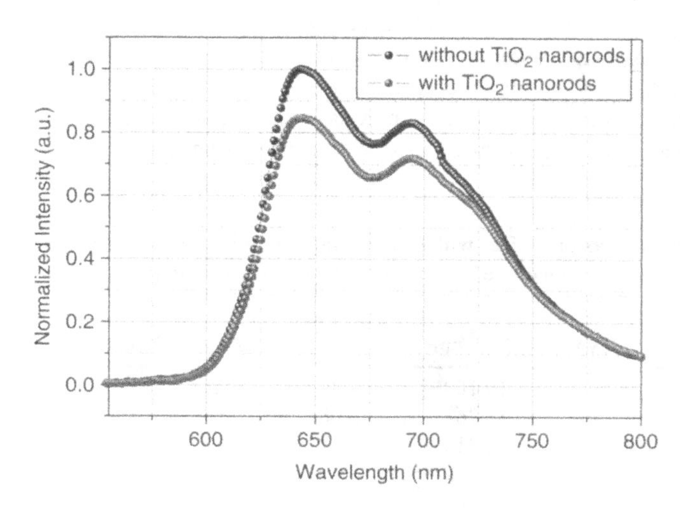

Figure 8.42. Photoluminescence (PL) spectra of the devices with and without TiO_2 nanorods [61].

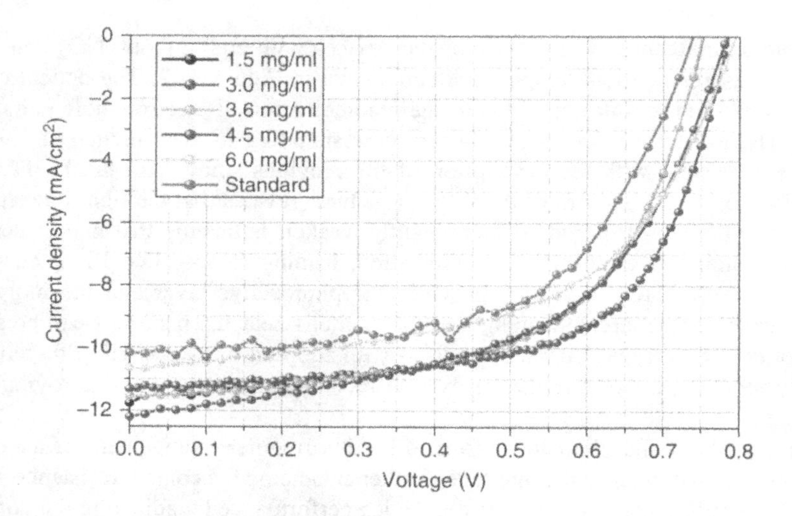

Figure 8.43. J–V curves of devices with different concentrations of TiO_2 nanorods suspended in the solution for deposition of the TiO_2 nanorod layer on ZnO film [61].

dent that the increase in J_{SC} directly contributes to the improvement of the *PCE*. However, as the concentration of TiO_2 solution increases to 6 mg/ml, large numbers of nanorods pile up unevenly, resulting in the increase of contact resistance and thus leading to a low photocurrent (J_{SC} = 10.59 mA/cm²).

In conclusion, the inverted-structure devices can be effectively improved by the sandwiched structure, which employs a pair of metal oxides to restrict the direction of carrier transportation, leading to effective suppression of current leakage. In particular, the photoactive layer is thickened due to the introduction of TiO_2 nanorods, leading to improved light harvest and photocurrent. Moreover, the charges in the photoactive layer are collected and transported via nanorods with relatively higher carrier mobility, which decreases the probability of charge recombination. According to the experi-

TABLE 8.5. Parameters of devices with different concentrations of TiO_2 nanorods suspended in the solution for deposition of the TiO_2 nanorod layer on ZnO film [61]

Concentration (mg/ml)	V_{OC} (V)	J_{SC} (mA/cm²)	Efficiency (%)	*FF* (%)	R_S (Ω-cm²)	R_{SH} (Ω-cm²)
0	0.74	10.16	4.40	58.5	3.18	599.1
1.5	0.76	11.76	5.08	56.8	1.22	892.8
3	0.78	11.31	5.61	63.5	2.17	664.9
3.6	0.76	11.54	5.10	58.3	1.28	740.7
4.5	0.78	12.06	5.03	53.3	2.72	325.2
6	0.8	10.59	4.67	55.1	4.24	468.1

ments, the best concentration of TiO_2 nanorod solution is 3mg/ml. Here, the multiple-layer structure of organic solar cells is demonstrated using low-cost solution processing. The advantages of paired blockers are shown in the organic solar cells with an inverted and sandwiched structure, achieving a high *PCE* of 5.61% using the P3HT organic material combined with inorganic oxides.

8.6 HYBRID SOLAR CELLS USING LOW-BANDGAP POLYMERS

8.6.1 Low-Bandgap Polymers with Sandwiched Structure

The same concept can be applied to low-bandgap polymers with sandwiched structure to obtain high efficiency and good stability. In a recent report [62], the low bandgap polymer PDTSTPD blended with $PC_{71}BM$ has been synthesized and used for solar cells. This polymer has a low bandgap of 1.69 eV, which enables light absorption up to the wavelength of 700 nm. Its LUMO level of –3.88 eV is only slightly above the LUMO level of the acceptor $PC_{71}BM$, –4.3 eV. Its HOMO level of –5.57 eV is also very close to the HOMO level of $PC_{71}BM$, –6.0 eV. The blend of PDTSTPD and $PC_{71}BM$ thus results in a high open-circuit voltage of 0.90 V and a calibrated PCE of 8.13% in the conventional structure. The corresponding J–V curve is shown in Figure 8.44. It is worth noting that the *PCE* obtained from the J–V curve might be different from that obtained from the EQE measurement. The J–V curve sometimes suffers from fluctuations in the light intensity of the solar simulator, so the corresponding *PCE* may not be as accurate as the one extracted from the EQE curve, which is usually much more consistent from one measurement to another. As a matter of fact, the *PCE* obtained from the J–V curve in Figure 8.44 is 8.8%, but the calibrated value of *PCE* from the insert of the EQE curve is more reliable.

For the device with the inverted structure using the above organic blend of PDTSTPD/$PC_{71}BM$ sandwiched by the oxides of ZnO_x and MoO_x, the open-circuit voltage and the *PCE* are still as high as 0.885 V and 6.7%, respectively, as shown by the J–V curve in Figure 8.45. Its device configuration is illustrated in Figure 8.46.

8.6.2 Improved Stability with Low-Bandgap Polymers in the Sandwiched Structure

Because the organic layer is protected by the oxide layer, its stability is significantly improved, similar to the previous experiments using the P3HT polymer. The device with the PDTSTPD/$PC_{71}BM$ sandwiched by the oxides of ZnO_x and MoO_x was placed in the atmosphere without encapsulation for over one month. The humidity level was 30%–60%. The *PCE* still retained 85% of the best value. The variation of the *PCE* with time is shown in Figure 8.47.

The stability can be further improved using air-stable conjugated polymers, as reported in the application to organic thin-film transistors (OTFTs) [63–66]. To possess sufficient stability toward oxidative doping under ambient conditions, conjugated

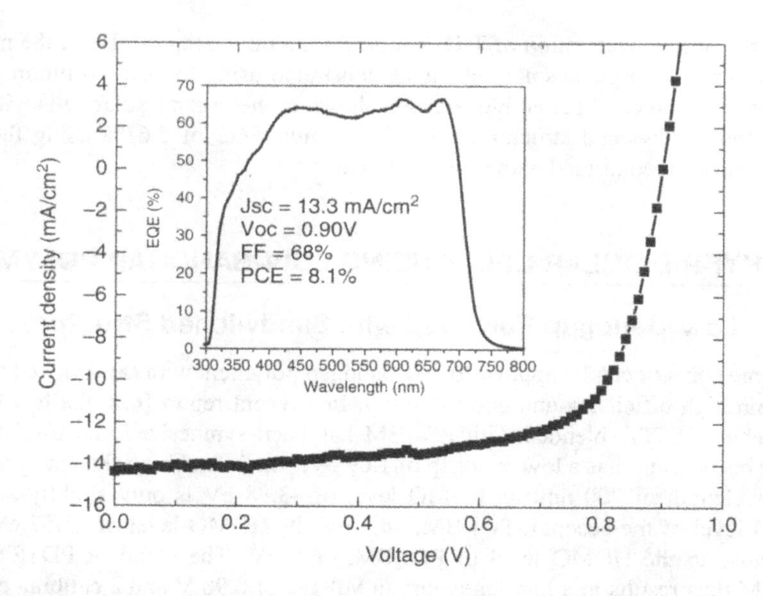

Figure 8.44. The current density-voltage (J–V) characteristics of the device with the low bandgap polymer PDTSTPD blended with $PC_{71}BM$. The insert is the corresponding external quantum efficiency [62].

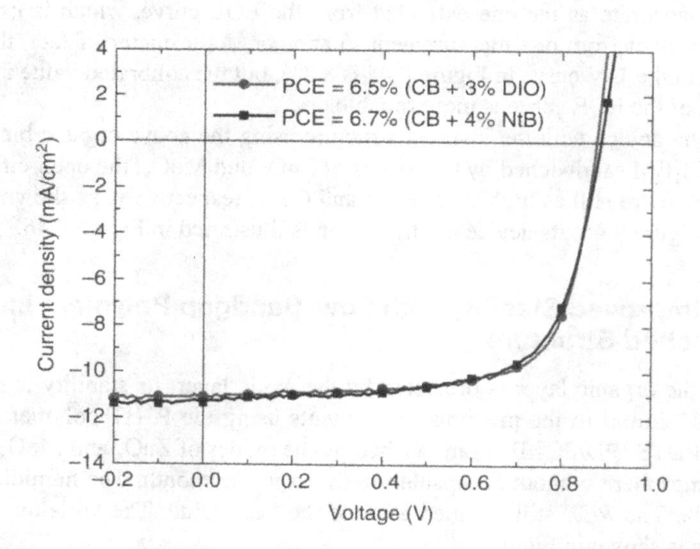

Figure 8.45. The current density–voltage (J–V) characteristics of the device with the low bandgap polymer PDTSTPD blended with $PC_{71}BM$ in the inverted structure [62].

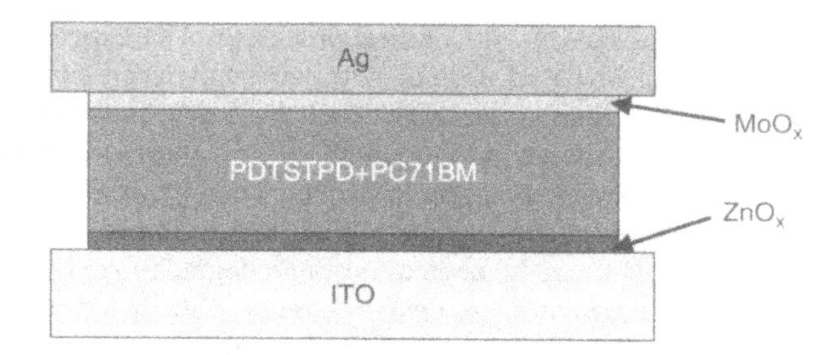

Figure 8.46. The schematic of the device with the blend of low bandgap polymer PDTSTPD and PC71BM sandwiched by the oxides of ZnOx and MoOx in the inverted structure [62].

polymers should have a higher ionization potential, corresponding to a lower HOMO level [67,68]. Conceptually, conjugated polymers with an HOMO level lower than –5.27 eV will possess oxidative stability. Thus, a highly air-stable polymer photovoltaic based on an alternating copolymer consisting of a thiophene–phenylene–thiophene (TPT) electron-donating unit and a 2,1,3-benzothia-diazole (BT) electron-accepting moiety (a-PTPTBT) [69,70] has been developed. The a-PTPTBT has a deep HOMO level of –5.4 eV and, therefore, is stable in air without being easily oxidized. This active layer (a-PTPTBT/$PC_{70}BM$) was sandwiched between and protected by a pair of metal oxide layers, ZnO and NiO [71]. Again, the ZnO layer was placed at the cathode side for hole blocking and the NiO layer was placed at the anode side for electron

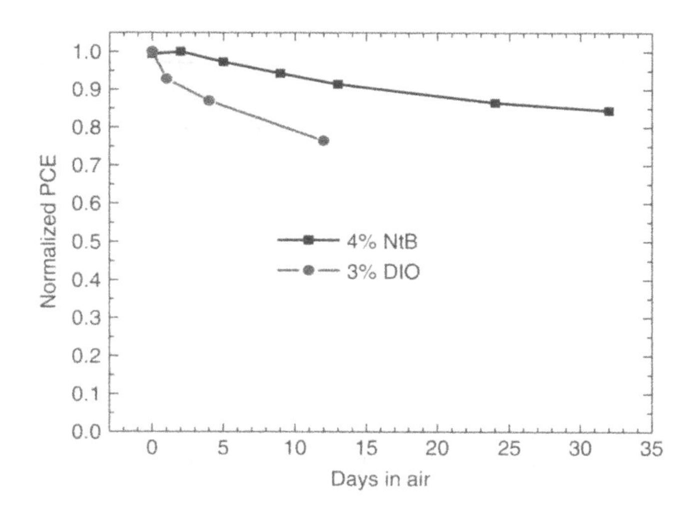

Figure 8.47. The variation of the *PCE* with the time for the device using the blend of low-bandgap polymer PDTSTPD and $PC_{71}BM$ sandwiched by the oxides of ZnO_x and MoO_x in the inverted structure [62].

blocking. Both were deposited using solution processes. The experiment discovered that the a-PTPTBT/$PC_{70}BM$-based devices with the sandwiched configuration of inverted structures have excellent air stability. Their *PCE* as a function of storage time under ambient conditions is shown in Figure 8.48. As revealed in this figure, the device reaches maximum efficiency at about the tenth day. Then its *PCE* remains at the best value for over 4520 hours (about 20 months) of storage under ambient conditions. Over such a long period of time, the device shows no degradation. The device was not encapsulated, either. This extraordinary device stability is attributed not only to the stable sandwiched structure that consists of both the inorganic and organic layers, but also to the intrinsically air-stable polymer material.

The device with the PDTSTPD/$PC_{71}BM$ blend should have similar stability to the one with a-PTPTBT/$PC_{70}BM$ because PDTSTPD also has a deep HOMO level of −5.57 eV, lower than −5.27 eV. However, the lifetime measurement of the device with the PDTSTPD/$PC_{71}BM$ blend is not as good. The reason might be due to the addition of NtB or DIO to the solvent during the process steps. The addition usually causes the morphology of the active blend to evolve with time because those added chemicals still have residues existing in the blend even after postannealing. Figure 8.47 reveals that the addition of 3% DIO leads to a worse stability than the addition of 4% NtB.

8.7 SI NANOWIRE–ORGANIC HYBRID SOLAR CELLS

The above organic–inorganic hybrid solar cells rely on organic polymers to absorb light. However, the most common polymer, P3HT, absorbs light with wavelength less than

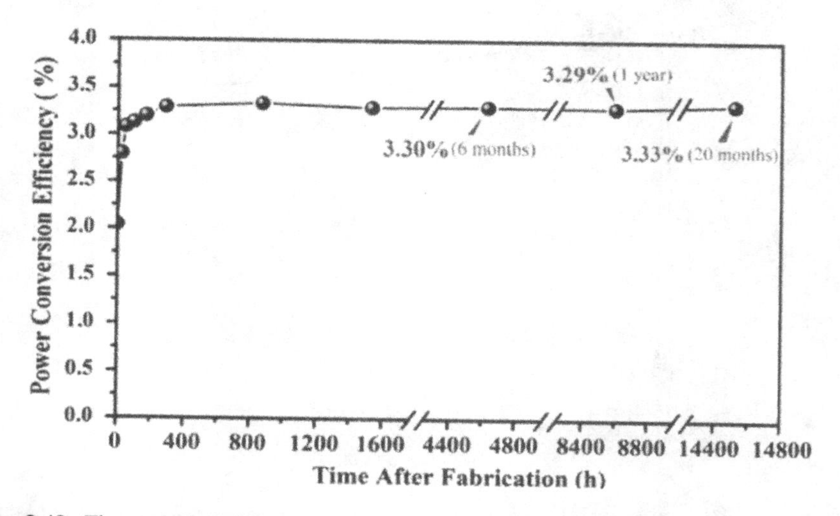

Figure 8.48. The variation of *PCE* with the time for a device using a blend of low-band-gap polymers a-PTPTBT and $PC_{70}BM$ sandwiched by the oxides of ZnO and NiO in the inverted structure.

650 nm. Recent efforts on the low-bandgap materials have increased the absorption range to 700 nm or longer [70, 72–76], but seldom over 800 nm. As indicated in Figure 2.16, the spectrum for wavelengths shorter than 800 nm contains less than 40% of total solar energy, so the polymer absorbs only a small portion of the sunlight. In comparison, the Si material with bandgap of 1.12 eV, corresponding to 1107 nm, can absorb about 70% of solar energy, so it is a much more efficient light absorber than polymers. In addition, the development of the electronics industry based on Si over decades has made Si a very mature and advanced material for vast applications. Therefore, it is also highly desirable to use Si for photovoltaics. Currently, crystalline Si solar cells are still the mainstream for sunlight harvesting in the photovoltaic industry. However, their fabrication involves high-cost and high-energy-consumption processes like diffusion, deposition, drying, etching, and so on. The reduction of fabrication cost and energy consumption is thus attractive. The combination of Si and organic materials can pave the way for this.

To date, researchers have demonstrated planar polymer/semiconductor heterojunction solar cells based on crystalline or amorphous silicon, such as poly-(CH3)3Si-cyclooctatetraene/n-Si [77], tetraphenylporphyrin/n-Si [78], 4-tri-N,N-diethylaniline/p-Si [79], poly(3-hexylthiophene)/a-Si [80], poly(3,4-ethylenedioxythiophene):poly(styrene-sulfonate)/n-Si [81], polyaniline/n-Si [82], and phthalocyanine/n-Si [83]. Unfortunately, with a planar polymer/silicon cell structure, the *PCE* is low. To improve the electron–hole separation and carrier collection, Si nanowire (SiNW) structures are used in solar cells. SiNWs give rise to an enormous increase of the p–n junction area and shorter carrier diffusion distance [84–88]. In addition, the nanowire structure can significantly reduce reflection [89–91] and induce strong light trapping between nanowires [92–94], enhancing light absorption. Recently, SiNWs were successfully combined with PEDOT to form solar cells with demonstrated *PCE* over 8% [95].

8.7.1 Fabrication of SiNWs

SiNWs could be fabricated using an aqueous, electroless etching method [96]. The fabrication steps are as follows. An n-type 1-10 Ω-cm Si (100) wafer is immersed in an aqueous solution of $AgNO_3$ and HF acid at room temperature. The concentrations of $AgNO_3$ and HF are 0.023 and 5.6 mol/L, respectively. In the solution, Ag^+ and Si have a redox reaction, so Ag^+ reduces to Ag, which deposits on the Si surface to form Ag nanoparticles. Then Si near the Ag nanoparticles oxidizes and becomes silicon dioxide (SiO_2), locating between the Ag and Si. The HF etches away the SiO_2. Because the electronegativity of Ag (1.9) is greater than that of Si (1.8), electrons of Si near the Ag nanoparticles move toward the Ag. This makes the Ag surface have a partially negative charge to attract Ag^+ in the etching solution. Because electrons gather near the surface, so Ag^+ reduces to Ag again. As a result, similar processes repeat. The chemical reaction is as follows:

$$Ag^+ + e^- \rightarrow Ag \tag{8-5}$$

$$Si + 2H_2O \rightarrow SiO_2 + 4H^+ + 4e^- \tag{8-6}$$

$$SiO_2 + 4HF \rightarrow H_2SiF_6 + 2H_2O \tag{8-7}$$

The reactions occur repeatedly. Because of the regular crystal orientation of the Si wafer, the reaction rate is much faster in the direction of [100] than the others [97]. Therefore, SiNW arrays normal to surface are formed on the wafer surface. Afterward, the arrays of SiNWs are immersed in a concentrated nitric acid bath to remove all Ag dentritic structures from the nanowire surfaces. Finally, the SiNWs are immersed in the buffered oxide etch (BOE) to remove the oxide layer from the SiNWs and form H-terminated bonds on the silicon surface.

8.7.2 The Fabrication of SiNW–Organic Hybrid Solar Cells

Poly(3,4-ethylenedioxy-thiophene):poly(styrenesulfonate) (Baytron P, H. C. Starck GmbH, Leverkusen, Germany) could be used to form SiNW/PEDOT heterojunctions. The PEDOT gel particles with a mean diameter of 80 nm are first dispersed in the aqueous solution. However, the SiNW surface has H-terminated bonds after being immersed in the BOE. Such a surface is hydrophobic, so PEDOT does not adhere to it. Thus, the SiNW surface should be modified to become hydrophilic. This could be done by placing the SiNWs in an environment with a relative humidity of 60% and 25°C for 2 hours to form a very thin layer of native oxide on the SiNW surface. Then it becomes hydrophilic and has a contact angle below 20°, so PEDOT can easily attach to it.

An ITO glass is spin-coated with a layer of PEDOT. The thickness of the wet PEDOT film is about 9 μm, which corresponds to the dry film with a thickness of 200 nm. Before the PEDOT thin film is dried, the top portion of the previously prepared SiNWs is immersed in the wet PEDOT thin film. Because the SiNWs fabricated by metal-assisted etching are vertically aligned and have uniform length, almost all SiNWs can be immersed in the wet PEDOT thin film. The wet PEDOT will automatically infiltrate into the spaces in the SiNWs with an effect similar to calligraphy, because of the hydrophilic property on the surface of the SiNWs. Afterward, the samples are annealed at 140°C for 10 min in an N2 environment to make the wet PEDOT dry. The dry PEDOT forms a compact film on the SiNW surface. The SiNWs are stuck onto the ITO glass via PEDOT. The solution process gives solar cells a cost benefit in comparison with the conventional vacuum process.

SEM imaging could be used to investigate the SiNW/PEDOT core-sheath structure. TEM images could be further taken to view the interface between the SiNW and the PEDOT. Figure 8.49(a) shows the side-view SEM images of the as-etched SiNWs without a PEDOT coating. As expected, they are vertically aligned on the silicon substrate. The top-view SEM image of the corresponding SiNWs is shown in Figure 8.49(b). This image shows that the size of SiNWs ranges from 20 to 289 nm. Figure 8.49(c) shows the side view of the SEM image for the SiNWs coated with PEDOT. It reveals that SiNWs are well covered by a layer of PEDOT. To closely examine the interface between the SiNW and PEDOT, TEM images are better. The SiNWs coated with PEDOT are detached from the ITO glass for TEM examination. Figure 8.49(d) shows that the SiNW is well surrounded by the PEDOT. Because the core of Si is single crystalline aligned to a zone axis, it has much stronger diffraction and thus darker

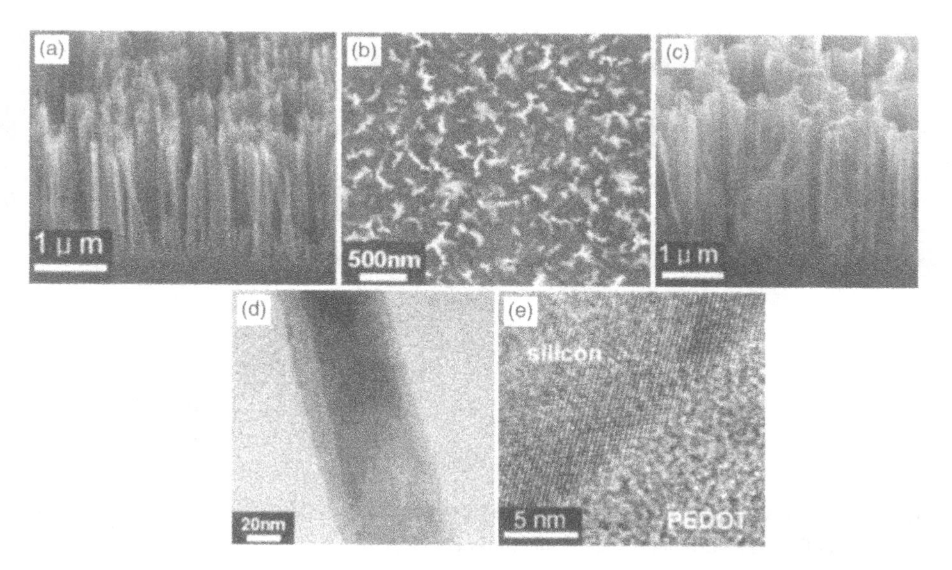

Figure 8.49. (a) Side-viewSEM image of the as-etched SiNWs. (b) Top-viewSEM image of the as-etched SiNWs. (c) SEM image of the SiNWs covered with the PEDOT thin film layer. (d) TEM image of the SiNW/PEDOT core-sheath structure. (e) TEM image of the SiNW/PEDOT interface [98].

contrast than that of the PEDOT sheath. This image shows that the thickness of PEDOT could be below 20 nm, indicating that the calligraphy effect works pretty well, so PEDOT has good adherence to the Si surface. A high-resolution TEM image is shown in Figure 8.49(e). It shows that the core is indeed Si with a crystalline structure, while the sheath is PEDOT without a crystalline structure. Also, there is no gap between the core Si and the sheath PEDOT.

Although the original PEDOT gel particles have a size of about 80 nm, they become lamellar particles after being dried. The stacks of the lamellae have a pancake-like morphology with a diameter of 20–25 nm and a thickness of 5–6 nm [99,100]. The high-resolution TEM image in Figure 8.49(e) shows that the PEDOT surrounding the SiNW consists of a few layers of lamellar particles to form a film with a thickness of about 20 nm.

8.7.3 Characteristics of SiNW–Organic Hybrid Solar Cells

The photovoltaic J–V characteristics are measured under an illumination intensity of 100 mW/cm^2 (AM1.5G). Figure 8.50 shows the measured J–V characteristics. The heterojunction of the Si and PEDOT could give rise to electrical rectification, as shown in the inset of the figure. The reverse-bias current is much smaller than the forward-bias current. The J–V characteristics of the SiNW/PEDOT cells clearly reveal a stable rectifying diode behavior. Therefore, PEDOT used here serves as the p-

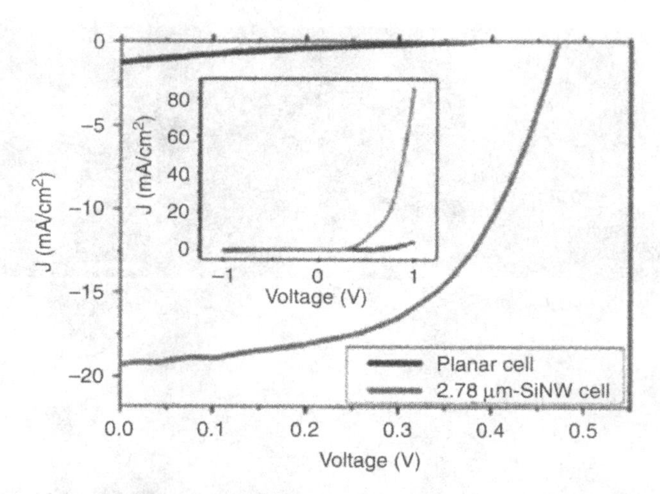

Figure 8.50. J–V characteristics of the SiNW–PEDOT cell and the planar Si/PEDOT cell. The inset shows the dark current–voltage characteristics [98].

type material. The n-type SiNW and p-type PEDOT together form a p–n junction with a depletion region that has a built-in field to separate the photo-generated electron–hole pairs.

Under AM1.5G light illumination, the SiNW–PEDOT hybrid solar cell with wire length of 2.78 μm shows a strong increase in short-circuit current density (J_{SC}), open-circuit voltage (V_{OC}), fill factor (*FF*), and *PCE*, as compared to the planar Si–PEDOT hybrid solar cell. The J_{SC} is improved from 1.27mA/cm² to 19.28 mA/cm², the V_{OC} from 0.34 to 0.47 V, and the *FF* from 18% to 61%, resulting in an improvement of the *PCE* from 0.08% to 5.09%. The reason for the increase in the short-circuit current is two-fold. First, the carrier diffusion distance from the core of the interface of the SiNW and PEDOT is only several tens of nanometers or less. In comparison, the corresponding distance may be several micrometers in the planar cells. Because the SiNW/PEDOT heterojunction is where the photo-generated electron–hole pairs dissociate and become individual electrons and holes, the short diffusion distance enhances the electron–hole separation. Thus, the carrier collection efficiency is greatly enhanced in the SiNW–PEDOT cells. Second, because of the light-trapping effect, the reflectance of the SiNW arrays is reduced to below 5% over the spectral range of 400–1100 nm. In contrast, the planar silicon surface exhibits over 30% of the reflectance. This means that the light trapping in the SiNW arrays increases the absorption of light, leading to the enhancement of the photocurrent.

The series resistance decreases from 60.42 Ωcm² for the planar cell to 1.47Ωcm² for the SiNW–PEDOT device. The reduction in the series resistance leads to improvement in the fill factor for the SiNW–PEDOT cells. This may be attributed to the increase of the heterojunction area with the SiNW structure so that the current density is greatly enhanced. The large series resistance may be also due to the bad contact between the planar surface and PEDOT because no special treatment is done for the planar Si surface.

8.7.4 The Influence of Si NW Length

The length of SiNW is found to have a significant influence on the performance of the SiNW–PEDOT hybrid solar cells [101]. A straightforward explanation is that long SiNWs will have a larger light trapping effect and, hence, absorb more light than short SiNWs, so the PCE will increase. Another set of SiNW–PEDOT hybrid solar cells was fabricated with the wire length as the control factor. Here PEDOT:PSS solution (Clevios™ P, H. C. V4 Stark GmbH, Leverkusen, Germany) is used to replace the previous one (Baytron P, H. C. Starck GmbH, Leverkusen, Germany) because its dry film has a 20 times larger conductivity, 200 S cm^{-1} vs. 10 S cm^{-1}.

Figure 8.51 shows the reflectance versus the wavelength for Si wafers with 0.37 μm, 2.15 μm, and 5.59 μm length of SiNWs, respectively. The reflectance of a planar wafer is also plotted for comparison. As expected, the planar wafer has a reflectance of over 35% for the wavelength range from 400 nm to 1100 nm and all SiNW arrays have much lower reflectance. Also, the reflectance decreases as the wire length increases, so SiNWs of 0.37 μm correspond to a higher reflectance than that of SiNWs with other lengths. However, the measured J–V curves under 1 sun 1.5 G illumination shown in Figure 8.52 reveals that the 0.37 μm SiNW–PEDOT:PSS solar cell has the highest PCE of 8.40%, highest J_{SC} of 24.24 mA cm^{-2}, and highest V_{OC} of 0.532 V. As the SiNW length extends to 5.59 μm, *PCE* is reduced from 8.40% to 3.76%, J_{SC} decays from 24.24 mA cm^{-2} to 13.06 mA cm^{-2}, and Voc decreases from 0.532 V to 0.435 V. Only the fill factor (*FF*) is preserved in the range between 63%

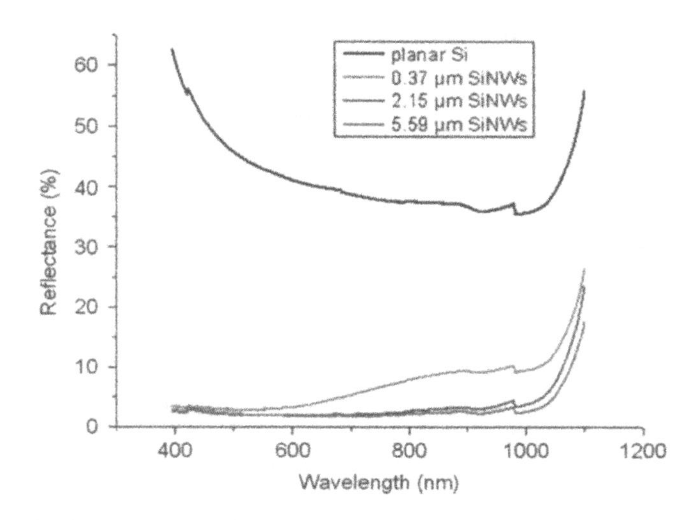

Figure 8.51. Reflectance of SiNWs in the wavelength ranging from 400 to 1100 nm [101].

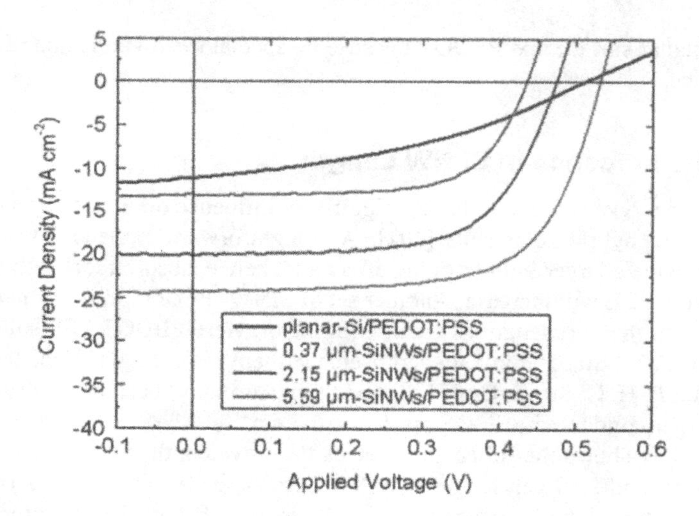

Figure 8.52. J–V curves of each SiNW device and planar Si device under the illumination condition of 1 sun AM1.5 G [101].

and 67%. These unexpected results stimulate us to further investigate the device characteristics.

Figure 8.53 shows the measured incident photon conversion efficiency (IPCE), including external quantum efficiency (EQE) and internal quantum efficiency (IQE). IQE is obtained from EQE divided by the transmittance of ITO/glass. Both EQE and IQE reveal that photons can be absorbed and transformed to current in the wavelength range from 400 nm to 1100 nm. The 0.37 μm SiNW–PEDOT:PSS solar cell has the best EQE and IQE compared to solar cells with different lengths of SiNWs. This is

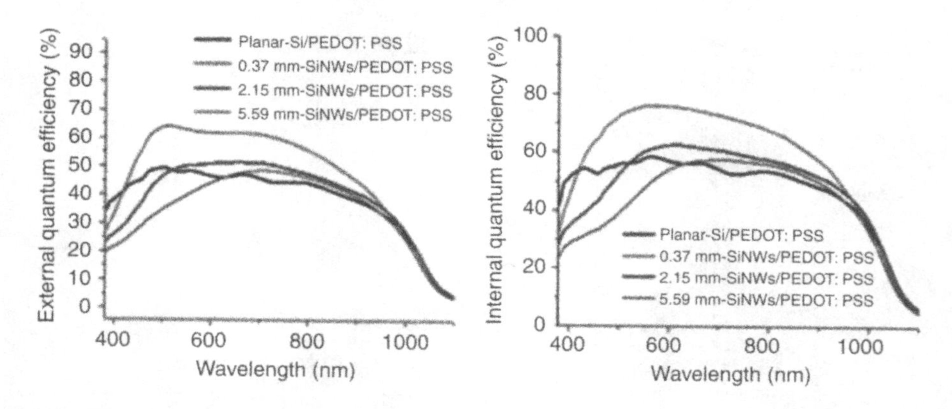

Figure 8.53. (Left) Incident photon conversion efficiency versus wavelength. (Right) Internal quantum efficiency versus wavelength [101].

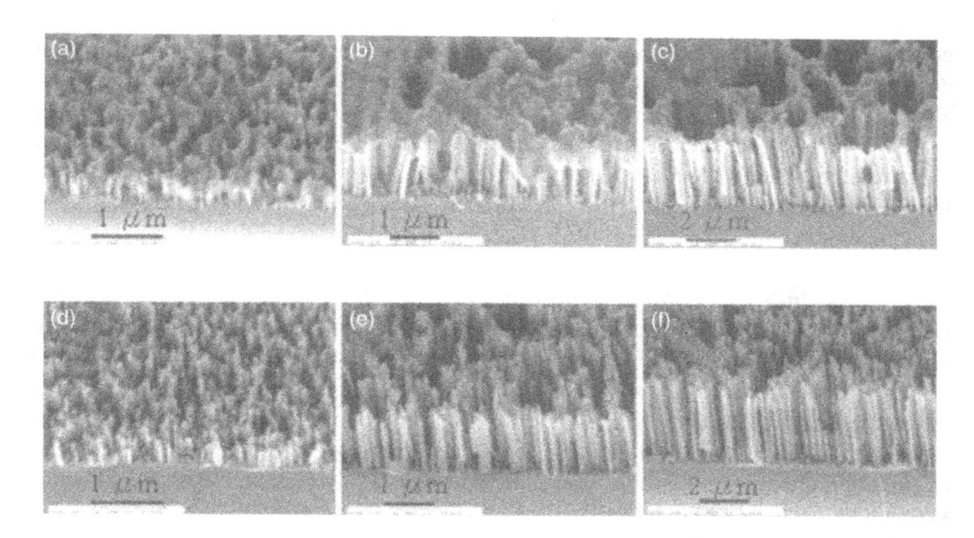

Figure 8.54. SEM images of SiNWs coated with PEDOT:PSS: (a) 0.37 mm SiNWs [101]; (b) 2.15 mm SiNWs [101]; (c) 5.59 mm SiNWs. SiNWs without being coated with PEDOT:PSS [101]; (d) 0.37 mm SiNWs; (e) 2.15 mm SiNWs; (f) 5.59 mm SiNWs.

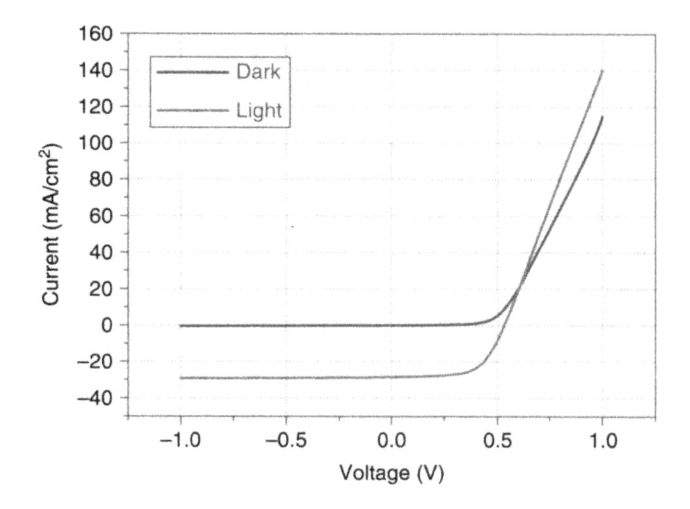

Figure 8.55. J–V curves of the SiNW–PEDOT solar cells with PEC of 9.45% under the illumination condition of 1 sun AM1.5 G.

consistent with the previously measured device paramaters of *PCE*, J_{SC}, and V_{OC}, but does not give us more hints to the explanation of better performance from shorter SiNWs.

SEM images are taken to see the detailed structure of the SiNW before and after being coated with PEDOT, shown in Figure 8.54. The images reveal two facts. First, the SiNWs are not well coated with PEDOT for the wire lengths of 2.15 and 5.59 μm. In comparison with the PEDOT:PSS (Baytron P, H. C. Starck GmbH, Leverkusen, Germany) that are able to well surround the SiNWs, the PEDOT:PSS (Clevios™ P, H. C. V4 Stark GmbH, Leverkusen, Germany) has a higher viscosity, so it has more difficulty penetrating into the spaces among SiNWs. As a result, the PEDOT:PSS used here are not able to infiltrate into the deep regions of the long SiNW structure. Second, the long SiNWs tend to aggregate at the top portion and form bundles. This further prohibits the infiltration of the PEDOT:PSS. These two factors make it hard to form a thorough SiNW–PEDOT p–n junction. To further improve the device performance, the aggregation phenomenon has to be eliminated and a highly conductive p-type polymer with a low viscosity has to be used.

As the deposition of PEDOT on SiNWs is improved, the power conversion efficiency can be further increased. *PCE* of 9.45% is obtained with the open-circuit voltage of 0.53 V, short-current density of 28.66 mA/cm^2, and fill factor of 62%. Its J–V curve and EQE versus wavelength are shown in Figures 8.55 and 8.56, respectively. The maximum EQE is only around 72%, occurring at wavelength of 500 nm. It indicates that there is still room for improvement. If the EQE is increased to near 100% for most of the wavelength range, the efficiency may be much more than 10%.

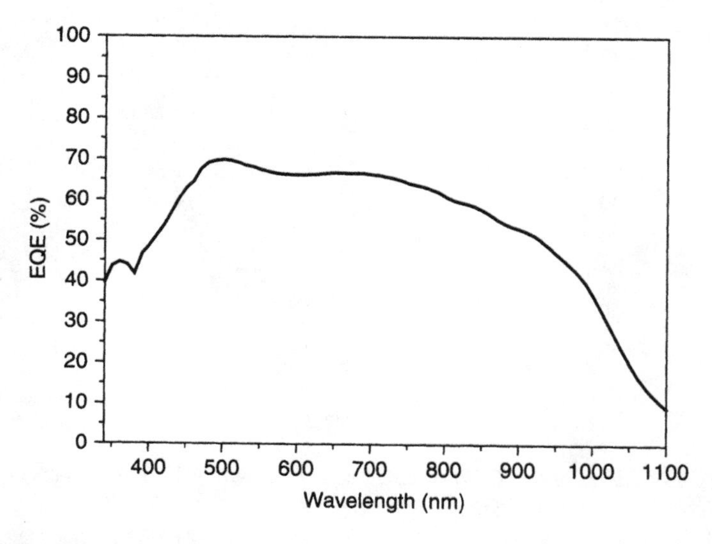

Figure 8.56. External quantum efficiency (EQE) versus wavelength of SiNW–PEDOT solar cells with *PEC* of 9.45%.

REFERENCES

1. M. Jørgensen, K. Norrman, and F. C. Krebs (2008), *Solar Energy Materials and Solar Cells, 92,* 686–714.
2. M. P. de Jong, L. J. van Ijzendoorn, and M. J. A. de Voigt (2000), *Applied Physics Letters, 77,* 2255–2257.
3. K. Kawano, R. Pacios, D. Poplavskyy, J. Nelson, D. D. C. Bradley, and J. R. Durrant (2006), *Solar Energy Materials and Solar Cells, 90,* 3520–3530.
4. M. O. Reese, A. J. Morfa, M. S. White, N. Kopidakis, S. E. Shaheen, G. Rumbles, and D. S. Ginley (2008), *Solar Energy Materials and Solar Cells, 92,* 746–752.
5. C.-Y. Chou, J.-S. Huang, C.-H. Wu, C.-Y. Lee, and C.-F. Lin (2009), *Solar Energy Materials and Solar Cells, 93,* 1608–1612.
6. M. S. White, D. C. Olson, S. E. Shaheen, N. Kopidakis, and D. S. Ginley (2006), *Applied Physics Letters, 89,* 143517.
7. K. Takanezawa, K. Tajima, and K. Hashimoto (2008), *Applied Physics Letters, 93,* 063308.
8. C. Tao, S. P. Ruan, G. H. Xie, X. Z. Kong, L. Shen, F. X. Meng, C. X. Liu, X. D. Zhang, W. Dong, and W. Y. Chen (2008), *Applied Physics Letters, 94,* 043311.
9. C. Tao, S. Ruan, X. Zhang, G. Xie, L. Shen, X. Kong, W. Dong, C. Liu, and W. Chen (2008), *Applied Physics Letters, 93,* 193307.
10. H. H. Liao, L. M. Chen, Z. Xu, G. Li and Y. Yang (2008), *Applied Physics Letters, 92,* 173303.
11. H.-H. Liao, L.-M. Chen, Z. Xu, G. Li, and Y. Yang (2008), *Applied Physics Letters, 92,* 173303.
12. T. H. Lowery, and K. S. Richardson (1987), *Mechanism and Theory in Organic Chemistry,* 3rd Edition, Harper & Row, New York.
13. L. J. A. Koster, V. D. Mihailetchi, and P. W. M. Blom (2006), *Applied Physics Letters, 88,* 093511.
14. Jing-Shun Huang (2010), "Development of Polymer Solar Cells with Organic-Inorganic Sandwiched Structures through Solution Processing," PhD dissertation, National Taiwan University, Taipei, Taiwan.
15. S. K. Hau, H. L. Yip, H. Ma, and A. K. Y. Jen (2008), *Applied Physics Letters, 93,* 233304.
16. M. D. Irwin, B. Buchholz, A. W. Hains, R. P. H. Chang, and T. J. Marks (2008), *Proceedings of the National Academy of Sciences of the United States of America, 105,* 2783–2787.
17. V. Shrotriya, G. Li, Y. Yao, C. W. Chu, and Y. Yang (2006), *Applied Physics Letters, 88,* 073508.
18. J. B. Kim, C. S. Kim, Y. S. Kim, and Y.-L. Loo (2009), *Applied Physics Letters, 95,* 183301.
19. A. Moliton, and J. M. Nunzi (2006), *Polymer International, 55,* 583–600.
20. J. Meyer, S. Hamwi, T. Bulow, H. H. Johannes, T. Riedl, and W. Kowalsky (2007), "Highly Efficient Simplified Organic Light Emitting Diodes," *Applied Physics Letters, 91,* 113506.
21. J. Meyer, T. Winkler, S. Hamwi, S. Schmale, H. H. Johannes, T. Weimann, P. Hinze, W. Kowlasky, and T. Riedl (2008), *Advanced Materials, 20,* 3839–3843.

22. J. Meyer, S. Hamwi, S. Schmale, T. Winkler, H. H. Johannes, T. Riedl, and W. Kowalsky (2009), *Journal of Materials Chemistry, 19,* 702–705.
23. M. J. Son, S. Kim, S. Kwon, and J. W. Kim (2009), *Organic Electronics, 10,* 637–642.
24. Y. K. Jeong, and G. M. Choi (1996), *Journal of Physics and Chemistry of Solids, 57,* 81–84.
25. N. Serin, T. Serin, Horzum, and Y. Celik (2005), *Semiconductor Science and Technology, 20,* 398–401.
26. G. Papadimitropoulos, N. Vourdas, V. E. Vamvakas, and D. Davazoglou (2006), *Thin Solid Films, 515,* 2428–2432.
27. G. B. Murdoch, M. Greiner, M. G. Helander, Z. B. Wang, and Z. H. Lu (2008), *Applied Physics Letters, 93,* 083309.
28. Jing-Shun Huang (2010), Chen-Yu Chou, and Ching-Fuh Lin, *IEEE Electron Device Letters, 31,* 332–334.
29. J. Y. Kim, S. H. Kim, H. H. Lee, K. Lee, W. L. Ma, X. Gong, and A. J. Heeger (2006), *Advanced Materials, 18,* 572–576.
30. F. C. Krebs (2008), *Solar Energy Materials and Solar Cells, 92,* 715–726.
31. D. C. Olson, S. E. Shaheen, R. T. Collins, and D. S. Ginley (2007), *Journal of Physical Chemistry C, 111,* 16670–16678.
32. K.-H. Tsai, J.-S. Huang, M.-Y. Liu, C.-H. Chao, C.-Y. Lee, S.-C. Hung, and C.-F. Lin (2009), *Journal of The Electrochemical Society, 156,* B1188–B1191.
33. G. Li, V. Shrotriya, J. S. Huang, Y. Yao, T. Moriarty, K. Emery, and Y. Yang (2005), *Nature Materials, 4,* 864–868.
34. X. N. Yang, J. Loos, S. C. Veenstra, W. J. H. Verhees, M. M. Wienk, J. M. Kroon, M. A. J. Michels, and R. A. J. Janssen (2005), *Nano Letters, 5,* 579–583.
35. L. M. Chen, Z. Hong, G. Li, Y. Yang (2009), *Advanced Materials, 21,* 1434.
36. M. Kakihana (1996), *J. Sol–Gel Sci. Technol., 6,* 7–55.
37. Ming-Yi Lin, Chun-Yu Lee, Shu-Chia Shiu, Ing-Jye Wang, Jen-Yu Sun, Wen-Hau Wu, Yu-Hong Lin, Jing-Shun Huang, and Ching-Fuh Lin (2010), *Organic Electronics, 11,* 1828–1834.
38 A. M. Shanmugharaj, S. Sabharwal, A. B. Majali, V. K. Tikku, A. K. Bhowmick (2002), *J. Mater. Sci., 37,* 2781–279.
39. L. W. Barbour, M. Hegadorn, J. B. Asbury (2007), *J. Am. Chem. Soc., 129,* 15884.
40. B. Pawelke, Cs. KoAsa, S. I. Chmela, W. D. Habicher (2000), *Polymer Degradation and Stability, 68,* 127–132.
41. A. Jagminas, J. Kuzmarskyte, G. Niaura (2002), *Appl. Surf. Sci., 201,* 129–137.
42. Y. C. Zhang, J. Y. Tang, G. L. Wang, M. Zhang, X. Y. Hu (2006), *J. Cryst. Growth, 294,* 278–282.
43. D. C. Olson, J. Piris, R. T. Collins, S. E. Shaheen, D. S. Ginley (2006), *Thin Solid Films, 496,* 26–29.
44. C. Y. Kuo, W. C. Tang, C. Gau, T. F. Guo, D. Z. Jeng (2008), *Applied Physics Letters, 93,* 033307.
45. Y. Y. Lin, T. H. Chu, C. W. Chen, W. F. Su (2008), *Applied Physics Letters, 92,* 053312.
46. W. U. Huynh, J. J. Dittmer, A. P. Alivisatos (2002), *Science, 295,* 2425–2427.

47. Q. Ahsanulhaq, A. Umar, and Y. B. Hahn (2007), *Nanotechnology, 18,* 115603.
48. M. N. R. Ashfold, R. P. Doherty, N. G. Ndifor-Angwafor, D. J. Riley, and Y. Sun (2007), *Thin Solid Films, 515,* 8679–8683.
49. S. Yamabi, and H. Imai (2002), *Journal of Materials Chemistry, 12,* 3773–3778.
50. R. A. Laudise, and A. A. Ballman (1960), *Journal of Physical Chemistry, 64,* 688–691.
51. Chen-Yu Chou (2009), "Organic-Inorganic Hybrid Solar Cells Based on ZnO Nanorods and Conjugated Polymer," MS thesis, National Taiwan University, Taipei, Taiwan, Chen-Yu Chou, Jing-Shun Huang, Chung-Hao Wu, Chun-Yu Lee, and Ching-Fuh Lin (2009), *Solar Energy Materials and Solar Cells, 93,* 1608–1612.
52. J.-S. Huang, C.-Y. Chou, and C.-F. Lin (2010), *Solar Energy Materials and Solar Cells, 94,* 182–186.
53. L. Vayssieres (2003), *Advanced Materials, 15,* 464–466.
54. J. M. Nápoles-Duarte, M. Reyes-Reyes, J. L. Ricardo-Chavez, R. Garibay-Alonso, and R. López-Sandoval (2008), *Physical Review B, 78,* 035425.
55. Y. Kim, S. Cook, S. M. Tuladhar, S. A. Choulis, J. Nelson, J. R. Durrant, D. D. C. Bradley, M. Giles, I. McCulloch, C. S. Ha, and M. Ree (2006), *Nature Materials, 5,* 197–203.
56. C. J. Brabec, A. Cravino, D. Meissner, N. S. Sariciftci, T. Fromherz, M. T. Rispens, L. Sanchez, and J. C. Hummelen (2001), *Advanced Functional Materials, 11,* 374–380.
57. D. C. Olson, Y. J. Lee, M. S. White, N. Kopidakis, S. E. Shaheen, D. S. Ginley, J. A. Voigt, and J. W. P. Hsu (2007), *Journal of Physical Chemistry C, 111,* 16640–16645.
58. D. C. Olson, S. E. Shaheen, R. T. Collins, and D. S. Ginley (2007), *Journal of Physical Chemistry C, 111,* 16670–16678.
59. D. C. Olson, Y. J. Lee, M. S. White, N. Kopidakis, S. E. Shaheen, D. S. Ginley, J. A. Voigt, and J. W. P. Hsu (2008), *Journal of Physical Chemistry C, 112,* 9544–9547.
60. Ing-Jye Wang ,Jing-Shun Huang, Yu-Hong Lin, and Ching-Fuh Lin (2010), *Solar Energy Materials & Solar Cells, 94,* 1681–1685.
61. Yu-Hong Lin (2010), "Study of Organic Polymer/Inorganic Semiconductor Hybrid Solar Cells in Inverted Structure," MS thesis, National Taiwan University, Taipei, Taiwan; Yu-Hong Lin, Po-Ching Yang, Jing-Shun Huang, Guo-Dong Huang, Ing-Jye Wang, Wen-Hao Wu, Ming-Yi Lin, Wei-Fang Su, and Ching-Fuh Lin (2011), *Solar Energy Materials and Solar Cells, 95,* 2511–2515.
62. T. Y. Chu, J. Lu, Y. Zhang, S. Wakim, J. Zhou, Z. Li, J. Ding, and Y. Tao (2011), "Highly Efficient Polymer Solar Cells Achieved by a Low Bandgap Copolymer with a Large Open Circuit Voltage," in *Proceedings of 26th European Photovoltaic Solar Energy Conference and Exhibition,* Hamburg, Germany, September; 1CO.9.2.
63. B. S. Ong, Y. Wu, P. Liu and S. Gardner (2011), *J. Am. Chem. Soc.,* 126, 3378–3379.
64. A. R. Murphy, J. Liu, C. Luscombe, D. Kavulak, J. M. J. Fréchet, R. J. Kline and M. D. McGehee (2005), *Chem. Mater., 17,* 4892–4899.
65. B. M. Medina, A. V. Vooren, P. Brocorens, J. Gierschner, M. Shkunov, M. Heeney, I. Mc-Culloch, R. Lazzaroni and J. Cornil (2007), *Chem. Mater., 19,* 4949–4956.
66. G. Lu, H. Usta, C. Risko, L. Wang, A. Facchetti, M. A. Ratner and T. J. Marks (2008), *J. Am. Chem. Soc., 130,* 7670–7685.
67. D. M. de Leeuw, M. M. J. Simenon, A. R. Brown and R. E. F. Einerhand (1997), *Synth. Met., 87,* 53–39.

68. N. Blouin, A. Michaud, D. Gendron, S. Wakim, E. Blair, R. Neagu-Plesu, M. Belletête, G. Durocher, Y. Tao and M. Leclerc (2008), *J. Am. Chem. Soc., 130,* 732–734.
69. S.–H. Chan, C–P. Chen, T.–C. Chao, C. Ting, C.–S. Lin and B.–T. Ko (2008), *Macromolecules, 41,* 5519–5526.
70. Y.–C. Chen, C.–Y. Yu, Y.–L. Fan, L.–I. Hung, C.–P. Chen, and C. Ting (2010), *Chem. Commun., 46,* 6503–6505.
71. Po-Ching Yang, Jen-Yu Sun, Shou-Yuan Ma, Yu-Min Shen, Yu-Hong Lin, Chih-Ping Chen, and Ching-Fuh Lin (2012), *Solar Energy Materials & Solar Cells, 98,* 351–356.
72. Eva Bundgaard and Frederik C. Krebs (2007), *Solar Energy Materials & Solar Cells, 91,* 954–985.
73. Renee Kroon, Martijn Lenes, Jan C. Hummelen, Paul W. M. Blom, and Bert de Boer (2008), *Polymer Reviews, 48,* 531–582.
74. S. H. Park, A. Roy, S. Beaupré, S. Cho, N. Coates, J. S. Moon, D. Moses, M. Leclerc, K. Lee, and A. J. Heeger (2009), *Nat. Photonics, 3,* 297–302.
75. Y. Liang, Z. Xu, J. Xia, S.-T. Tsai, Y. Wu, G. Li, C. Ray, and L. Yu (2010), *Adv. Mater., 22,* E135–E138.
76. T.-Y. Chu, J. Lu, S. Beaupre, Y. Zhang, J.-R. Pouliot, S. Wakim,J. Zhou, M. Leclerc, Z. Li, J. Ding, and Y. Tao (2011), *J. Am. Chem. Soc., 133,* 4250–4253.
77. M. J. Sailor, E. J. Ginaburg, C. B. Gorman, A. Kumar, R. H. Grubbs, and N. S. Lewis (1990), *Science, 249,* 1146–1149.
78. M. M. El-Nahass, H. M. Zeyada, M. S. Aziz, and M. M. Makhlouf (2005), *Thin Solid Films, 492,* 290–297.
79. M. M. El-Nahass, H. M. Zeyada, K. F. Abd-El-Rahman, and A. A. A. Darwish (2007), *Sol. Energy Mater. Sol. Cells, 91,* 1120–1126.
80. V. Gowrishankar, S. R. Scully, M. D. McGehee, Q. Wang, and H. M. Branz (2006), *Appl. Phys. Lett., 89,* 252102.
81. E. L. Williams, G. E. Jabbour, Q. Wang, S. E. Shaheen, D. S. Ginley, and E. A. Schiff (2005), *Appl. Phys. Lett., 87,* 223504.
82. W. Wang and E. A. Schiff, Appl. Phys. Lett., 2007, 91, 133504.
83. C. H. Lin, S. C. Tseng, Y. K. Liu, Y. Tai, S. Chattopadhyay, C. F. Lin, J. H. Lee, J. S. Hwang, Y. Y. Hsu, L. C. Chen, W. C. Chen, and K. H. Chen (2008), *Appl. Phys. Lett., 92,* 233302.
84. L. Tsakalakos, J. Balch, J. Fronheiser, B. A. Korevaar, O. Sulima, and J. Rand (2007), *Appl. Phys. Lett., 91,* 233117.
85. J. R., III Maiolo, B. M. Kayes, M. A. Filler, C. P. Putnam, M. D. Kelzenberg, H. A. Atwater, and N. S. J. Lewis (2007), *J. Am. Chem. Soc., 129,* 12346–12347.
86. A. P. Goodey, S. M. Eichfeld, K. K. Lew, J. M. Redwing, and T. E. J. Mallouk (2007), *J. Am. Chem. Soc., 129,* 12344–12345.
87. E. C. Garnett and P. J. Yang (2008), *Am. Chem. Soc., 130,* 9224–9225.
88. J. S. Huang, C. Y. Hsiao, S. J. Syu, J. J. Chao, and C. F. Lin (2009), *Sol. Energy Mater. Sol. Cells, 93,* 621–624.
89. L. Hu and G. Chen (2007), *Nano Lett., 7,* 3249–3252.
90. Y. F. Huang, S. Chattopadhyay, Y. J. Jen, C. Y. Peng, T. A. Liu, Y. K. Hsu, C. L. Pan, H. C. Lo, C. H. Hsu, Y. H. Chang, C. S. Lee, K. H. Chen, and L. C. Chen (2007), *Nat. Nanotechnol., 2,* 770–774.

91. V. Sivakov, G. Andra, A. Gawlik, A. Berger, J. Plentz, F. Falk, and S. H. Christiansen (2009), *Nano Lett., 9,* 1549–1554.
92. L. Tsakalakos, J. Balch, J. Fronheiser, M. Y. Shih, S. F. LeBoeuf, M. Pietrzykowski, P. J. Codella, B. A. Korevaar, O. Sulima, J. Rand, A. Davuluru, and U. J. Rapol (2007), *Nanophoton., 1,* 013552.
93. O. L. Muskens, J. G. Rivas, R. E. Algra, E. P. A. M. Bakkers, and A. Lagendijk (2008), *Nano Lett., 8,* 2638–2645.
94. O. L. Muskens, S. L. Diedenhofen, B. C. Kaas, R. E. Algra, E. P. A. M. Bakkers, J. G. Rivas, and A. Lagendijk (2009), *Nano Lett., 9,* 930–934.
95. Hong-Jhang Syu, Shu-Chia Shiu, and Ching-Fuh Lin (2011), in *Proceedings of 37th IEEE Photovoltaic Specialist Conference (IEEE PVSC 2011),* Seattle, Washington, USA, June 19–24, 2011.
96. K. Peng, J. Hu, Y. Yan, Y. Wu, H. Fang, Y. Xu, S. Lee, and J. Zhu (2006), *Adv. Funct. Mater., 16,* 387–394.
97. C-Y Chen, C-S Wu, C-J Chou, and T-J Yen (2008), *Adv. Mater., 20,* 3811.
98. Shu-Chia Shiu, Jiun-Jie Chao, Shih-Che Hung, Chin-Liang Yeh, and Ching-Fuh Lin (2010), *Chemistry of Materials, 22,* 3108–3113.
99. U. Lang, E. Muller, N. Naujoks, and J. Dual (2009), *Adv. Funct. Mater., 19,* 1215–1220.
100. A. M. Nardes, M. Kemerink, R. A. J. Janssen, J. A. M. Bastiaansen, N. M. M. Kiggen, B. M. W. Langeveld, A. J. J. M. van Breemen, and M. M. de Kok (2007), *Adv. Mater., 19,* 1196–1200.
101. Hong-Jhang Syu, Shu-Chia Shiu, and Ching-Fuh Lin (2012), *Solar Energy Materials & Solar Cells, 98,* 267–272.

EXERCISES

1. What are the disadvantages of the conventional structures of organic solar cells that use PEDOT and Al metal? Please explain the reasons.
2. What are the advantages of the inverted structures of organic solar cells?
3. What advantages do the sandwiched structures of organic solar cells have?
4. In the formation of organic solar cells with sandwiched structures, can the top oxide and the bottom oxide have the same band level? Why?
5. In the formation of organic solar cells with sandwiched structures, what condition is required to select the solvent used for the metal oxide to be deposited on the organic layer without damaging the organic layer?
6. Can NiO block the transportation of electrons toward the Ag electrode? Why?
7. Why does the insertion of a PCBM layer before the deposition of the P3HT:PCBM blend help the performance of the organic–ZnO nanorod hybrid solar cells?
8. How does the drying time of the P3HT:PCBM blend influence the performance of the organic–ZnO nanorod hybrid solar cells?
9. What are the possible ways to increase the drying time of the P3HT:PCBM blend?
10. How does the layer of the TiO_2 nanorods inserted between the ZnO layer and the active layer influence the performance of organic–inorganic hybrid solar cells?

11. Why does the devices with PV2000 gives a higher open-circuit voltage than those with a blend of P3HT:PCBM?
12. What conditions are required for a polymer to possess sufficient stability?
13. Compare the two materials of P3HT and PDTSTPD. Which one absorbs more light? Why?
14. Describe the mechanism of the aqueous electroless etching method that is used to form Si nanowires from the Si wafer.
15. Can the PEDOT solution adhere to the SiNW surface with H-terminated bonds? Why?
16. How will you modify the SiNW surface so that SiNW–PEDOT can form a nice core-sheath structure with good adhesion?
17. Why do the Si nanowire–PEDOT hybrid solar cells have better performance than the planar Si–PEDOT hybrid solar cells?
18. Does the Si with long Si nanowires absorb more light than the Si with short Si nanowires? Why?
19. Why do the Si nanowire–PEDOT hybrid solar cells with a short length have better performance than the Si nanowire–PEDOT hybrid solar cells with a long length?

9

OUTLOOK FOR HYBRID SOLAR CELLS

Wei-Fang Su

Hybrid solar cells are made from organic materials and inorganic materials. The organic materials are synthesized at relatively low temperature (< 200°C) in solution state and using abundant carbon-based raw materials, so they are low cost. For instance, polyethylene terephthalate (PET) (Eq. 4-1) is used for flexible substrates synthesized between 150–200°C, depending on the starting materials [1]. Poly (3-hexyl thiophene) (P3HT) (Eq. 4-2) is used for transistor and solar cells prepared at 66°C [2]. Poly(methyl methacrylate) (PMMA) (Eq. 4-3) is very useful in photoresist applications, which can be prepared below 80°C, depending on the reactants [1]. Inorganic nanomaterials (i.e., nanoparticles) are usually synthesized chemically using solution methods at relatively low temperature (< 300°C). For instance, the sol–gel technique is the most widely used approach to synthesize oxide-based semiconductor nanoparticles [3]. The sol–gel technique involves two step reactions: (1) hydrolysis of metal alkoxide to metal hydroxide at room temperature in acid or base catalyst, and then (2) the metal hydroxide is condensed into metal oxide by removing the water at moderate temperature, such as 90°C for titanium oxide. The sol–gel reactions can be expressed in the following equations. Thus, solution-based inorganic materials can be synthesized with energy conservation processes as compared to the high-vacuum and high-temperature processes commonly used for inorganic material processing.

Hydrolysis

$$M(OR)_n \xrightarrow{H^+/H_2O} M(OH)_{n-m}(OR)_m \qquad (9\text{-}1)$$

Organic, Inorganic, and Hybrid Solar Cells. By C.-F. Lin, W.-F. Su, C.-I Wu, and I-C. Cheng

Condensation

$$M(OH)_{n-m}(OR)_m \rightarrow H[O\text{—}M(OH)(OR)]_a\text{—}[O\text{—}M(OH)_2]_b\text{—}OR \quad (9\text{-}2)$$

$$H[O\text{—}M(OH)(OR)]_a\text{—}[O\text{—}M(OH)_2\text{—}O]_bR \rightarrow M_xO_y \quad (9\text{-}3)$$

Both organic material and inorganic materials used in the fabrication of hybrid solar cells are in solution states, so they can form thin films using conventional low-cost coating methods such as spin coating, spraying, dipping, and casting. The hybrid thin films combine the merits of organic materials and inorganic materials to be thermal stable, flexible and light weight. Therefore, the hybrid solar cell is the future for manufacturing long-life, flexible and light-weight solar cells.

Table 9.1 summarizes the state of the art of performances of different solar cells that were confirmed by the National Renewable Energy Laboratory (NREL) of the United States in 2010 [4]. The GaAs-based solar cell has the highest power conversion efficiency (27.6%) among the different types of solar cells due to GaAs being a direct-band-gap material with high absorption in the infrared region of the solar spectrum. However, the cost of GaAs solar cells is prohibitively high for daily electricity usage. Concentrated GaAs solar cells at 500X are being developed to reduce cost. The Si-based solar cell is less expensive with reasonable efficiency in the range of 12 to 20% and has been installed and used widely to generate renewable energy source of electricity in many countries, especially Germany and Japan. The CIGS thin-film solar cell has demonstrated impressive power conversion efficiency of close to 20%. However, the scale-up to large area CIGS solar cells has encountered many challenges due to the difficulties in controlling the stoichiometry of its four elements and the uniform crystallinity. Furthermore, indium is a rare element that is in limited supply.

At present, inorganic solar cells perform better than organic-based solar cells such as DSSC and polymer solar cells. However, the organic-based solar cells are catching up, with efficiency of more than 10%. The efficiency of DSSC mainly depends on the type of dye and electrolyte being used. Currently, using Ru-based dye together with a

TABLE 9.1. State of the art of performances of different solar cells under the global AM 1.5 spectrum (1000 W/cm^2) at 25°C, confirmed by NREL [4]

Type	*PCE* (%)	V_{OC} (V)	J_{SC} (mA/cm^2)	*FF* (%)	Area (cm^2)
GaAs	27.6 ± 0.8	1.10	29.6	84.1	0.9989
Si(poly C)	20.4 ± 0.5	0.66	38.0	80.9	1.0020
a-Si/μc-Si	11.9 ± 0.8	1.35	12.9	68.5	1.2270
CIGS	19.6 ± 0.6	0.71	34.8	79.2	0.9960
DSSC	10.4 ± 0.3	0.73	22.0	65.2	1.0040
Polymer	8.3 ± 0.3	0.82	14.5	70.2	1.0310
Hybrid	5.6 ± 0.1	0.79	11.3	63.0	0.0600

liquid electrolyte achieves the highest efficiency, but the Ru is an expensive rare element with limited supply. The liquid electrolyte of DSSC lacks long-term durability and safety. The polymer-based solar cell can solve the shortcoming of DSSC. Its development is progressing rapidly. Recently, the Mitsubishi Chemical Company announced that their polymer solar cell exhibited a cell efficiency of 9.2%. Konarka Technologies in Lowell, Massachusetts; Solarmer Energy Inc. in El Monte, California; and Heliatek in Dresden, Germany are now reporting cells with efficiencies greater than 8%. Many researchers in the field are confident that the figure could soon top 10% and possibly reach 15% [5]. Up to now, the power conversion efficiency of hybrid solar cells has reached about 6% [6–8]. The new concept of Si nanowire:polymer hybrid solar cells is being explored, with a promising power conversion efficiency over 8% (see Chapter 8, Section 8.7). The conventional Si solar cell is fabricated on wafers more than 100 micron thick due to the indirect-band-gap characteristic of Si and the current limitations of mechanical cutting technology, which results in a heavy, high-cost, and rigid solar cell. It is expected that progress will lead to 100 μm Si wafers being adapted in the year of 2020 [9]. The Si nanowire approach can solve the problems associated with the Si solar cells fabricated from thin Si wafer, and are also under active research [9,10]. Because Si nanowires could give rise to the light-trapping effect for increasing light absorption, the thickness of Si wafer can be greatly reduced, even though it is an indirect-band-gap semiconductor. We should be optimistic on pushing the use of thin Si wafers far before 2020.

The development of hybrid solar cells is relatively new and only started in the past few years. One can take the advantage of the progress in the polymer–fullerene solar cells and incorporate them into the hybrid solar cells to increase their efficiency. Highly ordered nanostructure inorganics are being proposed to enhance the transport of charge carriers and light trapping [11]. Interface engineering is also one of the key development areas to reduce the charge recombination and facilitate charge transport in hybrid solar cells [12]. With optimal nanostructure and interface engineering, and high light harvesting from both organic and inorganic materials, a power conversion efficiency of 15% can be reasonably expected for the hybrid solar cells.

In 2010, Nielsen and coworkers [13] have done very detailed performance and cost analyses of polymer–fullerene solar cells to determine the viability of the devices as commercial products for the period from 2010 to 2025. The results are summarized in Table 7.7. Efficiency is the essential parameter for solar cells in terms of maximizing energy production and minimizing cost. Low device efficiency demands a larger solar cell active area in order to produce the same energy output. A device efficiency of ~10% and a module efficiency of ~5% are regarded as critical values for commercial products. Currently, their main application is in consumer electronics with module efficiencies of 1–3% at a cost of 12 USD/Wp. The cost is expected to be lower than 0.5 USD/Wp and could be used as a renewable power source after 2018, with a life of 5–7 years and efficiency of 6–9%. From 2010 to 2011, within one year, the progress of polymer solar cells exceeded what Nielsen and coworkers discussed. The cell efficiency has reached almost 10% now; the cost will be reduced by half if the cost analysis is based on cell efficiency of 5%.

The progress in hybrid solar cells should be similar to that of polymer solar cells. Moreover, the hybrids are more thermally stable and environmentally benign than polymer solar cells in both forward and inverted types. Up to now, all the high-efficiency polymer solar cells (> 7%) have been forward types using air-sensitive photoactive layers and Al electrodes. An expensive multilayer hermetic packing is required for this type of polymer solar cell. The inverted hybrid solar cell uses a stable Ag electrode with an oxygen- and moisture-impermeable metal oxide electron transporting layer and hole transport layer to protect the photoactive layer, so essentially no packing is required for this type of solar cell or the packing can be a very simple and low-cost single-layer packing. That will reduce the cost of solar cells even further.

In summary, the hybrid solar cell will be the champion of long-life, low-cost, flexible solar cells. The dream of an environmentally benign, renewable, affordable energy source will come true when the cell efficiency of hybrid solar cells reaches 10% or higher.

REFERENCES

1. G. Odian (2004), *Principles of Polymerization,* 4th Ed., Wiley-Interscience.
2. T. M. Pappenfus, D. L. Hermanson, S. G. Kohl, J. H. Melby, L. M. Thoma, N. E. Carpenter, D. A. da Silva Filho, and J-L Bredas (2010), *Journal of Chemical Education, 87,* 522–525.
3. C. J. Brinker and G. W. Scherer (1990), *Sol–Gel Science: The Physics and Chemistry of Sol–Gel Processing,* Academic Press, NY.
4. M. A. Green, K. Emery, Y. Hishikawa, and W. Warta (2011), *Prog. Photovolt. Res. Appl., 19,* 84–92.
5. *News and Analysis in Science* (2011), 332(6027), 293.
6. Y. H. Lin, P. C. Yang, J. S. Huang, G. D. Huang, I. J. Wang, W. H. Wu, M. Y. Lin, W. F. Su, and C. F. Lin (2011), *Solar Energy Materials & Solar Cells, 95,* 2511–2515.
7. Y. J. Cheng, C. H. Hsieh, Y. He, C. S. Hsu, and Y. Li (2010), *Journal of the American Chemical Society, 132*(49), 17381–17383.
8. T. Y. Chu, J. Lu, Y. Zhang, S. Wakim, J. Zhou, Z. Li, J. Ding, and Y. Tao (2011), "Highly Efficient Polymer Solar Cells Achieved by a Low Bandgap Copolymer with a Large Open Circuit Voltage," in *Proceedings of 26th European Photovoltaic Solar Energy Conference and Exhibition,* Hamburg, Germany, September 2011; 1CO.9.2.
9. F. Dross, K. Baert, T. Bearda, J. Deckers, V. Depauw, O. El Daif, I. Gordon, A. Gougam, J. Govaerts, S. Granata, R. Labie, X. Loozen, R. Martini, A. Masolin, B. O'Sullivan, Y. Qiu, J. Vaes, D. Van Gestel, J. van Hoeymissen, A. V. Vanleenhove, K. Van Nieuwenhuysen, S. Venkatachalam, M. Meuris, and J. Poortmans (2011), "Crystalline Thin-Film Silicon: Where Crystalline Quality Meets Thin-Film Processing," in *Proceedings of 26th European Photovoltaic Solar Energy Conference and Exhibition,* Hamburg, Germany, September 2011; 3CP.1.5.
10. Shu-Chia Shiu, Shih-Che Hung, Hong-Jhang Syu, and Ching-Fuh Lin (2011), *Journal of the Electrochemical Society, 158*(2), D95–D98.
11. J. Weickert, R. B. Dunbar, H. C. Hesse, W. Wiedemann, and L. Schmidt-Mende (2011), *Adv. Mater., 23,* 1810–1828.

12. Y. C. Huang, J. H. Hsu, Y. C. Liao, W. C. Yen, S. S. Li, S. T. Lin, C. W. Chen, and W. F. Su (2011), *Journal of Materials Chemistry, 21*(12), 4450–4456.
13. T. D. Nielsen, C. Cruickshank, S. Foged, J. Thorsen, and F. C. Krebs (2010), *Solar Energy Materials & Solar Cells, 94,* 1553–1571.

EXERCISES

1. In the fabrication of solar cells, it has been found that cell efficiency is reduced when the area of a solar cell is increased. Explain why and provide solutions to solve the problem.
2. Nielsen and coworkers reported [10] that cell-module efficiency of 1–3% at a cost of 12 USD/Wp in 2010 will be reduced to lower than 0.5 USD/Wp at an efficiency of 6–9%. If we do the cost analysis from the efficiency only, the cost will be reduced from 12 USD/Wp to 2–4 USD/Wp rather than 0.5 USD/Wp. How would you explain the results of this analysis?
3. The long-term reliability of solar cells is hinged on the design and material of the packaging to enclose the solar cell and to protect it from hostile environments. Do a search on the state of the art in packaging design and materials used for silicon solar cells. Would they be suitable for the packaging of forward P3HT–PCBM solar cells or invert P3HT–PCBM solar cells? Explain your answers.

INDEX

www.ingramcontent.com/pod-product-compliance
Lightning Source LLC
Chambersburg PA
CBHW070833040225
21358CB00001B/7

9781118168530